ESTIMATING IN BUILDING CONSTRUCTION

THIRD EDITION

ESTIMATING IN BUILDING CONSTRUCTION

Frank R. Dagostino

Registered Architect
Architectural and Construction Consultant

Dean, Industrial and Engineering Technology
Trident Technical School
Charleston, SC

PRENTICE HALL, Englewood Cliffs, New Jersey 07632

LIBRARY OF CONGRESS
Library of Congress Cataloging-in-Publication Data

Dagostino, Frank R.
    Estimating in building construction / Frank R. Dagostino. -- 3rd
ed.
        p.   cm.
    Includes index.
    ISBN 0-13-289737-7
    1. Building--Estimates.   I. Title.
TH435.D18 1989
692'.5--dc19   88-6734
                                                           CIP

Editorial/production supervision: Claudia Citarella
Manufacturing buyer: Bob Anderson

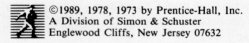
©1989, 1978, 1973 by Prentice-Hall, Inc.
A Division of Simon & Schuster
Englewood Cliffs, New Jersey 07632

Printed in the United States of America

10 9 8 7 6 5 4 3 2

ISBN   0-13-289737-7

Prentice-Hall International (UK) Limited, *London*
Prentice-Hall of Australia Pty. Limited, *Sydney*
Prentice-Hall Canada, Inc., *Toronto*
Prentice-Hall Hispanoamericana, S.A., *Mexico*
Prentice-Hall of India Private Limited, *New Delhi*
Prentice-Hall of Japan, Inc., *Tokyo*
Simon & Schuster Asia Pte. Ltd., *Singapore*
Editora Prentice-Hall do Brasil, Ltda., *Rio de Janeiro*

CONTENTS

PREFACE   xiii

CHAPTER 1   INTRODUCTION TO ESTIMATING   1

1-1   General Introduction, 1
1-2   Estimating Opportunities, 2
1-3   The Estimator, 5
1-4   Quantity Surveying, 6
1-5   Types of Bids, 7
1-6   Cost Analysis Methods, 8
1-7   The Contract Documents, 9
1-8   Bidding Information, 10
1-9   Availability of Contract Documents, 11
1-10   Sources of Estimating Information, 12
1-11   Computers, 13
       Review Questions, 13

CHAPTER 2   CONTRACTS, BONDS, AND INSURANCE   14

2-1   The Contract System, 14
2-2   Types of Agreements, 15
2-3   Lump-Sum Agreement (Stipulated-Sum), 16
2-4   Unit Price Agreement, 16
2-5   Cost-Plus-Fee Agreements, 17
2-6   Agreement Provisions, 18
2-7   Bonds, 19
2-8   Bid Bond, 20

2–9   Performance Bond, 20
2–10   Labor and Material Bond, 21
2–11   Miscellaneous Bonds, 21
2–12   Obtaining Bonds, 21
2–13   Insurance, 24
       Review Questions, 25

CHAPTER 3   SPECIFICATIONS   26

3–1   Specifications, 26
3–2   Construction Specifications Institute (CSI), 27
3–3   Invitation to Bid (Advertisement for Bids), 30
3–4   Instructions to Bidders (Information for Bidders), 30
3–5   Bid (Proposal) Forms, 33
3–6   Form of Owner-Contractor Agreement, 34
3–7   General Conditions, 34
3–8   Supplementary General Conditions, 34
3–9   Technical Trade Sections, 35
3–10   Alternates, 36
3–11   Addenda, 36
3–12   Errors in the Specifications, 37
       Review Questions, 38

CHAPTER 4   THE ESTIMATE   40

4–1   Organization, 40
4–2   Daily Log, 41
4–3   Notebook, 41
4–4   To Bid or Not to Bid, 43
4–5   The Estimate, 43
4–6   Site Investigation, 45
4–7   Subcontractors, 46
4–8   Materials, 47
4–9   Workup Sheets, 48
4–10   Summary Sheet, 50
4–11   Errors and Omissions, 51
       Review Questions, 51

CHAPTER 5   OVERHEAD AND CONTINGENCIES   52

5–1   Overhead, 52
5–2   General Overhead (Indirect), 52
5–3   Job Overhead (General Conditions, Direct Overhead), 54
5–4   Scheduling, 56
5–5   Contingencies, 61
5–6   Checklist, 67
       Review Questions, 69

CHAPTER 6   LABOR   70

6–1   General, 70
6–2   Unions—Wages and Rules, 71
6–3   Field Reports, 71
6–4   Pricing Labor, 72

6–5   Labor Diagrams, 72
       Review Questions, 73

CHAPTER 7   EQUIPMENT   75

7–1   General, 75
7–2   Operation Costs, 76
7–3   Depreciation, 77
7–4   Interest, 78
7–5   Ownership Costs, 78
7–6   Rental Costs, 79
7–7   Miscellaneous Tools, 80
7–8   Cost Accounting, 80
7–9   Mobilization, 81
7–10   Checklist, 81
       Review Questions, 81

CHAPTER 8   EXCAVATION   83

8–1   General, 83
8–2   Specifications, 83
8–3   Soil, 84
8–4   Excavation—Unit of Measure, 84
8–5   Equipment, 84
8–6   Earthwork—New Site Grades; Rough Grading, 85
8–7   Perimeters and Areas, 91
8–8   Topsoil Removal, 92
8–9   General Excavation, 94
8–10   Special Excavation, 105
8–11   Backfilling, 108
8–12   Excess and Borrow, 114
8–13   Spreading Topsoil, Finish Grade, 115
8–14   Landscaping, 115
8–15   Pumping (Dewatering), 116
8–16   Rock Excavation, 116
8–17   Subcontractors, 117
8–18   Overhead and Profit, 117
8–19   Excavation Checklist, 117
8–20   Piles, 118
8–21   Pile Checklist, 120
8–22   Asphalt Paving, 120
       Review Questions, 122

CHAPTER 9   CONCRETE   125

9–1   Concrete Work, 125
9–2   Estimating Concrete, 126
9–3   Reinforcing, 130
9–4   Vapor Barrier, 135
9–5   Accessories, 136
9–6   Concrete Finishing, 138
9–7   Curing, 141
9–8   Transporting Concrete, 143
9–9   Forms, 144
9–10   Form Liners, 147
9–11   Checklist, 148
       Review Questions, 150

CHAPTER 10 MASONRY 151

10-1 General, 151
10-2 Specifications, 151
10-3 Labor, 152
10-4 Bonds (Patterns), 152
10-5 Concrete Masonry, 154
10-6 Specifications—Concrete Masonry, 155
10-7 Estimating—Concrete Masonry, 157
10-8 Clay Masonry, 160
10-9 Specifications—Brick, 162
10-10 Estimating Brick, 162
10-11 Specifications—Tile, 165
10-12 Estimating Tile, 165
10-13 Stone Masonry, 166
10-14 Specifications—Stone, 166
10-15 Estimating Stone, 166
10-16 Mortar, 167
10-17 Accessories, 168
10-18 Reinforcing Bars, 177
10-19 Cleaning, 177
10-20 Equipment, 178
10-21 Cold Weather, 178
10-22 Subcontractors, 179
10-23 Checklist, 179
        Review Questions, 180

CHAPTER 11 WATERPROOFING AND DAMPPROOFING 181

11-1 Waterproofing, 181
11-2 Membrane Waterproofing, 181
11-3 Estimating Membrane Waterproofing, 182
11-4 Integral Method, 184
11-5 Metallic Method, 184
11-6 Dampproofing, 184
11-7 Painting Method, 185
11-8 Estimating Painting, 185
11-9 Parging (Plastering), 186
11-10 Estimating Parging, 186
11-11 Checklist, 186
        Review Questions, 187

CHAPTER 12 STRUCTURAL SYSTEMS 188

12-1 Precast Concrete, 188
12-2 Specifications, 189
12-3 Estimating, 190
12-4 Precast Tees, 190
12-5 Precast Slabs, 193
12-6 Precast Beams and Columns, 194
12-7 Miscellaneous Precast, 196
12-8 Precast Costs, 197
12-9 Precast Checklist, 197
12-10 Structural Steel, 198

12-11 Structural Steel Framing, 198
12-12 Steel Joists, 199
12-13 Steel Decking, 203
12-14 Miscellaneous Structural Steel, 205
12-15 Steel Erection Subcontractors, 205
12-16 Steel Checklist, 205
12-17 Wood Systems, 206
12-18 Wood Trusses, 206
12-19 Laminated Beams and Arches, 208
12-20 Wood Decking, 209
12-21 Plywood Systems, 209
12-22 Wood Checklist, 210
       Review Questions, 212

CHAPTER 13   WOOD FRAME CONSTRUCTION   213

13-1   Frame Construction, 213
13-2   Board Measure, 214
13-3   Floor Framing, 215
13-4   Wall Framing, 232
13-5   Ceiling Assembly, 247
13-6   Roof Assembly, 249
13-7   Trusses, 255
13-8   Insulation, 255
13-9   Trim, 256
13-10 Labor, 257
13-11 Checklist, 258
       Review Questions, 258

CHAPTER 14   DRYWALL AND WETWALL CONSTRUCTION   260

14-1   General, 260
14-2   Supporting Construction, 260
14-3   Suspended Ceiling Systems, 264
14-4   Types of Assemblies, 265
14-5   Wallboard Types, 266
14-6   Drywall, 267
14-7   Column Fireproofing, 268
14-8   Accessories, 269
14-9   Wetwall Construction, 273
14-10 Plaster, 274
14-11 Lath, 277
14-12 Accessories, 280
14-13 Checklist, 280
       Review Questions, 281

CHAPTER 15   WINDOWS, CURTAIN WALLS, DOORS, AND HARDWARE   282

15-1   Window and Curtain Wall Frames, 282
15-2   Accessories, 285
15-3   Doors, 287
15-4   Fire-Rated Doors, 289

15–5 Acoustical Doors, 290
15–6 Overhead Doors, 291
15–7 Folding Doors and Partitions, 291
15–8 Prefitting and Machining(Premachining), 291
15–9 Prefinishing, 292
15–10 Door Frames, 293
15–11 Hardware, 294
15–12 Accessories, 295
15–13 Checklist for Doors and Frames, 295
15–14 Glass, 296
Review Questions, 297

CHAPTER 16 FINISHES 298

16–1 Flooring, 298
16–2 Wood Flooring, 298
16–3 Resilient Flooring, 299
16–4 Carpeting, 303
16–5 Tile, 305
16–6 Painting, 306
16–7 Checklist, 308
Review Questions, 309

CHAPTER 17 ROOFING 310

17–1 Roofing, 310
17–2 Roof Areas, 310
17–3 Shingles, 311
17–4 Built-up Roofing, 313
17–5 Corrugated Sheets (Including Ribbed V-Beam), 316
17–6 Metal Roofing, 316
17–7 Slate, 316
17–8 Tile, 317
17–9 Sheets, Tile, and Shingles Formula, 317
17–10 Liquid Roofing, 318
17–11 Flashing, 318
17–12 Insulation, 318
17–13 Roofing Trim, 319
17–14 Labor, 319
17–15 Equipment, 319
17–16 Checklist, 320
Review Questions, 320

CHAPTER 18 ELECTRICAL 321

18–1 Electrical Work, 321
18–2 Single Contracts, 322
18–3 Separate Contracts, 323
18–4 Coordination Requirements, 323
18–5 Checklist, 325
Review Questions, 325

CHAPTER 19 PLUMBING 326

19–1 Plumbing Work, 326
19–2 Single Contracts, 326

19–3   Separate Contracts, 328
19–4   Coordination Requirements, 328
19–5   Checklist, 329
Review Questions, 330

CHAPTER 20   HEATING, VENTILATING, AND AIR CONDITIONING   331

20–1   HVAC Work (Heating, Ventilating, and Air Conditioning), 331
20–2   Single Contracts, 331
20–3   Separate Contracts, 332
20–4   Coordination Requirements, 332
20–5   Checklist, 334
Review Questions, 335

CHAPTER 21   PROFIT   336

21–1   Profit, 336

APPENDIX A   DRAWINGS AND OUTLINE SPECIFICATIONS OF A SMALL COMMERCIAL BUILDING   338

APPENDIX B   COMMON TERMS USED IN THE BUILDING INDUSTRY   343

APPENDIX C   AIA GENERAL CONDITIONS AND STANDARD FORM OF AGREEMENT   352

APPENDIX D   CONVERSIONS   385

APPENDIX E   DRAWINGS AND OUTLINE SPECIFICATIONS OF A SMALL COMMERCIAL BUILDING   390

APPENDIX F   SMALL COMMERCIAL BUILDING   395

APPENDIX G   DRAWINGS AND OUTLINE SPECIFICATIONS OF A RESIDENCE   397

APPENDIX H   DRAWINGS AND OUTLINE SPECIFICATIONS OF A RESIDENCE   402

INDEX   408

PREFACE

The third edition has been updated to include new cost data for the material and labor examples for the commercial building and residence used as examples in the text. It still contains an example of a small residence estimate and commercial building in a step-by-step takeoff procedure.

This book has been designed to give a thorough understanding of the interrelationships between contract documents and the quantity takeoff of the materials, and to impart an understanding of the factors that affect the time required to do each portion of the work.

The first four chapters deal in depth with bidding procedures, contract documents, contracts, bonds, and insurance. The remaining chapters follow a format of briefly describing the materials and how the different types may affect the bid, installation procedures as they may affect the bid, the unit of measure of the work, estimating the quantity of materials, and the relationship of the specifications.

A small commercial building and residence are used as examples throughout the text. In addition, small commercial buildings and residences have been included in the appendices to be used to gain estimating experience.

Frank R. Dagostino

ESTIMATING
IN BUILDING
CONSTRUCTION

1
INTRODUCTION TO ESTIMATING

1-1 GENERAL INTRODUCTION

Building construction estimating is the determination of probable construction costs of any given project. There are many items that influence and contribute to the cost of a project; each item must be compiled and analyzed. Since the estimate is prepared before the actual construction of the project, a great deal of study must be put into the construction documents; and this makes estimating one of the most important phases of any contractor's business.

In most instances, it is necessary to submit an estimate on the cost of the project, often in competitive bidding against other firms. The competition in construction bidding is keen, sometimes with ten or more firms bidding on a single project. In order to stay in business, a contractor must be low bidder on a certain number of projects; yet their prices cannot be so low that it is impossible to make a profit on them.

Since the estimate is prepared from the working drawings and specifications of a building, the ability of the estimator to visualize all of the different phases of construction becomes a prime ingredient in successful bidding.

The working drawings usually contain information relative to design, location, dimensions, and construction of the project, while the specifications are a written supplement to the drawings and include information pertaining to materials and workmanship. The working drawings and specifications *must* be considered together when an estimate is being prepared. The two complement each other, often there is an overlap in the information they convey. The bid submitted must be based on the drawings *and* specifications. You, as the estimator, are responsible for everything contained in the specifications as well as for that which is covered on the drawings. Read everything thoroughly; recheck all items whenever it is necessary. The plans must be measured and checked carefully, and all listings of quantities and materials done as clearly and accurately as possible. Each item must be listed with as much information about it as is reasonably possible.

Estimating the cost of a project is a process subject to many variables that may affect the actual construction of the project. These include weather, transportation, soil conditions, labor strikes, material availability, and subcontractors available. Regardless of the variables involved, the estimator must strive to prepare as accurate an estimate as possible. Estimating is not a guessing game. Carefully organized work, based on the estimator's best judgment and records of past projects completed, will result in accurate bids.

1-2 ESTIMATING OPPORTUNITIES

For anyone who is not aware of the many opportunities in the estimating field this section will review some of the areas in which a knowledge of estimating is necessary. Generally, a knowledge of the procedures for estimating is required by almost everyone involved in or associated with the field of construction. From the estimator, who may be involved solely with the estimating of quantities of materials and pricing of the project using the computer to increase speed and providing accurate data from previous projects, to the carpenter, who must order the material required to build the framing for a home, this knowledge is needed to do the best possible job at the most favorable cost. This includes the project designer, drafters, engineers, contractors, subcontractors, material suppliers, and material representatives.

Architectural offices. Basically the architectural office will require estimates at three stages: preliminary (rough square foot or cubic foot costs), cost evaluation during drawing preparation (usually more accurate square foot or cubic foot costs), and a final estimate (usually based on material and installation costs, to be as accurate as possible).

In large offices the estimating may be done by an estimator hired primarily to do all required estimating. In many offices, the estimating may be done by the chief drafter, head or lead architect, or perhaps someone else in the office who has developed the skills required to do the required estimating. Many ambitious drafters, with a basic knowledge of estimating have worked themselves into the position of doing the estimating for the office.

Engineering offices. The engineering offices involved in the design of construction projects include civil, structural, mechanical (plumbing, heating, air conditioning), electrical, and soil analysis. All of these engineering design phases require preliminary estimates, estimates while the drawings are being prepared, and the final estimates as the drawings are complete. As in architectural offices, the estimating may be done by someone who primarily does the estimating or by a head drafter, or engineer in charge of the project.

General contractors (U.S.A.). Typically, the general contractor makes *detailed* estimates (discussed in Sec. 1-6) which are used to determine what the company would charge to do the work required.

The estimator will have to take off the quantities (amounts) of each material, determine the cost to furnish (buy and get to the site) and install each material in the project, assembly of the bids (prices) of subcontractors, as well as determining all of the costs of insurance, permits, office staff, and the like. In smaller companies one person may do all of the estimating, while in larger companies several people may be used to negotiate a final price with an owner or to provide a competitive bid. Many times, the contractor's business involves providing assistance to the owners beginning with the planning stage and continuing through the actual construction of the project (commonly called design and build contractors). In this type of business the estimators will also provide preliminary estimates and then update them periodically until a final price is set.

Contractors (in countries or situations where a quantity survey (Sec. 1–4) is provided, such as Canada, parts of Europe, and highway projects in the U.S.A.). Estimating for projects with quantity surveys involves reviewing the specifications for the material requirements and the drawings for a review of the type of construction used and assemblies of materials used. Then the estimator will spend part of the time getting prices from subcontractors and material suppliers and the rest of the time deciding on how the work may be most economically accomplished.

The estimator may still have to do a quantity take off for negotiated bid projects (when the contractor and owner come to an agreement on the amount of money). In addition, if they are involved in design-build projects there will have to be preliminary as well as final estimates.

Subcontractors. These individuals, companies, or corporations are hired by the general contractor to do a particular portion of the work on the project. Subcontractors are available for all of the different work required to build any project and include excavation, concrete, masonry (block, brick, stone), interior partitions, drywall, wetwall, acoustical ceilings, painting, steel and precast concrete, erection, windows and metal and glass curtainwalls, roofing, flooring (resilient, ceramic and quarry tile, carpeting, wood, terrazzo), interior wall finishes such as: wallpaper, wood paneling, sprayed-on finishes, and the list continues on to include all materials and finishes required.

The subcontractor carefully checks the drawings and specifications and submits a price to the construction companies who will be bidding on the project. The price given may be a unit or lump-sum price. Unit prices are those the contractors bid on as to what they would charge per unit (such as: per square foot, per block, per thousand brick, per cubic yard of concrete), for example, the bid might be $2.70 per lineal feet of concrete curbing. Even subcontractors who bid on a unit basis will do at least a rough quantity takeoff so they can have an idea of what is involved in the project, at what stages they would be needed, how long it would take to complete their work, and how many workers and equipment will be required. The lump-sum bid would be a cost to install, or furnish and install, a portion of work, for example, the bid might state "agrees to furnish and install all Type I concrete curbing for a sum of $1,875.00."

Each subcontractor will need someone (or several people) to check specifications, review the drawings, determine the quantities required, and put the proposal together. It may be a full-time estimating position or part of the duties assumed, perhaps in addition to purchasing materials, helping to schedule the projects, working on shop drawings required, being the company representative who visits contractors, architects, engineers, etc. to promote the use of the product being installed or the use of their firm as the installer.

Material suppliers. Suppliers submit bids to the contractors (and subcontractors) to supply the materials required for the construction of the project. Virtually every material used in the project will be estimated by a material supplier and be bid as unit or lump-sum prices. Estimators will have to check the specifications and drawings to be certain that the materials (the materials they produce or sell) will meet all of the requirements of the contract, determine about how much of the material is required and when during the construction of the project the materials will be required.

Large material suppliers may use a full-time estimator or it may be a part of someone's responsibilities. Quite often this estimating is done by the manufacturer's representative.

Manufacturers' representatives. These persons represent certain material or product suppliers or manufacturers. They spend at least part of their time visiting members of the construction industry (architects, engineers, contractors, subcontractors, owners, developers, etc.) to be certain they are aware that the material is available, and what its uses and approximate costs are. In a sense they are salespeople, but their services and the expertise they develop in their product lines make a good manufacturer's representative welcome in the industries offices not as salespersons but as needed sources of information concerning the materials and products they represent. Representatives may work for one company or they may represent two or more.

Manufacturer's representatives will generally do quantity takeoffs of projects which use the materials they represent. They will also carefully check the specifications and drawings to be certain the materials meet all requirements. If some aspect of the specifications or drawings tends to exclude their product, or if they feel there may be a mistake or misunderstanding in these documents, they may call the architect-engineers and discuss it with them. In addition, many times they will be involved in working up various cost analyses of what the materials or products cost installed and in working to devise new uses for the materials, alternate construction techniques, and even the development of new products.

Project management. These companies specialize in providing professional assistance in planning the orderly construction of a project while keeping an accurate and updated financial statement of the project. Such companies are often hired by the owner to coordinate a large project, and many construction firms also make use of their services. Among the various types of owners are private individuals, companies, corporations, municipal government agencies (such as public works and engineering departments), and various public utility companies.

Both the firms involved in project management, as well as someone on the staff of the owner being represented, must be knowledgeable in estimating and scheduling of a project.

Government. When government is involved in any phase of construction, personnel with experience in construction and estimating are required. Included are local, state or province, and nationwide agencies including: those involved in highways, roads, sewage treatment, schools, courthouses, nursing homes, hospitals, and single- and multiple-family dwellings financed, or qualifying for financing, by the government.

Employees may be involved in preparing or assisting to prepare preliminary and final estimates; reviewing estimates from architects, engineers, and contractors; the design and drawing of the project and preparation of the specifications.

In addition, where the quantity survey method of bidding is used (Sec. 1-4), the quantity survey of materials may be done by the governmental agency or department (or by a professional quantity surveyor or firm).

Professional quantity surveyors. Professional quantity surveyors are discussed in Sec. 1-4 and require estimators to make unit quantity takeoffs of materials required to build a project. Such individuals and firms are available to provide this service to all who have need of it, including governmental agencies.

Free-lance estimators. Such estimators will do a material takeoff of a portion or entire project for whomever may want them done. This estimator may work for the owner, architect, engineer, contractor, subcontractor, material supplier, or manufacturer. In some areas the estimator will do a material takeoff of a project being competitively bid and then sell the quantity list to one or more contractors who intend to submit a bid on the project.

Many times this type of estimator will even work with individual subcontractors on the preparation of the unit or lump-sum proposal that will be sent out to the contractors. This type of business may be one person or a company with several employees. The employees may be full time; often part-time help is also used during busy seasons.

Many times a talented individual has a combined drafting and estimating business. Part of the drafting business may include any shop drawings (drawings showing sizes of materials and installation details) for subcontractors, material suppliers, and manufacturers' representatives.

Residential construction. Estimators are also required for all of the contractors, material suppliers, and manufacturers' representatives, and most of the subcontractors involved in residential construction. From the designer who plans the house and the drafter who draws the plans and elevations to the carpenters who put up the rough framing and the roofers who install the roofing material, a knowledge of estimating is necessary.

The designer and drafter should plan and draw the house plans using standard material sizes whenever possible (or being aware of it when they are not using standard sizes). In addition, they will need to give preliminary and final estimates to the owner. Workers need to have a basic knowledge of estimating so they can be certain that adequate material has been ordered and will be delivered by the time it is needed.

1-3 THE ESTIMATOR

As one of the most important employees in the contractor's organization, the estimator should be carefully selected. Most estimators work first doing quantity takeoff; as they develop experience and judgment, they develop into estimators. The following abilities are most important to the success of an estimator. They should be more than just read through. Any weaknesses affect the estimator's ability to produce complete and accurate estimates. If individuals lack any of these abilities, they must first be able to admit it and second begin to acquire the abilities they lack. Those with actual field experience who are subsequently trained as estimators are often most successful in this field.

To be able to do quantity takeoffs, the estimator:

1. Must be able to read and measure plans.

2. Must have a knowledge of mathematics that includes addition, subtraction, multiplication, division, and an understanding of the decimal system. Most measurements and computations are made in lineal feet, square feet, square yards, cubic feet, and cubic yards. The quantities are usually multiplied by a unit price to calculate material costs.

3. Must have the patience and ability to do careful, thorough work.

To be an estimator you must go a step further, you:

1. Must be able to visualize the project from looking at the drawings, through the various phases of construction, and you must provide the solutions to various problems, such as where the contractor will place equipment and materials.

2. Must have enough construction experience to possess a good knowledge of job conditions, including methods of handling materials on the job, the most economical methods of construction, and the problems of handling labor. With this experience, estimators will be able to construct the job in their minds and thus get the most accurate estimate on paper.

3. Must have sufficient knowledge of labor performances and operations so that you can convert them into costs on a project. You must understand how much work can be accomplished under given conditions. Experience in construction and a study of projects that have been completed are required to develop this ability.

4. Must have the ability to keep a fund of information on costs of all kinds, including those of labor, material, overhead, equipment, as well as knowledge of the availability of all the required items.

5. Must be able to meet bid deadlines and still remain calm. Even in the rush of last-minute phone calls and the competitive feeling, which seems to electrify the atmosphere just before the bids are due, you must "keep your cool."

People cannot be taught experience and judgment, but they can be taught an acceptable method of preparing an estimate, items included in the estimate, calculations required, and how to make them. They can also be warned against possible errors and alerted to certain problems and dangers, but the practical experience and use of good judgment required cannot be taught and must be obtained through their own experiences.

How closely the estimated cost will agree with the actual cost will depend, to a large extent, on the estimators' skill and judgment. Their skill enables them to use accurate estimating methods, while their judgment enables them to visualize the construction of the project throughout the stages of construction.

1-4 QUANTITY SURVEYING

In Canada, parts of Europe, and most road construction projects in the United States of America, the quantities of materials required on the project are determined by a professional quantity surveyor. These quantities are then made available to all bidders on the project and the general contractor's estimator will work up a price for each of the materials required. Part of a typical quantity survey is shown in Fig. 1–1.

No.	Quantity	Unit	Item
241	715	l.f.	Curb, type A
243	75	l.f.	Curb, type A
0873A	12,500	each	8" concrete block
0873B	5,280	each	12" concrete block
0873C	3,700	each	4" concrete block

Figure 1-1 Quantity Survey

In this method of bidding, the contractors are all bidding based on the same quantities and the contractor's estimator spends time developing the unit prices which will be bid. For example, the bid may be $47.32 per cu. yd. of concrete. Since all of the contractors are bidding on the same quantities, they will work on keeping the cost of purchasing and installing the materials as low as possible.

As the project is actually built, the actual number of units required is checked against the original number of units on which the estimates were made. For example, in Fig. 1-1, the original quantity survey called for 715 lineal feet of concrete curbing. If 722 l.f. were actually installed, then the contractor would be paid for the additional 7 l.f. If 706 l.f. were used, then the owner would pay only for the 706 l.f. installed and not the 715 in the original quantity survey. This type of adjustment is quite common. When errors do occur and there is a large difference between the original quantity survey and the actual number of units, an adjustment is made. Small adjustments are usually made at the same unit rate that the contractor bid. Large errors may require that the unit price be renegotiated.

Professional quantity surveyors prepare the quantities when they are used outside the U.S.A. In the U.S.A., the highway quantity surveys are done by estimators within the government department, such as the local, state, or federal department of transportation.

NOTE: For general construction work in the United States of America each construction firm makes its own estimate of the amount of materials required to build the project. This means that the quantities may vary considerably from firm to firm and chances for error are increased.

1-5 TYPES OF BIDS

Basically, there are two bidding procedures by which the contractor gets to build a project for an owner:

1. Competitive bidding
2. Negotiated bidding.

Competitive bidding involves each contractor submitting lump-sum bids (prices) in competition with other contractors to build the project. In most cases the lowest lump-sum bidder is awarded the contract to build the project, as long as the bid form and proper procedures have been followed and it is a reputable firm. (Refer also to Sec. 3-3.) Most commonly, the bids must be delivered to the person or place specified by a time stated in the Instruction to Bidders (Secs. 1-8 and 3-4).

Negotiated bidding involves the contractor working with the owner (or through the owner's architect-engineer) to arrive at a mutually acceptable price for the construction of the project. It often involves negotiations back and forth on materials used, sizes, finishes, and similar items which affect the price of the project. Owners may negotiate with as many contractors as they wish. This type of bidding is often used when owners know what contractor they would like to build the project, in which case a competitive bidding would waste time. The biggest disadvantage of this arrangement is that the contractor may not feel the need to work quite as hard to get the lowest possible prices as when a competitive bidding process is used.

Cost-plus agreements, also used in many projects, have the owner pay the contractor all of the costs of construction plus some kind of percentage or fee. This type of arrangement is thoroughly discussed in Sec. 2–5. Both competitive and negotiated bids require extensive material, labor, and equipment estimates. The cost-plus arrangement may not require so detailed a estimate if it is for a small project, such as a residence. However, on large projects the contractor may be required to submit detailed estimates for the work as the project progresses.

1–6 COST ANALYSIS METHODS

In competitive bidding, when you are submitting a proposal on a project, you want to be low bidder and still make a profit, and you can't afford to take chances with shortcuts. In order to obtain accurate estimates, you must take the extra time to figure the work in as much detail as is required to provide accurate material and labor costs. Shortcuts may be acceptable when preparing preliminary estimates, but you should never use them when you are involved in competitive bidding.

1. *Detailed Method.* The detailed method includes determination of the quantities and costs of everything required to complete the work including materials, labor, equipment, insurance, bonds, and overhead, as well as an estimate of profit. This is the only method that should be used for competitive bidding. Each item of the project should be broken down into its parts and estimated. Each piece of work has a distinct labor requirement that should be estimated. Then, during, and after the job, it is possible to compare estimated and actual costs of doing the work.

   In this manner the operation in question can be analyzed with an eye to cost control on future projects. This type of estimate also requires that costs be specifically allowed for the various types of equipment that may be required on the job.

   The detailed estimate must establish the estimated quantities and costs of material, the time required for and costs of labor, the equipment required and its cost, the items required for overhead and the cost of each item, and the percent of profit desired considering the investment, the time to complete, and the complexity of the project.

2. *Preliminary Methods: Area and Volume Methods.* The *volume method* involves computing the number of cubic feet contained in the building and multiplying that volume by an assumed cost per cubic foot. Using the *area method,* you compute the square footage of the building and multiply that area by an assumed cost per square foot. Both methods require skill and experience in adjusting the unit cost to the varying conditions of each project. The chance for error is high. As a note of warning, it should be added that most of the unit prices published in

manuals do not include the site improvements, landscaping, utilities outside the building, unusual foundations, or cost of the land or furnishings.

In using published unit costs, select a project most nearly like the one you are estimating with regard to size, proposed use, site and soil conditions, mechanical conditions, and climate. Always add a contingency allowance of 5 to 10 percent.

These methods are preliminary (approximate) methods, and are used by architects and engineers for first consultations with clients and by contractors to guide their thinking in the first approach to a probable cost of a building. Contractors will also compare the approximate methods with the final detailed bid and, if a large discrepancy exists, they usually recheck the bid.

In using the area and volume methods, you *must* keep comparative data completely up to date in your office. At the end of each job you should update all records. Computers are especially useful for such tasks.

1-7 THE CONTRACT DOCUMENTS

The bids submitted for any construction project are based on the contract documents. If estimators are to prepare a complete and accurate estimate, they must become familiar with all of the documents. The documents are listed and briefly described in this article. Further explanations of the portions and how to bid them are contained in later chapters.

The Contract Documents comprise the *Owner-Contractor Agreement,* the *General Conditions of the Contract,* the *Supplementary General Conditions,* and the *Drawings* and *Specifications,* including all *Addenda* incorporated in the documents before their execution. All of these taken together form the *contract.*

Working drawings. Those actual plans (drawings, illustrations) from which the project is to be built. They contain the dimensions and locations of building elements and materials required and delineate how they fit together.

General conditions. These define the rights, responsibilities, and relations of all parties to the construction contract.

Supplementary general conditions. *(Special Conditions)* Since conditions vary by locality and project, these are used to amend or supplement portions of the General Conditions.

Specifications. *Specifications* are written instructions concerning project requirements that describe the quality of materials to be used and the results to be provided by the application of construction methods.

Agreement. The document that formalizes the construction contract, it is the basic contract and incorporates by reference all of the other documents and makes them part of the contract; it states the contract sum.

Addenda. The *addenda* statements or drawings that modify the basic contract documents after they have been issued to the bidder, but prior to the taking of bids. They may provide clarification, correction, or changes in the other documents.

1-8 BIDDING INFORMATION

There are several sources of information pertaining to the projects available for bidding. Public advertising (Advertisement for Bids) is required for many public contracts. The advertisement is generally placed in newspapers, trade magazines, and journals, and a notice is posted in public places. Private owners often advertise in the same manner to attract a large cross section of bidders (Fig. 1-2). Included in the advertisement is a description of the nature, extent, and location of the project; the owner; availability of bidding documents; bond requirements; and the time, manner, and place that the bids will be received.

FRANK R. DAGOSTINO

1098 BRIGHTWOOD LANE

HARTFORD, CONNECTICUT

ADVERTISEMENT TO BIDDERS

Sealed proposals covering the Construction, Plumbing, Electrical, Heating and

Ventilating for the new two (2) story building for Design Unlimited, Inc. in

accordance with the Contract Documents.

Sealed proposals will be received by the Architect until 2:00 pm on October 11,

19—— when they will be publicly opened and read.

Bidders must provide bid security, in the amount of 10% of the bid. Bid

security may be a bid bond or certified check.

Drawings and specifications may be obtained from the Architect for a deposit of

$120.00 for each set. Deposits are refundable to all bidders.

Figure 1-2 Advertisement to Bidders

Reporting services, such as the F. W. Dodge Company, provide additional bidding information to the bidder. The *Dodge Reports* are issued for particular, defined localities throughout the country, and separate bulletins are included that announce new projects within the defined area, as well as provide a constant updating on jobs previously reported. The updating may include a listing of bidders, low bidders, awards of contracts, or abandonment of projects. In short, the updates provide information that is of concern to the contractors.

Other reporting services are the Associated General Contractors and local building contractor groups. They generally perform the same type of service as the *Dodge Reports* but are not quite as thorough nor as widespread. In most locales, the reporting services provide plan rooms where interested parties may review the drawings and specifications of current projects. While most prime contractors will obtain several sets of contract documents for bidding, the various subcontractors and material suppliers make extensive use of such plan rooms.

1-9 AVAILABILITY OF CONTRACT DOCUMENTS

There is usually a limit on the number of sets of contract documents a prime contractor may obtain from the architect-engineer, and this limitation is generally found in the *Invitation to Bid* or *Instructions to Bidders*. It is partly due to the restriction on the number of sets of contract documents that the reporting services, which make the documents available to all interested parties, provide the plan rooms. Subcontractors, material suppliers, and manufacturers' representatives can usually obtain prints of individual drawings and specifications sheets at cost upon request to the architect-engineer, but it should be noted that the cost is rarely refundable.

The architect-engineer will require a deposit for each set of contract documents obtained by the prime contractors. The deposit, which acts as a guarantee for the safe return of the contract documents, usually ranges from $10.00 to over $150.00 per set and is usually refundable. It should be realized that the shorter the bidding period, the greater the number of sets that will be required. Also, a large complex job requires extra sets of contract documents to make an accurate bid.

In order to obtain the most competitive prices on a project, a substantial number of subcontractors and material suppliers must bid the job. In order to obtain the most thorough coverage, there should be no undue restrictions on the number of sets of contract documents available. If this situation occurs, it is best to call the architect-engineer and discuss the problem.

During the bidding period, be certain that the contract documents are kept together. Never lend out portions of the documents. This practice will eliminate subcontractors' and material suppliers' complaints that they did not submit a complete proposal since they lacked part of the information required for a complete bid. The best method available to prime contractors to avoid this problem is to set aside space in their offices where the subcontractors and materials suppliers estimators may work. This will also encourage the most competitive prices since, when experienced estimators are working from an incomplete set of documents, they will protect themselves in their proposals from the unknown; in contrast, inexperienced estimators may not realize that portions are missing and thus may submit a proposal that they will later try to void. While it is true that the lowest prices are desired, it must be remembered at all times that it is in everyone's best interest for all parties concerned to make a profit. In this manner the contract documents never leave the contractor's office and are available to serve a large number of bidders efficiently.

The F. W. Dodge Company also provides a service called SCAN, which provides microfilm copies of the contract documents to subscribing members (Fig. 1-3). Primarily, large projects are covered by this service, but for subcontractors, materials

Figure 1-3 SCAN

suppliers, and manufacturers' representatives this provides an invaluable service since they are able to keep the microfilm as a record of what they were bidding and are able to work on their proposal at their convenience without waiting for other people to finish with the documents in the plan room. It provides a convenient, complete service that is adaptable to most offices.

1–10 SOURCES OF ESTIMATING INFORMATION

For matters relevant to estimating and costs, the best source of information is your past experience. This is why a careful accurate accounting system combined with accuracy in field reports is so important. All the information relating to the job can be kept track of and thus will be available for future reference. Many firms feed this information into computers, which the company may own, lease, or share time on. In this manner the information is more readily available for future reference.

There are several "guides to construction cost" manuals available. However, a word of extreme caution is offered regarding the use of the manuals. They are only *guides:* the figures should *never* be used to prepare an actual estimate. The manuals may be used as a guide in checking current prices and should enable the estimator to follow a more uniform system and save valuable time. The actual pricing in the manuals is most appropriately used in helping architects check approximate current prices and facilitate their preliminary (which in itself is only approximate) estimate accuracy.

In most areas the local construction firms have grouped together to form associations with the aim of promoting the industry. Many of these associations hold seminars that deal with the problems of construction today (including use of computers, how to obtain bonding, cold weather concrete, or other pertinent local, regional, or national projects). Some have advancement programs whereby employees of member firms are given the opportunity to attend classes (often free) in conjunction with local colleges. If such an organization is located in or near your area, become a part of it.

Technical information can often be obtained from technical or trade organizations. A list of many of these organizations and the addresses of their home offices should be kept on file. Many have regional offices throughout the world, and a note to the home office will give you the appropriate address.

1-11 COMPUTERS

The amazing world of computers can assist the estimator in several ways. The computer may be used to analyze labor and material costs from past projects, project labor and material costs to future projects, and facilitate information retrieval systems and building product data retrieval.

Computer estimating still involves the estimator taking off the dimensions and compiling material data. This information can then be fed to the computer to obtain the labor, equipment, and material costs that may be used to compile the final estimate. To perform this type of function, the computer must be programmed to proceed through the required steps, and appropriate unit costs must be stored in it. The computer will generate only the material that it is programmed to deliver; no less, no more. A mistake in the program or an incorrect symbol fed in, and incorrect information will be returned. The computer is one of the finest tools available to the modern construction office, but it can generate no more than it is programmed to do. It cannot exercise judgment or experience on any given item. It should be used in conjunction with the experienced judgment of estimators; it cannot replace them.

REVIEW QUESTIONS

1. What is the difference between doing quantity takeoffs and doing a full estimate?
2. Why is the detailed method of estimating the only method which is used for competitive bidding?
3. What are the preliminary methods of estimating? How may they be of help to the general construction estimator?
4. What are the contract documents? Why are they important?
5. Why is it important to bid only from a full set of contract documents?
6. What variables may affect the actual cost of construction in contrast to the way it was bid?

2

CONTRACTS, BONDS, AND INSURANCE

2-1 THE CONTRACT SYSTEM

Contracts may be awarded either by a single contract for the entire project or by separate contracts for the various phases required for completion of the project. The *single contract* comprises all of the work required for the completion of a project and is the responsibility of a single prime contractor. This centralization of responsibility provides that one of the distinctive functions of the prime contractor is to direct and coordinate all parties involved in completing the project. All of the subcontractors (including mechanical and electrical) and material suppliers involved in the project are responsible directly to the prime contractor, who in turn is responsible directly to the owner. The prime contractor is responsible for work completed in accordance with the contract documents, completion of the work on schedule, and payment of all construction costs related to the project.

Under the system of *separate contracts* the owner signs separate agreements for the construction of various portions of a project. The separate awards are often broken into the following phases:

1. General Construction
2. Plumbing
3. Heating (Ventilating, Air Conditioning)
4. Electrical
5. Sewage Disposal (if applicable)
6. Elevators (if applicable)

7. Specialities
8. Other.

In this manner the owner retains the opportunity to select the contractors for the various important phases of the project. Also, the responsibility for the installation and operation of these phases is directly between the owner and contractors rather than through the general contractor; and the owner or agents representing the owner must provide the coordination between the contractors.

There is disagreement in the construction industry itself as to which system is better. There are states that require by law the award of separate contracts when public money is involved. Most general contractor organizations favor single contracts, but in contrast, most large specialty contract groups favor separate contracts. However, in most instances the owner must decide. Under the single contract the prime contractor will include a markup on the subcontracted work for the coordination of the project. Often it is this markup that encourages the owner to use separate contracts. If the general contractor does not assume the responsibility for management and coordination of the project, someone else will have to or the result may be utter confusion. For an added fee many architects will provide this service.

On occasion the owners will attempt to coordinate the project. However, unless they have a good deal of construction experience, their efforts will prove futile as they attempt to keep each contractor's work on schedule and to get each contractor to coordinate all phases of the work.

2-2 TYPES OF AGREEMENTS

The *owner-contractor agreement* formalizes the construction contract. It incorporates, by reference, all of the other contract documents. The owner selects the type of agreement that will be signed; it may be a standard form of agreement such as the American Institute of Architects (AIA) or other organizations provide their members or a form prepared for a particular project.

The agreement generally includes a description of the project and contract sum. Other clauses pertaining to alternates accepted, completion date, bonus and penalty clauses, and any other items that should be amplified are included.

It is suggested that before the contractor actually signs a contract, a financial check of the owners should be made, not only for their ability to pay, but also for their reputation as prompt payers.

If the contractor has any questions as to the owner's ability to pay during the bidding period, a preliminary check should be made at that time.

No contract should ever be signed until it is checked by the lawyers of all parties. Each party's lawyers will normally give attention only to matters that pertain to their client's welfare. All contractors should employ the services of a lawyer who understands construction agreements and will take the time to review them carefully.

Types of agreements generally used are as follows:

1. Lump-Sum Agreement (Stipulated-Sum, Fixed-Price) (Sec. 2-3)
2. Unit-Price Agreement (Sec. 2-4)
3. Cost-Plus-Fee Agreements (Sec 2-5).

2-3 LUMP-SUM AGREEMENT (STIPULATED-SUM)

The contractor agrees to construct the project, in accordance with the contract documents, for a set price arrived at through competitive bidding or negotiation. The contractor agrees that the work will be satisfactorily completed, regardless of the difficulties encountered. This type of agreement (Fig. 2–1 and Appendix C) provides the owner advance knowledge of construction costs. The accounting process is simple and it creates centralization of responsibility in single-contract projects.

. . . agrees to build the project in accordance with the contract documents herein described for the lump sum of $275,375.00 . . . .

Figure 2-1   Lump-Sum Agreement

It is also flexible with regard to alternates and changes required on the project. However, there are some disadvantages as well. The contractor must complete the work at a guaranteed price even though the costs involved may be more than estimated. Other disadvantages include the possibility that unscrupulous contractors may attempt to obtain maximum profit by cutting the quality of work and materials on the project, and the fact that, based on the lowest competitive bid, an incompetent contractor may be selected.

Because of the very nature of the lump-sum price, it is important that the contractor be able to understand accurately and completely the work required on the project at the time of bidding.

2-4 UNIT PRICE AGREEMENT

In a *unit price agreement* the contractor bases the bid on estimated quantities of work and on completion of the work in accordance with the contract documents. The quantity takeoff is made by the government agency for highway projects in the U.S.A. and by professional quantity surveyors in Canada and Europe. Some design-build contractors in the U.S.A. also make use of this type of contract. Part of a typical quantity survey is shown in Fig. 2–2. The number on the left ties the material in the quantity takeoff with the specifications, providing easy cross-reference.

No.	Quantity	Unit	Item
0532	12,228	Each	Brick, face
0721	150	c.y.	Concrete footings
0722	85	c.y.	Concrete slabs

Figure 2-2   Typical Quantity Survey

All bidders will base their bids on the quantities listed and will bid on a unit price basis (or a price per cubic yard, lineal foot, square foot, square, etc.). In

Fig. 2–3 the illustration shows the tabulated bid (of the top three bidders) for a portion of the quantities.

No.	Quantity	Unit	Item	1	2	3
0532	12,228	Each	Brick, face	$ 0.52	0.49	0.54
0721	150	c.y.	Concrete footings	92.00	87.00	94.50
0722	85	c.y.	Concrete slabs	96.25	112.00	104.00

Figure 2-3   Tabulated Bid for Quantity Survey

Payments are made based on the price that the contractor bids for each unit of work, and field checks with measurements of work actually completed must be made by a field crew which represents the owner. This means that neither the owner nor the contractor will know the exact cost of the project until its completion. The biggest advantages of the unit price agreement are that:

1. It allows the contractors to spend most of their time working on pricing the labor and materials required for the project while checking for the most economical approach to handle the construction process.
2. Under lump-sum contracts each contractor does a quantity takeoff, which increases the chances for quantity errors considerably. There is less of a chance for error in the unit price bidding since all contractors use the same quantities and can devote more time to pricing. Of course, errors in pricing can and do happen in both types of bidding.

Unit pricing and quantity surveying are also discussed in Sec. 1–4.

2–5 COST-PLUS-FEE AGREEMENTS

In *cost-plus-fee agreements* the contractor is reimbursed for the construction costs as defined in the agreement. However, the contractor is not reimbursed for all items, and complete understanding of reimbursable and nonreimbursable items is required. This agreement is often used when speed and high quality have a higher priority than the desire for the lowest possible price. It is often used when construction must begin before the drawings and specifications are completed. Extensive accounting is required, and there should be an advance understanding of what type of accounting methods shall be followed. Work out in advance all details concerning record keeping, purchasing, and the procedure that all concerned will follow.

There are a variety of types of fee arrangements, any of which may be best in a given situation. The important point is that whatever the arrangement, it must be clearly understood by all parties, not only the amount of the fee, but also how it will be paid to the contractor throughout the period of the contract.

Percentage fee.   The owner has the opportunity to profit if prices go down and changes in the work may be readily made.

The major disadvantage is that the fee increases with construction costs, so there is no desire on the contractor's part to keep costs low.

Fixed fee.   The advantages include the owner's ability to reduce construction time by beginning construction before the drawings and specifications are completed, and removal of the temptation for the contractor to increase costs or cut quality while maintaining a professional status. Also, changes in the work are readily made. Among the disadvantages are that the exact cost of the project is not known in advance, extensive accounting is required, and that keeping costs low depends on the character and integrity of the contractor.

Fixed fee with guaranteed cost.   Advantages are that a guaranteed maximum cost is assured to the owner; it generally provides an incentive to contractors to keep the costs down if they will share in any savings. Again, the contractor assumes a professional status. Disadvantages include the fact that drawings and specifications must be complete enough to allow the contractor to set a realistic maximum cost. Extensive accounting is required as in all cost-plus agreements.

Sliding scale fee.   This provides an answer to the disadvantages of the percentage fee, since as the cost of the project goes up, the percent fee of construction goes down. The contractor is motivated to provide strong leadership so that the project will be completed swiftly at low cost. Disadvantages are that the costs cannot be predetermined, extensive changes may require modifications of the scale, and that extensive accounting is required.

Fixed fee with a bonus and penalty.   The contractor is reimbursed the actual cost of construction plus a fee. A target cost estimate is set up and, if the cost is less than the target amount, the contractor receives a bonus of a percentage of the savings. If the cost goes over the target figure, there is a penalty (reduction of percentage).

2-6 AGREEMENT PROVISIONS

While the exact type and form of agreement may vary, there are certain provisions that are included in all of them. Contractors (as well as their lawyers) must check each item carefully before signing the agreement.

Scope of the work.   The project, drawings, and specifications are identified; the architect is listed. The contractor agrees to furnish all material and perform all of the work for the project in accordance with the contract documents.

Time of completion.   Involved in this portion is the starting and completion time. Starting time should never precede the execution date of the contract. The completion date is expressed either as a number of days or as a specific date. If the number of days is used, it should be expressed in calendar days and not working days to avoid subsequent disagreements as to the completion date. Any liquidated damages or penalty and bonus clauses would usually be included here; they should be clearly written and understood by all parties concerned.

The contract sum.   Under a lump-sum agreement the *contract sum* is the amount of the accepted bid or negotiated amount. The accepted bid amount may be adjusted by the acceptance of alternates or by minor revisions that were negotiated with the contractor after receiving the bid. In agreements that involve cost-plus there are generally articles concerning the costs for which the owner reimburses the contractor. Customarily, not all costs are paid by the contractor and reimbursed by the owner; reimbursable and nonreimbursable items should be listed. The contractor should be certain that all costs incurred in the construction are included somewhere. Also, in cost-plus-fee agreements, the exact type of compensation should be stipulated.

Progress payments. Prime items include due dates for payments, retained percentage, work in place, and stored materials.

The *due date* for payments is any date mutually acceptable to all concerned. The maximum time the architect can hold the contractor's Application for Payment and how soon after the architect makes out the Certificate of Payment the owner must pay the contractor should be spelled out. There should also be some mention of possible contractor action if these dates are not met. Generally, the contractor has the option of stopping work. Some contracts state that if the contractor is not paid when due, the owner must also pay interest at the legal rate in force in the locale of building.

Retained percentage. It is customary for the owner to withhold a certain percentage of the payments. This is referred to as *retainage* and is protection for the owner to assure completion of the contract. The most typical retainage is 10 percent, but other percentages are also used. On some projects this retainage is continued through the first half of the project, but not through the last half.

Schedule of values. The contractor furnishes the architect-engineer with a statement that shows the amount allowed for each division of the work. Often, these are broken down into the same divisions as the specifications.

Work in place and stored materials. The *work in place* is usually geared as a percentage of the work to be completed. The amounts allowed for each division in the schedule of values are used as the base amounts due on each item. The contractor may also receive payment for materials stored on the site or some other location agreed upon in writing. The contractor may have to present proof of purchase, bills of sale, or other assurances to receive payment for materials stored off the job site.

Acceptance and final payment. This sets up a time for final payment to the contractor. When the final inspection, certification of completion, acceptance of the work, and issuance of the final payment are completed the contractor will receive the *final payment,* which is the amount of retainage withheld throughout the construction period. Many agreements are set up so that if full completion is held up through no fault of the contractor, the architect can issue a certificate for part of the final retainage.

2-7 BONDS

Often referred to as *surety bonds,* bonds are written documents that describe the conditions and obligations relating to the agreement. (In law a *surety* is one who guarantees payment of another party's obligations.) The bond is not a financial loan, but acts as an endorsement of the contractor. Under the terms of the bond, the owner is compensated for actual damages sustained from any default of the contractor. The bond guarantees that the contract documents will be complied with, and all costs relative to the project will be paid. If the contractor is in breach of contract, the surety must complete the terms of the contract and pay all costs incurred up to the face of the contract.

A corporate surety who specializes in contract bonds is most commonly used by contractors. The owner will reserve the right to approve the surety company and form of bond, since the bond is worth no more than the company's ability to pay. To eliminate the risk of nonpayment, the contract documents will on occasion require that the bonds be obtained from one specified company. To contractors this may mean doing business with an unfamiliar company, and they may be required to submit financial reports, experience records, projects, in progress and completed, as well as other material, which could make for a long delay before the bonds are approved. It is up to the owner to decide whether the surety obtained by the contractor is acceptable or to specify a company. In the latter case, the contractor has the option of complying with the contract documents or not submitting a bid on the project.

There is no standard form of surety bond that is applicable to every project. *Statutory bonds* are bond forms that conform to a particular governing statute; they vary from one jurisdiction to another. *Nonstatutory bonds* are the bonds used where a statutory form is not required. There is no standard form of bond nationally accepted, and the customary bond forms used by the surety companies are generally employed.

2-8 BID BOND

The *bid bond* insures that if a contractor is awarded the bid within the time specified, the contractor will enter into the contract and provide all other specified bonds. If they fail to do so without justification, the bond shall be forfeited to the owner. The amount forfeited shall in no case exceed the amount of the bond or the difference between the original bid and the next larger bid that the owner may in good faith accept.

These bonds are usually provided by the contractor's surety free or for a small annual service charge of from $25.00 to $100.00. The usual contract requirements for bid bonds specify that they must be 5 to 10 percent of the bid price, but higher percentages are sometimes used. Contractor should inform the surety company once the decision to bid a project is made, especially if it is a larger amount than they usually bid or if they already have a great deal of work. Once a surety writes a bid bond for a contractor, they are obligated to provide the other bonds required for the project. This is why surety companies may do considerable investigation of contractors before they will write a bid bond for them, particularly if it is a company with which they have not done business before or that has never had a bid bond before.

2-9 PERFORMANCE BOND

The *performance bond* guarantees the owner that, within limits, the contractor will perform all work in accordance with the contract documents and that the owner will receive the project built in substantial agreement with the documents. It protects the owner against default on the part of the contractor up to the amount of the bond penalty. The warranty period of one year is usually covered under the bond also.

The contractor should check the documents to see if this bond is required and in what amount and must also make the surety company aware of all requirements. Most commonly these bonds must be made out in the amount of 100 percent of the contract price.

The rates vary according to the classification of work being bid on. The standard classification breakdown is shown in Table 2-1. If the work required on the project comes under more than one classification, whichever premium rate is the highest is the one used. Almost all general construction work on buildings rates a "B" classification.

The premium rates are subject to change without notice, and it is possible to get lower rates from "preferred companies" if the contractor is acceptable to the company.

2-10 LABOR AND MATERIAL BOND

The *labor and material bond* also referred to as a *payment bond* guarantees the payment of the contractor's bill for labor and materials used or supplied on the project. It acts as protection for the third parties and the owner, who is exempted from any liabilities in connection with claims against the project.

In public works, the language of the required statute will determine whether a specific item of labor or material is covered.

Claims must be filed in accordance with the requirements of the bond used. Most often there is the limitation included in the bond that the claimant must give written notice to any two of three parties, general contractor, owner, or surety, within 90 days after the last day the claimant performed any work on the project or supplied materials to it.

2-11 MISCELLANEOUS BONDS

Subcontractor bonds. *Performance, Labor,* and *Payment bonds* are those which the subcontractors must supply to the prime contractor. They protect the prime contractor against financial loss and litigations due to the default by a subcontractor. Since these bonds vary considerably, prime contractors may require use of their own bond forms or reserve the right to approval of both the surety and form of bond.

License or permit bond. This is the bond that is required of the prime contractor where a state law or municipal ordinance requires a contractor's license or permit. The bond guarantees compliance with statutes and ordinances.

Lien bond. The bond is provided by the prime contractor and indemnifies the owner against any losses resulting from liens filed against the property.

2-12 OBTAINING BONDS

The surety company will thoroughly check a contractor out before it furnishes a bid bond. The surety checks such items as financial stability, integrity, experience,

Table 2-1 Construction Contract Classifications
For Surety Bond Premium Rates

CLASSIFICATION			
A-1	A	B	MISCELLANEOUS CONTRACTS
Doors	Airfield grading	Air conditioning	Ash removal
Fire alarms	Airfield surfacing	Airport hangers	Automatic sprinkler
Floodlights	Airfield runways	Aqueducts	system
Floors, wood and	Ceilings, metal or	Breakwaters	Bridges
composition	acoustical tile	Buildings, erection	Buildings, pre-
Gas tanks	Ducts, under-	and repair	fabricated
Guard rails	ground	Canals	Culverts
Iron work, orna-	Elevators	Dams	Demolition
mental	Glazing	Dikes	Draying
Lock gates	Golf courses	Ditches	Dredging
Machinery	Greenhouses	Docks	Garbage removal
Metal windows	Landscaping	Electrical work	Grade eliminations
Parking meters	Levees	Excavation	Hauling
Pipelines, oil or	Millwork	Filling stations	Highways
gas	Murals	Foundations	Maintenance
Police alarms	Painting	Gas mains	Overpasses
Radio towers	Parking areas	Gas piping	Roads
Refrigerating	Piping, high	Grain elevators	Shipbuilding
plants	pressure	Gunite contracts	Shoring
Signal systems,	Playgrounds	Heating and	Sidewalks, curb and
railroad	Revetments	ventilating	gutter
Signs	Roadbeds, railroad	Incinerators	Street paving
Stack rooms	Roofing	Jetties	Structural iron and
Stand pipes	Sandblasting	Locks	steel
Steel shutters	Stone furnishing	Masonry	Test borings
Street lighting	Storage tanks,	Piers	Timber cutting
Tanks, gas	metal	Piling	Underpasses
Thermostat equip-	Tennis courts	Pipelines, water	Viaducts
ment	Transmission	Plants, power	
Towers, water	lines	Plants, sewage-	
Track laying	Waterproofing	disposal	
Traffic control	Wind tunnels	Plastering	
Weatherstripping		Plumbing	
Window cleaning		Seawalls	
		Sewers	
		Spillways	
		Stone setting	
		Subways	
		Terminals	
		Tile	
		Tunnels	
		Waterworks	
		Wells	
		Wharves	

equipment, and professional ability of the firm. The contractor's relations with sources of credit will be reviewed as will financial statements current and past. At the end of the surety company's investigations, it will establish a maximum bonding capacity for that particular contractor. The investigation often takes time to complete, so contractors should apply well in advance of the time at which they desire bonding—waits of two months are not uncommon.

Each time the contractor requests a bid bond for a particular job, the application must be approved. If the contractor is below the work load limit and there is nothing unusual about the project, the application will be approved quickly. But if the maximum bonding capacity is approached, or if the type of construction is new to the particular contractor or is not conventional, a considerably longer time may be required.

The surety puts the contractor through investigations before giving a bond for a project to be sure that the contractor is not overextended. In order to be successful, the contractor requires equipment, working capital, and an organization; and none of these should be spread thin. The surety thus checks the contractor's availability of credit so that, if already overextended, the contractor will not take on a project that is too big. The surety will want to know if the contractor has done other work similar to that about to be bid upon. If so, the surety will want to know how large a project it was. The surety will encourage contractors to stay with the type of work they have the most experience in. The surety may also check progress payments and amount of work to be subcontracted.

If the surety refuses the contractor a bond, the contractor must first find out why and then attempt to demonstrate to the surety that the conditions questioned can be resolved. The contractor must remember that the surety is in business to make money and can only do so if the contractor is successful. The surety is not going to take any unnecessary chances in the decision to bond a project. At the same time, some surety companies are more conservative than others. If you feel the company is too conservative or not responsive enough to your needs as a contractor, shop around, talk with other sureties, and try to find one that will work with your organization.

If you, the contractor, are approaching a surety for the first time, pay particular attention to what services the company provides. Some companies provide a reporting service, which includes projects being bid and low bidders. Also, when a contractor is doing public work, the company can find out when the contractor can expect to get payment and in what stage it is at a given time. Select the company that seems to be the most flexible in its approach and offers the greatest service to the contractor.

2-13 INSURANCE

The contractor must provide insurance for the business in addition to the insurance required by the contract documents. The contractor's selection of an insurance broker is of utmost importance since the broker must be familiar with construction requirements and problems. The broker also must protect the contractor against the wasteful overlapping of protection, and yet there can be no gaps in the insurance coverage that might cause the contractor serious financial loss.

Copies of the insurance requirements in the contract documents should be forwarded immediately to the insurance broker. The broker should be under strict instructions from the contractor that all insurance must be supplied in accordance with the contract documents. The broker will then supply the cost of the required insurance to the contractor for consideration in the bidding proposal.

Insurance is not the same as a bond. With insurance policies, a financial responsibility is assumed for a specified loss or liability.

In addition to the insurance required by the contract documents, the contractor also has insurance requirements. By law, the contractor is required to maintain such insurance as workman's compensation, unemployment, social security, and motor vehicle insurance. Other insurances that are usually carried include fire, liability, accident, life, hospitalization, and business interruption.

No attempt will be made in this book to describe all of the various types of insurance that are available. A few of the most common types are described.

Workmen's compensation insurance. This policy provides the benefits required by law and covers employees killed or injured during the course of work. The rates charged for this insurance vary in different states, for different types of work, and for different contractors, depending on their own work records with regard to accidents. Workers should be classified correctly to keep rates as low as possible. Rates are charged as a percentage of payroll and vary considerably. The rates may range from less than 1 percent to over 30 percent, depending on the location of project and the type of work being performed. The cost of the policy is paid in full by the contractor.

Builder's risk fire insurance. This insurance protects projects under construction against direct loss due to fire and lightning. The insurance should cover temporary structures, sheds, materials, and equipment stored at the site. The cost usually ranges from $0.40 to $1.05 per $100 of valuation, depending on the project location, type of construction assembly, and the company's past experience with a contractor. If desirable, the policy may be extended to all direct loss causes by windstorms, hail, explosion, riot, civil commotion, vandalism, and malicious mischief. Also available are endorsements that cover earthquakes and sprinkler leakage.

Other policies which fall under the category of "Project and Property Insurance" are:

1. Fire insurance on the contractor's buildings
2. Equipment insurance
3. Burglary, theft, robbery insurance
4. Fidelity insurance—protects the contractor against loss caused by any dishonesty on the part of employees.

REVIEW QUESTIONS

1. Compare the advantages and disadvantages of separate versus single contracts.
2. Under separate contracts what are the responsibilities of each contractor?
3. List and briefly define the types of agreements for payment which may be used.
4. In your opinion which type agreement is best and why?
5. What are the important provisions which are included in the agreement?

6. What is retainage? Why is it used?

7. Why is it important that the estimator read the contract agreement over and understand it?

8. What are bid bonds? Are they required with all bid proposals?

9. What are performance bonds? Are they required with all bid proposals?

10. How may bonds be obtained?

11. How does insurance differ from bonds?

12. Under what circumstances may a bid bond be forfeited and what happens when it is? How does this affect the general contractor who forfeited the bond?

3

SPECIFICATIONS

3-1 SPECIFICATIONS

Specifications, as defined by the AIA (American Institute of Architects), are the written descriptions of materials, construction systems, and workmanship. The AIA further states that the quality of materials and the results to be provided by the application of construction methods are the purpose of the specifications.

All of the above is true, however, occasionally the specifications are geared to protect the architect and owner from legal suits and extra cost charges. This is one of the reasons that many contractors prefer not to do work with certain architect-engineers. The architect-engineer is supposed to show impartiality toward the owner and contractor, but unfortunately there are a few whose order of priority seems to be the architect-engineer first, the owner second, and the contractor a distant third. Of course, this is not true of the majority of architects, but it is suggested that the contractor avoid bidding on projects produced by such firms unless fully aware of what is involved.

Remember, the bid you, as contractor, submit is based on the drawings *and* specifications. You are responsible for everything contained in the specifications as well as what is covered on the drawings. Read them thoroughly; review them when it is necessary. There is a tendency to read only the portions of the specifications that refer to materials and workmanship. However, you are also responsible for anything said in the Proposal (or Bid Form), the Information to Bidders, the General Conditions, and the Supplementary General Conditions.

There is a tendency among estimators to simply skim over the specifications. Reading the average set of specifications is time consuming, but many important items are mentioned only in the specifications and not at all on the drawings. Since the specifications are part of the contract documents, the general contractor is responsible for all of the work and material mentioned in them.

The bound volume of specifications contains items ranging from the types of bonds and insurance required to the type, quality, and color of materials used on the job. A thorough understanding of the materials contained in the specifications may make the difference between being low bidder and not. On one job one of the estimators included in the bid an amount slightly over $2,000 for the cost of a performance bond. The sad part was that a performance bond was not required. By carefully reading the specifications, the bid would have been over $2,000 less for this one item. It makes you wonder what else may have been inadvertently put in (or left out).

There is no question that skimming the specifications is risky. Either your bids will be too high since you must allow for contingencies or too low from not allowing for all items in the required construction of the project.

Another example of sloppy specification analysis pertains to items that might be special. This is especially true if they are only slightly different from the standard product. An estimator called the supplier of masonry units and asked for a quote on standard units. The supplier asked what job the bid request was for and then commented that the other firms had asked for quotes on a colored masonry unit. Upon checking the specifications, it was learned that a colored unit was required. This type of mistake happens far too frequently.

The bound volume of specifications is generally presented in the following sequence:

1. Invitation to Bid (Advertisement for Bidders), Fig. 3–1
2. Instructions to Bidders, Fig. 3–2
3. Bid (or Proposal) Forms, Fig. 3–3
4. Form of Owner-Contractor Agreement
5. Form of Bid Bond
6. Forms of Performance Bonds
7. Alternates
8. General Conditions of the Contract
9. Supplementary General Conditions
10. Technical Trade Sections.

When one is using separate contracts or when engaged on a large job, it is common to use separate bound volumes of specifications for the mechanical and electrical trades.

3-2 CONSTRUCTION SPECIFICATIONS INSTITUTE (CSI)

The Construction Specifications Institute (CSI) has developed a specifications format which divides the major areas involved in building construction into 16 divisions and each division is subdivided into specific areas. For example, Division 8 covers doors and windows while the subdivision 08500 deals specifically with metal windows. This CSI format is shown in Fig. 3–4. This format has found wide acceptance in the entire construction industry including many architectural offices and sources of material, product, and design information such as: *Architectural Graphic*

INVITATION FOR BIDS

Sealed proposals will be received by the Architect, Frank R. Dagostino.

Location of Bid Opening: Office of the Architect, 1098 Brightwood Lane, Hartford Connecticut.

Bids will be received up to 2:00 P.M. Oct. 11, 19– –, for the erection and completion of:

OFFICE BUILDING FOR
DESIGNS UNLIMITED, HARTFORD, CONN.

Separate bids will be received for the following portions of the work and the deposits as noted:

General Construction	$120.00
Plumbing	$ 50.00
Heating and A/C	$ 50.00
Electrical	$ 50.00

Prime bidders may obtain one complete set of bid documents on a deposit basis. Full deposit for one set of bid documents will be refunded to those who submit a bona fide proposal for the prime contract and who return the documents in usable condition within 10 days after Bid Date.

Additional prints of drawings and specifications for prime bidders and all prints of drawings and specifications for other parties will be issued on a cost of reproduction (nonrefundable) as follows:

Drawings	$ 1.50 per sheet
Complete set of specifications	$20.00
Individual sheets of specifications	$ 0.30
Handling	$ 5.00
Postage (actual cost)	

Plans and specifications for the proposed work will be in file in the offices of the Associated General Contractors and the F. W. Dodge Corporation.

Bidders' proposals for the work must be accompanied by a paid bond for 10% of the amount of bid or a certified check made payable to the Owner. As a further alternate, bidders may submit a cash bid deposit in the amount of 10% of their bid. Said deposit may, at the discretion of the Owner, be retained in the event of failure of successful bidder to execute contract and give satisfactory surety bond as required within 10 days after award of contract.

In consideration of the Owner receiving his bid, each bidder agrees that no bid may be withdrawn after scheduled closing time for receipt of bids for a period of 30 days. It is the intention of the Owner to return all bid deposits, except the three lowest to be held until contracts have been awarded or deferred but in no event longer than 30 days.

A Performance Bond and a Labor and Material Payment Bond will be required for all contractors in an amount equal to 100% of the contract price, guaranteeing faithful performance of the contract and payment of all persons supplying labor and/or materials for the construction of the project.

Bids will be opened publicly and read in the presence of those interested. The Owner reserves the right to reject any or all bids, or to accept the lowest legal bid deemed in the best interest of the Owner, and to waive informalities.

Contracts may be awarded on the basis of separate proposals, combined proposals, and/or alternates, whichever is to the best advantage of the Owner. The award of each contract will be made to the lowest responsible bidder as soon as practicable, provided that in the selection of equipment or materials, a contract may be awarded to a responsible bidder other than the lowest in the interest of standardization or ultimate economy if such is clearly evident.

Figure 3-1 Invitation to Bid

Standards and *Sweet's File*. The CSI format also ties in easily with computer programming and cost accounting practices. While it is not necessary to memorize the major divisions of the format, their constant use will commit them to your memory.

DEFINITIONS

All definitions in the General Conditions, AIA Document A201, apply to the Information for Bidders.

Owner is Design Unlimited, Inc, Hartford, Conn.
Architect is Frank R. Dagostino, Hartford, Conn.
Engineer is Frank R. Dagostino, Hartford, Conn.
Contractor will be the successful bidder with whom a contract is signed by the Owner.

RESPONSIBILITY

By making a bid bidders represent that they have read and understand the bidding documents. It is also each bidder's responsibility to visit the site and become familiar with any local conditions which may affect the work.

BIDDING REQUIREMENTS

Submit all bids on prepared forms and at the time and place specified under "PROPOSALS."

All blanks pertinent to the contract proposal must be filled in.

BID SECURITY

Bid security, in the amount of 10% of the bid, must accompany each proposal. Bid bonds and certified checks, made payable to the Owner, are acceptable bid security. Bid security will be returned to all unsuccessful bidders after award of contract has been made by the Owner. If the successful bidder fails to enter into contract within ten (10) days after receiving written notice of the award of contract, that contractor's bid security shall be forfeited to the Owner.

PERFORMANCE BOND

The Owner reserves the right to require the bidder to furnish bonds covering the faithful performance of the Contract and the payment of all obligations arising from the performance of the Contract in such form and amount as the Owner may require.

In the event the Owner elects to require said executed bond, the Owner shall cause the Contract sum to be increased by the cost to the Contractor of said bond.

AWARD OR REJECTION OF BID

The Owner reserves the right to reject any or all bids and to waive any informality or irregularity in any bid received. The Owner reserves the right to reject a bid if the data is not submitted as required by the bidding documents, if no bid security has been furnished, or if the bid is in any way incomplete or irregular.

POST-BID INFORMATION

Within seven (7) days after the Contract signing, the Contractor shall submit an itemized statement of costs for each major item of work and a list of the proposed subcontractors, for approval.

PROPOSALS

Sealed proposals will be received by the Architect at the Architect's office, 1098 Brightwood Lane, Hartford, Conn., until 2:00 p.m. on October 11, 19--.

All proposals together with the bid security must be submitted in sealed envelope marked "PROPOSAL" and bear the title of the work and the name of the bidder.

All bidders are invited to be present at the bid opening. Proposals will be publicly opened and read aloud at that time.

Figure 3-2 Instructions to Bidders

SUBSTITUTIONS

All bids must be based on the materials and equipment described in the bidding documents.

Requests for material substitution must be submitted to the Architect, in writing, not less than ten (10) days prior to the bid date. Any approval of proposed substitution will be set forth in an addendum.

INTERPRETATIONS

All requests for interpretation or correction of any ambiguity, inconsistency, or error in the bidding documents must be made in writing not less than seven (7) days prior to the bid date.

Interpretations and corrections will be issued as an addendum by the Architect. Only written interpretations and corrections are binding. Interpretations and corrections are not binding when made verbally and not spelled out in writing.

Figure 3-2 *(cont.)*

3–3 INVITATION TO BID (ADVERTISEMENT FOR BIDS)

In public contracts, the public agencies must conform with regulations that relate to the method they use for advertising for bids. Customarily, notice of proposed bidding is posted in public places and by advertising for bids in newspapers, trade journals, and magazines. Where the advertisement is published, how often and over what period of time it will be published, vary considerably according to the jurisdictional regulations. An estimator must not be bashful. If you are interested in a certain project, never hesitate to call and ask when it will be bid, when and where the agencies will advertise for bids, or any other information that may be of importance.

Generally, the advertisement describes the location, extent, and nature of the work. It will designate the authority under which the project originated. Concerning the bid, it will give the place where bidding documents are available and list the time, manner, and place where bids will be received. It will also list bond requirements and start and completion dates of work.

In private contracts, owners often do not advertise for bidders. They may choose to negotiate with a contractor of choice, put the job out to bid on an invitation basis, or put the project out for competitive bidding. If the owner puts a job out for competitive bidding, the architect-engineer will call the construction reporting services, who will pass the information on to their members or subscribers.

3–4 INSTRUCTIONS TO BIDDERS (INFORMATION FOR BIDDERS)

Instructions to Bidders is the document that states the procedures to be followed by all bidders. It states in what manner the bids must be delivered, the time, date, and location of bid opening, and whether it is a public opening. (Bids may either be

GENERAL CONSTRUCTION PROPOSAL FORM
DESIGNS UNLIMITED INC.
Hartford, Connecticut

Proposals are due on October 11, 19--, at the Architect's office, 1098 Brightwood Lane, Hartford, Connecticut.

Having carefully read the Information for Bidders, General Conditions, Supplementary General Conditions, the complete General Construction Specifications Divisions 1 through 17, Drawings A/101 through A/108, HVAC/1 through HVAC/3, PL/1 through PL/4 and E/1 through E/4, all dated September 29, 19--, for the new building for Designs Unlimited, and any addenda issued on the work, the undersigned submits proposals to furnish all labor and materials for the complete General Construction Work as follows:

BASE BID: _____ ($ _____) Dollars

_____

ALTERNATE PROPOSALS
Alternate No. One _____ ($ _____) Dollars
Alternate No. Two _____ ($ _____) Dollars

In accordance with the provision of the Information for Bidders, the estimated date of substantial completion is _____ .

ADDENDA: Receipt of Addendum No. _____
_____ is hereby acknowledged.

The Undersigned certifies that the Contract Documents have been carefully examined and that the site of the work has been personally visited and that the amount and nature of the work to be done is understood and that at no time will a misunderstanding of the Contract Documents be pleaded.

ACCEPTANCE OF PROPOSAL: The Undersigned understand(s) that the Owner reserves the right to accept or reject any or all Proposals, but that if written notice of the acceptance of the above Proposal is mailed, telegraphed, or delivered to the Undersigned within thirty (30) days after the formal opening of bids, or any time thereafter before this proposal is withdrawn, the Undersigned will enter into, execute, and deliver a contract within five (5) days after the date of said mailing, telegraphing, or delivering of such notice.

The above proposals are submitted by:

Firm: _____

By: _____

Titles: _____

Address: _____

Dated: _____ , 19____

Note:

Contractor to submit his Proposal in duplicate and retain the triplicate.

Figure 3-3 Bid or Proposal Form

opened publicly and read aloud or opened privately.) The Instructions to Bidders states where the drawings and specifications are available and the deposit required. It also lists the form of Owner-Contractor agreement to be used, bonds required, times of starting and project completion, and any other bidder requirements.

Each set of Instructions to Bidders is different, so read them all carefully. In reading the Instructions to Bidders, you should especially note several items.

BIDDING REQUIREMENTS, CONTRACT FORMS, AND CONDITIONS OF THE CONTRACT

00010	PRE-BID INFORMATION
00100	INSTRUCTIONS TO BIDDERS
00200	INFORMATION AVAILABLE TO BIDDERS
00300	BID FORMS
00400	SUPPLEMENTS TO BID FORMS
00500	AGREEMENT FORMS
00600	BONDS AND CERTIFICATES
00700	GENERAL CONDITIONS
00800	SUPPLEMENTARY CONDITIONS
00850	DRAWINGS AND SCHEDULES
00900	ADDENDA AND MODIFICATIONS

Note: Since the items listed above are not specification sections, they are referred to as "Documents" in lieu of "Sections" In the Master List of Section Titles, Numbers, and Broadscope Explanations.

SPECIFICATIONS

DIVISION 1—GENERAL REQUIREMENTS

01010	SUMMARY OF WORK
01020	ALLOWANCES
01025	MEASUREMENT AND PAYMENT
01030	ALTERNATES/ALTERNATIVES
01040	COORDINATION
01050	FIELD ENGINEERING
01060	REGULATORY REQUIREMENTS
01070	ABBREVIATIONS AND SYMBOLS
01080	IDENTIFICATION SYSTEMS
01090	REFERENCE STANDARDS
01100	SPECIAL PROJECT PROCEDURES
01200	PROJECT MEETINGS
01300	SUBMITTALS
01400	QUALITY CONTROL
01500	CONSTRUCTION FACILITIES AND TEMPORARY CONTROLS
01600	MATERIAL AND EQUIPMENT
01650	STARTING OF SYSTEMS/COMMISSIONING
01700	CONTRACT CLOSEOUT
01800	MAINTENANCE

DIVISION 2—SITEWORK

02010	SUBSURFACE INVESTIGATION
02050	DEMOLITION
02100	SITE PREPARATION
02140	DEWATERING
02150	SHORING AND UNDERPINNING
02160	EXCAVATION SUPPORT SYSTEMS
02170	COFFERDAMS
02200	EARTHWORK
02300	TUNNELING
02350	PILES AND CAISSONS
02450	RAILROAD WORK
02480	MARINE WORK
02500	PAVING AND SURFACING
02600	PIPED UTILITY MATERIALS
02660	WATER DISTRIBUTION
02680	FUEL DISTRIBUTION
02700	SEWERAGE AND DRAINAGE
02760	RESTORATION OF UNDERGROUND PIPELINES
02770	PONDS AND RESERVOIRS
02780	POWER AND COMMUNICATIONS
02800	SITE IMPROVEMENTS
02900	LANDSCAPING

DIVISION 3—CONCRETE

03100	CONCRETE FORMWORK
03200	CONCRETE REINFORCEMENT
03250	CONCRETE ACCESSORIES
03300	CAST-IN-PLACE CONCRETE
03370	CONCRETE CURING
03400	PRECAST CONCRETE
03500	CEMENTITIOUS DECKS
03600	GROUT
03700	CONCRETE RESTORATION AND CLEANING
03800	MASS CONCRETE

DIVISION 4—MASONRY

04100	MORTAR
04150	MASONRY ACCESSORIES
04200	UNIT MASONRY
04400	STONE
04500	MASONRY RESTORATION AND CLEANING
04550	REFRACTORIES
04600	CORROSION RESISTANT MASONRY

DIVISION 5—METALS

05010	METAL MATERIALS
05030	METAL FINISHES
05050	METAL FASTENING
05100	STRUCTURAL METAL FRAMING
05200	METAL JOISTS
05300	METAL DECKING
05400	COLD-FORMED METAL FRAMING
05500	METAL FABRICATIONS
05580	SHEET METAL FABRICATIONS
05700	ORNAMENTAL METAL
05800	EXPANSION CONTROL
05900	HYDRAULIC STRUCTURES

DIVISION 6—WOOD AND PLASTICS

06050	FASTENERS AND ADHESIVES
06100	ROUGH CARPENTRY
06130	HEAVY TIMBER CONSTRUCTION
06150	WOOD-METAL SYSTEMS
06170	PREFABRICATED STRUCTURAL WOOD
06200	FINISH CARPENTRY
06300	WOOD TREATMENT
06400	ARCHITECTURAL WOODWORK
06500	PREFABRICATED STRUCTURAL PLASTICS
06600	PLASTIC FABRICATIONS

DIVISION 7—THERMAL AND MOISTURE PROTECTION

07100	WATERPROOFING
07150	DAMPROOFING
07190	VAPOR AND AIR RETARDERS
07200	INSULATION
07250	FIREPROOFING
07300	SHINGLES AND ROOFING TILES
07400	PREFORMED ROOFING AND CLADDING/SIDING
07500	MEMBRANE ROOFING
07570	TRAFFIC TOPPING
07600	FLASHING AND SHEET METAL
07700	ROOF SPECIALTIES AND ACCESSORIES
07800	SKYLIGHTS
07900	JOINT SEALERS

DIVISION 8—DOORS AND WINDOWS

08100	METAL DOORS AND FRAMES
08200	WOOD AND PLASTIC DOORS
08250	DOOR OPENING ASSEMBLIES
08300	SPECIAL DOORS
08400	ENTRANCES AND STOREFRONTS
08500	METAL WINDOWS
08600	WOOD AND PLASTIC WINDOWS
08650	SPECIAL WINDOWS
08700	HARDWARE
08800	GLAZING
08900	GLAZED CURTAIN WALLS

DIVISION 9—FINISHES

09100	METAL SUPPORT SYSTEMS
09200	LATH AND PLASTER
09230	AGGREGATE COATINGS
09250	GYPSUM BOARD
09300	TILE
09400	TERRAZZO
09500	ACOUSTICAL TREATMENT
09540	SPECIAL SURFACES
09550	WOOD FLOORING
09600	STONE FLOORING
09630	UNIT MASONRY FLOORING
09650	RESILIENT FLOORING
09680	CARPET
09700	SPECIAL FLOORING
09780	FLOOR TREATMENT
09800	SPECIAL COATINGS
09900	PAINTING
09950	WALL COVERINGS

DIVISION 10—SPECIALTIES

10100	CHALKBOARDS AND TACKBOARDS
10150	COMPARTMENTS AND CUBICLES
10200	LOUVERS AND VENTS
10240	GRILLES AND SCREENS
10250	SERVICE WALL SYSTEMS
10260	WALL AND CORNER GUARDS
10270	ACCESS FLOORING
10280	SPECIALTY MODULES
10290	PEST CONTROL
10300	FIREPLACES AND STOVES
10340	PREFABRICATED EXTERIOR SPECIALTIES
10350	FLAGPOLES
10400	IDENTIFYING DEVICES
10450	PEDESTRIAN CONTROL DEVICES
10500	LOCKERS
10520	FIRE PROTECTION SPECIALTIES
10530	PROTECTIVE COVERS
10550	POSTAL SPECIALTIES
10600	PARTITIONS
10650	OPERABLE PARTITIONS
10670	STORAGE SHELVING
10700	EXTERIOR SUN CONTROL DEVICES
10750	TELEPHONE SPECIALTIES
10800	TOILET AND BATH ACCESSORIES
10880	SCALES
10900	WARDROBE AND CLOSET SPECIALTIES

DIVISION 11—EQUIPMENT

11010	MAINTENANCE EQUIPMENT
11020	SECURITY AND VAULT EQUIPMENT
11030	TELLER AND SERVICE EQUIPMENT
11040	ECCLESIASTICAL EQUIPMENT
11050	LIBRARY EQUIPMENT
11060	THEATER AND STAGE EQUIPMENT
11070	INSTRUMENTAL EQUIPMENT
11080	REGISTRATION EQUIPMENT
11090	CHECKROOM EQUIPMENT
11100	MERCANTILE EQUIPMENT
11110	COMMERCIAL LAUNDRY AND DRY CLEANING EQUIPMENT
11120	VENDING EQUIPMENT
11130	AUDIO-VISUAL EQUIPMENT
11140	SERVICE STATION EQUIPMENT
11150	PARKING CONTROL EQUIPMENT
11160	LOADING DOCK EQUIPMENT
11170	SOLID WASTE HANDLING EQUIPMENT
11190	DETENTION EQUIPMENT
11200	WATER SUPPLY AND TREATMENT EQUIPMENT
11280	HYDRAULIC GATES AND VALVES
11300	FLUID WASTE TREATMENT AND DISPOSAL EQUIPMENT
11400	FOOD SERVICE EQUIPMENT
11450	RESIDENTIAL EQUIPMENT
11460	UNIT KITCHENS
11470	DARKROOM EQUIPMENT
11480	ATHLETIC, RECREATIONAL AND THERAPEUTIC EQUIPMENT
11500	INDUSTRIAL AND PROCESS EQUIPMENT
11600	LABORATORY EQUIPMENT
11650	PLANETARIUM EQUIPMENT
11660	OBSERVATORY EQUIPMENT
11700	MEDICAL EQUIPMENT
11780	MORTUARY EQUIPMENT
11850	NAVIGATION EQUIPMENT

DIVISION 12—FURNISHINGS

12050	FABRICS
12100	ARTWORK
12300	MANUFACTURED CASEWORK
12500	WINDOW TREATMENT
12600	FURNITURE AND ACCESSORIES
12670	RUGS AND MATS
12700	MULTIPLE SEATING
12800	INTERIOR PLANTS AND PLANTERS

DIVISION 13—SPECIAL CONSTRUCTION

13010	AIR SUPPORTED STRUCTURES
13020	INTEGRATED ASSEMBLIES
13030	SPECIAL PURPOSE ROOMS
13080	SOUND, VIBRATION, AND SEISMIC CONTROL
13090	RADIATION PROTECTION
13100	NUCLEAR REACTORS
13120	PRE-ENGINEERED STRUCTURES
13150	POOLS
13160	ICE RINKS
13170	KENNELS AND ANIMAL SHELTERS
13180	SITE CONSTRUCTED INCINERATORS
13200	LIQUID AND GAS STORAGE TANKS
13220	FILTER UNDERDRAINS AND MEDIA
13230	DIGESTION TANK COVERS AND APPURTENANCES
13240	OXYGENATION SYSTEMS
13260	SLUDGE CONDITIONING SYSTEMS
13300	UTILITY CONTROL SYSTEMS
13400	INDUSTRIAL AND PROCESS CONTROL SYSTEMS
13500	RECORDING INSTRUMENTATION
13550	TRANSPORTATION CONTROL INSTRUMENTATION
13600	SOLAR ENERGY SYSTEMS
13700	WIND ENERGY SYSTEMS
13800	BUILDING AUTOMATION SYSTEMS
13900	FIRE SUPPRESSION AND SUPERVISORY SYSTEMS

DIVISION 14—CONVEYING SYSTEMS

14100	DUMBWAITERS
14200	ELEVATORS
14300	MOVING STAIRS AND WALKS
14400	LIFTS
14500	MATERIAL HANDLING SYSTEMS
14600	HOISTS AND CRANES
14700	TURNTABLES
14800	SCAFFOLDING
14900	TRANSPORTATION SYSTEMS

DIVISION 15—MECHANICAL

15050	BASIC MECHANICAL MATERIALS AND METHODS
15250	MECHANICAL INSULATION
15300	FIRE PROTECTION
15400	PLUMBING
15500	HEATING, VENTILATING, AND AIR CONDITIONING (HVAC)
15550	HEAT GENERATION
15650	REFRIGERATION
15750	HEAT TRANSFER
15850	AIR HANDLING
15880	AIR DISTRIBUTION
15950	CONTROLS
15990	TESTING, ADJUSTING, AND BALANCING

DIVISION 16—ELECTRICAL

16050	BASIC ELECTRICAL MATERIALS AND METHODS
16200	POWER GENERATION
16300	HIGH VOLTAGE DISTRIBUTION (Above 600-Volt)
16400	SERVICE AND DISTRIBUTION (600-Volt and Below)
16500	LIGHTING
16600	SPECIAL SYSTEMS
16700	COMMUNICATIONS
16850	ELECTRIC RESISTANCE HEATING
16900	CONTROLS
16950	TESTING

Figure 3-4   CSI Specifications Format

The following comments are presented to help guide your future analysis of this type of document. The Instructions to Bidders must be checked thoroughly for each project.

Proposals.   Be sure to check exactly where the bids are being received. Be sure to check each addendum to see if the time or location has been changed. It is rather embarrassing (as well as unprofitable) to wind up in the wrong place at the right time, or the right place at the wrong time.

Did you hear the one about the estimator who never thoroughly read the instructions to bidders and assumed the bid opening was at the architect's office (as was customary for this particular architect)? Imagine the horror when the bids were due at 2:00 P.M. and at 1:30 the estimator pulled into a near-empty parking lot. A quick check in the architect's office revealed that the bids were being opened at the owner's office, which was located about 15 minutes away. Fortunately, the bid arrived in time, but there was no happy soap opera ending since it was not the low bid anyway. It would have been no joke had the bid arrived late, since it is extremely rare for the architect-engineer to accept a bid that is even a minute late.

Commencement and completion.   Work on the project shall commence within a specified period after execution of the contract. The contractor will have to determine the number of calendar days to complete the project. Be careful to be realistic in your completion date, most contractors have a tendency to be overly optimistic with work schedules. At the same time, don't be too conservative either, since it may cause the owner to express concern over your ability to expedite the work.

Responsibility of bidders.   Read these thoroughly—they indicate that you should carefully check to be sure you have received *all* of the drawings and specifications. They clearly put the burden on you to visit the site.

Award or rejection of bid.   The owner reserves the right to:

1. reject any or all bids.
2. accept a bid that is not the lowest.
3. reject any proposal not prepared and submitted in accordance with the contract documents.

These are common stipulations; in effect, owners may contract to whichever bidder they select. Needless to say, it causes some hard feelings when it happens.

3–5 BID (PROPOSAL) FORMS

A prepared *proposal form* is included in the bound set of specifications, and the contractor must use this form to present a bid. By using a prepared bid form, the owner can evaluate all bids on the same basis.

The proposal form stipulates the price for which the contractor agrees to perform all of the work described in the contract documents. It also assures that if the owner accepts the proposal within a certain time period, the contractor must enter into an agreement or the owner may keep the bid security as liquidated damages.

The proposal must be submitted according to the requirements of the instructions to bidders. Any deviation from these requirements for the submission of the proposal may result in the proposal being rejected.

Be sure you fill in all blanks of the proposal, acknowledge receipt of all addenda, submit required number of copies of the proposal, supply proper bid security, and be at the right place at the correct time. Countless bids have never been opened because they were delivered a few minutes late.

3-6 FORM OF OWNER-CONTRACTOR AGREEMENT

This portion of the specifications spells out exactly what type or form of agreement between owner and contractor will be used. The agreement may be a standard form published by the AIA in which case reference may be made to it and a copy may not be actually included. Various agencies of the government that are controlling the work usually have their own forms of agreement; the same is true of many corporations. In these cases, a copy of the agreement is included in the specifications. If the contract agreement is unfamiliar to you, make sure your lawyer reviews it before you submit your bid. If the form of agreement is unacceptable to you, the contractor, you may prefer not to bid that particular project. Types of agreements are discussed in Chapter 2.

3-7 GENERAL CONDITIONS

The *General Conditions* assembled and published by the American Institute of Architects are the most commonly used standard form. They have found wide acceptance throughout the industry. Many branches of government, as well as large corporations, have also assembled their own versions of general conditions. Contractors must read each article carefully and make appropriate notes as they proceed. In some cases it is best for the contractor to give a copy of the general conditions to a lawyer for review and comment. In the event the contractor decides it is not in their best interests to work under the proposed set of general conditions, the architect-engineer should be informed of the reasons and ask if the architect would consider altering the conditions. If not, the contractor may decide it is best not to bid a particular project. Typically, all general conditions will include, in one form or another, the 14 topics included in the AIA General Conditions. A copy of the AIA General Conditions is included in Appendix C. The General Conditions clearly spell out the rights and responsibilities of all the parties. Obviously, the most stringent demands are placed on the contractors since they are entrusted with the responsibility of actually building the project.

3-8 SUPPLEMENTARY GENERAL CONDITIONS

The *Supplementary General Conditions* of the specifications amends or supplements portions of the General Conditions. It is through these supplemental conditions that the General Conditions are geared to all of the special requirements of geography, local requirements, and individual project needs. Part of the supplemental conditions annul or amend the articles in the General Conditions, while the remaining portion adds extra articles.

You must carefully check the supplementary conditions to add. Each set of Supplementary General Conditions is different; they cannot be standardized.

Items that may be covered in this section include insurance, bonds, and safety requirements. Also included may be comments concerning:

1. Pumping and shoring
2. Dust control
3. Temporary offices
4. Temporary enclosures
5. Temporary utilities
6. Temporary water
7. Material substitution
8. Soil conditions
9. Signs
10. Cleaning
11. Shop drawings—drawings that illustrate how specific portions of the work shall be fabricated and/or installed
12. Surveys.

As you review the supplementary general conditions you will note that many of the requirements included in them will cost money and an amount will have to be included in the estimate to cover those costs. Once you are actually doing the estimate it will be necessary to carefully go through the entire supplementary general conditions, note all items which must be covered in the bid, and decide on how much to allow for each item.

3-9 TECHNICAL TRADE SECTIONS

The *technical trade sections* are generally put in the same order as the work will be done in the field. There is an attempt to subdivide them so that they correspond with union jurisdictions, but often this is impossible. These specifications include the type of materials required and the result that is expected. Frequently, the specifications also stipulate the method that must be used to obtain the result. When a particular method is specified, the contractor should base bids accordingly.

The material portions of the specifications usually mention the physical properties, performance requirements, handling, and storage requirements. Often, specific brands or types of material are listed as the standard of quality required. Sometimes two or three acceptable brands are specified, and the contractor has a choice of which to supply. If the contractor wishes to substitute another manufacturer's materials, it must be done in accordance with the contract documents.

Results that may be specified include items such as the texture of the material, appearance, noise reduction factors, allowable tolerances, heat loss factors, and colors.

3-10 ALTERNATES

In many projects the owner requests prices for alternate methods or materials of construction (Fig. 3–5). These alternates are generally spelled out on a separate listing in the specifications and they are listed on the proposal form. The alternates may be either an *add price* or a *deduct price,* which means that you either add the price to the base bid or you deduct it from the base bid. Be sure the price you include for any alternates is complete and includes all taxes, overhead, and profit. When an owner has a limited budget, the system of alternates allows a choice of what to spend money on.

1C—ALTERNATES

ALTERNATE NO. ONE

The Contractor shall state in the Proposal Form the amount to be added or deducted from the base bid should he furnish and apply two (2) coats of Toch Brothers Masonry Paint, or approved equal, or externally exposed concrete block, and eliminate the Tonecrete on the concrete block.

ALTERNATE NO. TWO

The Contractor shall state in the Proposal Form the amount to be added or deducted from the base bid should he furnish and apply paint (as specified in Section 11C) in lieu of Cold Glazed Cement on gypsum board.

Figure 3-5   Alternates

Since lump-sum contracts are awarded on the basis of the total base bid, plus or minus any alternates accepted, there is always concern that the owner will select alternates in a way that will help a particular contractor become low bidder. This concern has become so great that some contractors will not bid projects with a large number of alternates. In order to relieve the contractor of this concern, many architects include in the contract documents the order of acceptance of the alternates. Alternates deserve the same estimating care and consideration as the rest of the project; do not rush through them or leave them until the last minute.

3-11 ADDENDA

The time period after the basic contract documents have been issued to the bidders and before the bids are due is known as the *bidding period.* Any amendments, modifications, revisions, corrections, and explanations issued by the architect-engineer during the bidding period are effected by issuing the *addenda* (Fig. 3–6). The statements, and any drawings included, serve to revise the basic contract documents. They notify the bidder of any corrections in the documents, interpretations required, and any additional requirements as well as other similar matters. The addenda must be in writing.

Since the addenda become part of the contract documents, it is important that all prime contractors promptly receive copies of them. Many architect-engineer offices send copies to all parties who have plans and specifications (including the

FRANK R. DAGOSTINO
1098 BRIGHTWOOD LANE
HARTFORD, CONNECTICUT

ADDENDUM NO. 1
DESIGNS UNLIMITED INC.
Oct. 1, 19--

GENERAL

Bid date has been revised to 2:00 p.m. October 18, 19--. Bid will be received at the Architect's office, 1098 Brightwood Lane, until that time. Bids will be publicly opened and read.

SPECIFICATIONS

1C Article "ALTERNATE NO. ONE"—Delete in full as written and substitute the following therefore:

The Contractor shall state in his Proposal Form the amount to be added to or deducted from his Base Bid should he furnish and install all materials as shown on drawings AL/1 and in accordance with the other bidding documents.

3B Article "MATERIALS"—Amend by adding the following paragraph thereto as follows:

All ends of the double tee planks which rest on steel beams shall have a bearing plate 1/8 x 2-3/4 x 5 inches attached to the reinforcing bar and welded, on the job, to the steel beam.

Figure 3-6 Addendum

planning rooms). The addenda are also of concern to the subcontractors, material suppliers, and manufacturers' representatives who are preparing proposals for the project, since the revisions may affect their bids.

The receipt of the addenda for a particular project by the reporting service will be noted in the information they send to their subscribers. It is suggested that the contractor call the architect-engineer's office once several days before bids are to be received, and again the day before, to check that all addenda have been received. Most proposals have a space provided in which the contractor must list the addenda received. Failure to fill this space in may result in the bid being disqualified.

3–12 ERRORS IN THE SPECIFICATIONS

Ideally, the final draft of the technical trade sections should be written concurrently with the preparation of the working drawings. The specification writer and production staff should keep each other posted on all items so that the written and graphic portions of the document complement and supplement each other.

Many architect-engineers have been very successful in achieving this difficult balance. Unfortunately, there are still offices that view specification writing as dull and dreary. Because of this feeling, they sometimes assign a person not sufficiently skilled to this extremely important task. Also, many times the specifications are put off until the last minute and then rushed so that they are published before they have been proofread. Some architect-engineers brush the errors off that arise with "we'll pick it up in an addendum."

Another practice that results in errors is the use of the specifications from one job on a second job. This involves the cutting up of an old specification, cutting a portion of the old specification out, inserting portions to cover the new job and having it typed up. Inadvertently, some item is usually left out in such cases, for instance, an allowance for brick when the job has a natural stone veneer.

The real question is what to do when you find such an error. If the error is discovered early in the bidding period and you do not need an immediate answer, note the error on a sheet specifically for that purpose and keep a list of all errors and omissions on the sheet. You are strongly urged to keep one sheet solely for the purpose of errors and omissions so that you can find them when they are needed. Most specifications require that all requests for interpretations must be made in writing and state how many days before the date set for opening bids they must be received. Check for this (often in Instructions to Bidders), and note it on your errors and omissions sheet. It is further stipulated in the specifications that all interpretations shall be made in writing in the form of addenda and sent to all bidders. In actual practice it is often accepted that estimators will telephone the architect-engineer's office and request clarifications (interpretations really). If the interpretation will materially affect your bid, be certain you receive it in writing and avoid problems later on.

If there are contradictions between the drawings and the specifications, attempt to get them resolved as early in the bidding period as possible.

In keeping a list of all discrepancies (errors) and any items you don't thoroughly understand, be sure you have notations about where on the drawings and specifications the problems occur. In this manner, when you ask for clarification, it is relatively easy to explain exactly what you want. Don't call the architect-engineer for each problem individually—call for a few at a time. Often the questions answer themselves as you become more familiar with the drawings and specifications. When calling architect-engineers, remember to be courteous. Everyone makes mistakes, and being courteous will help keep them "on your side." Besides, the information may have been there; you simply overlooked it. Don't wait until bids are due to call with questions, since verbal interpretations often will not be given, and even if they are, the person who knows the answers may not be available at the time. No matter what the project, type of estimate, or anything else, the keynotes to success are COOPERATION and ORGANIZATION.

REVIEW QUESTIONS

1. What items are contained in the specifications?
2. Why is it important for the estimator to study the specifications?
3. Why is it important to deliver the proposal at the proper time and place?
4. Where in the contract documents would you refer to determine whether bid or performance bonds are required?
5. If the proposal forms are opened by the architect-engineer and one of the blanks is not filled in, what may happen to the proposal?
6. How do Supplementary General Conditions differ from General Conditions?

7. What information is contained in the technical trade sections of the specifications?

8. What is an alternate? How is it handled in the bidding process?

9. What is an addendum? When is it used? Why is it important that all subcontractors and suppliers be familiar with it?

10. How should errors in the specifications be handled by the estimator?

4

THE
ESTIMATE

4-1 ORGANIZATION

The organization that should be maintained by the estimator cannot be overemphasized. The organization required includes the maintaining of complete and up-to-date files; it must include a complete breakdown of costs for each of the projects, both of work done by company forces and of work done by subcontractors. The information should include quantities, materials prices, labor conditions, costs, weather conditions, job conditions, delays, plant costs, overhead costs, and salaries of foremen and superintendents. All data filed must be filed in an orderly manner. Some large firms microfilm their reports and file them for future reference; other large firms put the information into a computer for quick recall. Still other firms file their data in a convenient, orderly method for quick reference. The suggested type of data filing is the outline set up for this purpose by the Construction Specifications Institute (CSI). As an added advantage this will key your files to the suggested CSI *Uniform System for Construction Specifications,* which is gaining wide acceptance among architects.

The estimate of the project being bid must be systematically done, neat, clear, and easy to follow. The estimator's work must be kept organized to the extent that in an unforeseen circumstance (such as illness or automobile accident), someone else may step in, complete the estimate, and submit a proposal on the project. If the estimator has no system, if the work done cannot be read and understood, then there is no possible way that anyone can pick up where the original estimator left off. The easiest way for you to judge the organization of a particular estimate is to ask yourself if someone else could pick it up, review it and the contract documents, and be able to complete the estimate. Ask yourself: Are the numbers labeled? Are calculations labeled? Where did the numbers come from? What materials are you estimating?

4-2 DAILY LOG

One of the most important responsibilities of anyone involved in the construction process (or any other business) is the orderly planning of what must be done and the accumulation of information of what was done. One of the most effective, informal approaches to the accumulation of this information is the daily log (Fig. 4-1).

The daily log helps plan each day, it lists (in a few words) work to be done, phone calls and personal visits that are made and received, as well as providing a convenient place to keep track of important future dates. Questions that arise concerning the past and the future, business trips, project visits, and meetings with people are only a few examples of the information the log would provide you with. If you had made a call on Wednesday and were told the person was away until the following Tuesday, you would note in the log for the following Tuesday to call him again.

The next question is what to include in the log: the answer is anything that concerns you in your work should be included. The log would contain phone calls made and received (include number and with whom you spoke), trips, work scheduled to be completed each day, bid time and date, meetings (include where and with whom), addenda received, drawings and specifications received, and similar items.

The size of the log used depends on where it will be used. The most common type is about 5 × 8 inches so that it may be conveniently carried with you when you are out of the office. In this manner dates of future meetings can be immediately checked, and the miscellaneous items to be completed during the day are not forgotten. The log should be ruled and have the day and date at the top of each page. A typical day as recorded in a daily log is shown in Fig. 4-1.

No one really wants to bother keeping a daily log. It's been said to be a bother and to serve no real purpose. True, at first the daily log seems a chore, but only until a person is accustomed to keeping it up to date. If you work in an office where no one else keeps a daily log, you can expect a few chuckles at first, but it never takes long before the jokers end up asking for information you have in your log: When did we bid a particular job? Have you noted on what date I went to a particular meeting? and so on. And be sure to keep the old daily log available for reference. Just recently it was necessary for me to check back to see who attended a particular meeting that was held three years ago. So, *start your daily log today* and begin to develop the good habit of keeping a daily log, not just as an estimator but whatever job you have.

4-3 NOTEBOOK

A notebook should be kept of each estimate prepared. The notebook should be broken down into several areas: the workup sheets, summary sheets, errors and omissions sheets, the proposals received from subcontractors, proposals received from material suppliers and manufacturers' representatives, and notes pertaining to the project. Also, a listing of all calls made to the architect-engineer should be kept

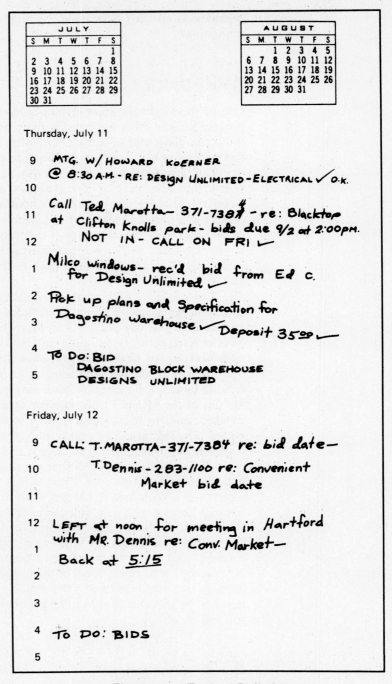

Figure 4-1 Typical Daily Log

together specifying who called, who was contacted at the architect-engineer's office, the date of the call, and what was discussed. The notebook should be neat, and easy to read and understand.

4-4 TO BID OR NOT TO BID

It is impossible for a contractor to submit a proposal for every project that bids are being taken on. Through personal contact and the reporting services, the contractor finds out what projects are out for bid and then must decide which projects to submit a proposal for. Many factors must be considered, for example the type of construction involved compared to the type of construction the contractor is usually involved in, the location of the project, the size of the project in terms of total cost and in relation to bonding capacity, the architect-engineer, the amount of work currently under construction, the equipment available, and whether qualified personnel are available to run the project.

There are also certain projects that a contractor is not allowed to submit a proposal on. The owners may accept proposals only from contractors who are invited to bid; other projects may have certain conditions pertaining to work experience or years in business that must be met.

4-5 THE ESTIMATE

Once you, the contractor, have decided to bid on a particular project, arrange to pick up or have sent to you the number of contract documents required for the estimators to prepare a proposal. A detailed estimate will be prepared for the project and a proposal will be made based on this estimate.

The estimator should proceed with the estimate in a manner that will achieve the greatest accuracy and completeness possible. In school 90 percent accuracy rates an A; in estimating 90 percent accuracy could not be tolerated. The accuracy required must be in the range of 98 to 99 percent for all major items on the estimate.

Listed below are the steps in working up a detailed estimate. These steps should form a basis for your estimating. Read and understand them.

1. Carefully check the drawings *AND* the specifications to be sure that you have everything, including all addenda. Not all architectural and engineering offices number their drawings in the same manner so sometimes there can be confusion as to whether you have all the drawings. Always take the time to number them since a missing drawing could result in ultimate bankruptcy. Architectural drawings are usually prefixed with the letter "A." If we assume there are 23 architectural drawings then they should be numbered "Sheet A-1 of 23," "Sheet A-2 of 23," and so on through the 23 drawings. (Beware—sometimes there will be a sheet A-1 and a sheet A-1A. This can occur if a sheet is added just before bidding time.)

   Structural drawings may be prefixed with the letter "S-1 of ____," or they may be included with the architectural drawings.

   Mechanical drawings may be prefixed with "M-1 of ____," "P-1 of ____," "HVAC-1 of ____," "E-1 of ____."

   On some jobs there are no prefixes before the numbers. Look over the drawings and see how the architect-engineer numbered them. Usually in the front of the specifications there is a list of all the drawings included in the set. Check this list against the drawings you received. If there are any discrepancies, check with the architect-engineer and complete your set. Follow the same procedure with the specifications: Check the list in the front of the specifications against what was received.

2. Scan the drawings to get a "feel" for the building. How large is it? What shape is it? What are the principal materials? Pay particular attention to the elevations. At this step it is important that the estimator understand the project. Make a mental note of exterior finish materials, amount of glass required, and any unusual features.

3. Review the floor plans, again getting the "feel" of the building. The estimator should begin to note all unusual plan features of the building. Look it over, follow throughout the rooms starting at the front entrance. Again, make a mental note of what type of walls are used, note whether enlarged floor plans showing extra dimensions or whether special room layouts are required.

4. Begin to examine the wall sections for a general consideration of materials, assemblies, and makeup of the building. Take special note of any unfamiliar details and assemblies; lightly circle them with a red pencil so you can readily refer to them.

5. Review the structural drawings. Note what type of systems are used, what equipment will be required. Once again, if the structural system is unusual the estimator should make at least a mental note to spend extra time on this area.

6. Review the mechanical drawings paying particular attention to how they will affect the general construction, underground work requirements, outlet requirements, chases in walls, and other items of this sort. Even under separate contracts, the mechanical portions must be checked.

7. *Specifications*—Remember that the bid you submit is based on the drawings *and* specifications. You are responsible for everything contained in the specifications as well as for what is covered on the drawings. Read and study them thoroughly, and review them when necessary. Take notes on all unusual items contained in the specifications.

8. *Visit the Site*—After making a preliminary examination of the drawings and specifications, the estimator should visit the site. The visit should be made by the estimator or other experienced person and not by a novice. The information that is obtained will influence the bidding of the project. Refer to Sec. 4–6 for the items to examine on the site.

9. *Staff Consultation*—Even though estimators must rely on their own experience in construction, it is imperative that they create and maintain a close liaison with the office personnel and field superintendents. After you, the estimator, have become familiar with the drawings and specifications, call a meeting with the people who would most likely hold the key supervisory positions if you are the successful bidder. Be sure to allow these people time, before the meeting, to become familiar with the project. During this meeting the project would be discussed in terms of construction methods that could be followed, the most desirable equipment to use, the time schedules to be followed, and personnel needed on the project.

10. Checking carefully through the General Conditions and Supplementary General Conditions, make a list of all items contained in the specifications which will affect the cost of the project.

11. Send a copy of all insurance requirements for the project to your insurance company and all bonding requirements for the project to your bonding company.

12. The estimator may now begin to take off the quantities required. Each item must be accounted for, and the estimate itself must be as thorough and complete as possible.

   The items should be listed in the same manner and with the same units of measure in which the work will be constructed on the job. Whenever possible, the estimate should follow the general setup of specifications. This work is done on a workup sheet. As each item is estimated, the type of equipment to be used for each phase should be listed. The list will vary depending on equipment owned and what is available for rent. Prices on equipment to be purchased or rented must be included.

13. At the time the estimator is preparing the quantity takeoff on workup sheets, there are several other items to be taken care of:

   a. Sending out postcards to notify subcontractors, material suppliers, and manufacturers' representatives that the company is preparing a proposal for the project and asking them if they intend to submit bids on the project.

   b. The portions of the specifications concerning insurance and bonds required should be copied and sent to the appropriate insurance and bond companies.

   c. Begin to make a list of all items of overhead that must be included in the project. This will speed up the future pricing of this item.

14. The information on the workup is carried over to the summary sheet. Work *carefully;* double-check all figures. If possible have someone go over the figures with you.

   The most common error is the misplaced decimal point. Other common errors include:

   a. errors in addition, subtraction, multiplication, and division

   b. omission of items such as materials, labor, plant, or overhead

   c. errors in estimating the length of time required to complete the project

   d. errors in estimating wastage

   e. errors in estimating quantities of materials

   f. errors in transposing numbers from one sheet to another.

15. Having priced everything, make one last call to the architect-engineer's office to check the number of addenda issued to be sure you have received them all. Double-check the time, date, and place that bids are being received. Double-check that all of the requirements for submission of the proposal have been followed; be sure the proposal is complete.

4-6 SITE INVESTIGATION

It is required by the contract documents that the contractor visit the site. The importance of the visit and the items to be checked vary, depending on the type of project and its location. As a contractor expands to relatively new and unfamiliar areas, the importance of the preliminary site investigation increases as does the list of items that must be checked. Examples of the type of information that should be collected are:

1. Site access
2. Availability of utilities (electric, water, telephone)
3. Site drainage
4. Transportation facilities
5. Any required protection or underpinning of adjacent property
6. A rough layout of the site locating proposed storage trailer and equipment locations
7. Subsurface soil conditions (bring a post hole digger to check this)
8. Local ordinances and regulations, and note any special requirements (permits, licenses, barricades, fences)
9. The local labor situation and local union rules
10. Availability of construction equipment rentals, the type and conditions of what is available as well as the cost
11. Prices and delivery information from local material suppliers (request they send you proposals for the project)
12. The availability of local subcontractors (note their names, addresses, and what type of work they usually handle)
13. The conditions of the roads leading to the project, low bridges, and load limits on roads or bridges
14. Housing and feeding facilities if workers must be imported
15. Banking facilities.

4-7 SUBCONTRACTORS

A *subcontractor* is a separate contractor, who is hired by the prime contractor to perform certain portions of the work. Specialty work, such as pile driving, elevator installation, plumbing, heating, and electrical work, is the most common type of work subcontracted out. Often, less specialized work, such as drywall construction, masonry work, and flooring, is also subcontracted out.

The amount of work that the prime contractor will subcontract out varies from project to project. Some federal and state regulations limit the proportion of a project that may be subcontracted out, but this is rarely the case in private work.

If subcontractors will be used, the contractor must be certain to notify them early in the bidding period so they have time to prepare a complete, accurate proposal. If rushed, there is a tendency for the subcontractor to bid high just for protection against what might have been missed.

The use of subcontractors can be economical, but it does not mean that an estimate need not be done for a particular portion of work. Even if the estimator intends to subcontract the work out, an estimate of the work must be prepared. It is possible that the estimator will not receive proposals for a project before the bid date and will have to use an estimated cost of the work in totaling the proposal. All subcontractors' proposals are compared with the estimator's price

(Fig. 4–2); it is important that a subcontractor's price is neither too high nor too low. If either situation exists, the estimator should call the subcontractor and discuss the proposal with him.

McBill Precast
1215 Briarhill Rd.
Hartford, Connecticut
October 1, 19--

Ace Construction
501 Hightower St.
Hartford, Connecticut

RE: Designs Unlimited
Bids Due: Oct. 11, 19--

Gentlemen:

We propose to furnish and install all precast double tees for the above mentioned project for the lump sum of: $13,250.00

Furnish double tee fillers (for ends of tees, literature enclosed). $3.75 each

All materials are bid in accordance with the contract documents, including Addendum #1.

Sincerely,

Charles McBill

Figure 4-2 Subcontractor's Proposal

The subcontractor's proposal is often phoned in to the general contractor's office at the last minute. This is done because of the subcontractor's fear that the contractor will tell other subcontractors the proposal price and encourage lower bids. This practice is commonly referred to as "bid peddling," and it is discouraged. Even when proposals are sent in writing, many subcontractors will call at the last minute and revise their price. This often leads to confusion and makes it difficult for the estimator to analyze all bids carefully.

In checking subcontractor proposals, note especially what is included and what is left out. Each subsequent proposal may add or delete items. Often the proposals set up certain conditions, such as use of water, heat, or hoisting facilities. The estimator must compare each proposal and select the one that is the most economical.

All costs must be included somewhere. If the subcontractor does not include an item in the proposal, it must be considered elsewhere.

4–8 MATERIALS

For each project being bid the contractor will request quotations from materials suppliers and manufacturers' representatives for all materials required. While on occasion a manufacturer's price list may be used, it is more desirable to obtain written quotations that spell out the exact terms of the freight, taxes, time required for delivery, the materials included in the price, and the terms of payment (Fig. 4–3). The written proposals should be checked against the specifications to make certain that the specified material was bid.

```
                                              United Block Company
                                              712 Charles Blvd.
                                              Hartford, Connecticut
                                              October 1, 19--

Ace Construction                              RE: Designs Unlimited
501 Hightower St.
Hartford, Connecticut

Dear Sir:

We are pleased to quote you on the materials required for the above referred
project.

All materials quoted meet the requirements of the drawings and specifications.

        8 x 8 x 16      Concrete Block          $1.18
        8 x 4 x 16      Concrete Block          $0.82
        8 x 6 x 16      Concrete Block          $0.98
        8 x 12 x 16     Concrete Block          $1.57
        Mortar, per bag                         $5.25
        Brick, Tru-red range                    $295.00 per M

All prices delivered to jobsite, plus sales tax.

Delivery: 24 hour notice.

Sincerely,
```

Figure 4-3   Manufacturer's Quotation

All material costs you enter on your workup and summary sheets (Sec. 4-9) must be based on delivery to the job site, not including tax. This means the total will take into account all necessary freight, storage, transportation insurance, and inspection costs. The taxes should be added on the summary sheet. Remember that the sales tax that must be paid on a project is the tax in force in the area in which the project is being built; sometimes cities have different rates than the county they are in. Take time to check.

4–9 WORKUP SHEETS

The estimator uses two basic sheets in his estimate, the *workup sheet* and the *summary sheet*. The workup sheet is used to make calculations and sketches, and to generally "work up" the cost of each item. Material and labor costs should always be estimated separately. Labor costs vary more than material costs, and the labor costs will vary in different stages of the project. For example, a concrete block will cost less for its first three feet than for the balance of its height and the labor cost goes up as the scaffold goes up—yet material costs remain the same.

When beginning the estimate on workup sheets, the estimator must be certain to list the project name and location, the date that the sheet was worked on, and the estimator's name. All sheets must be numbered consecutively, and when completed, the total number of sheets noted on each sheet (e.g., if the total number were 56, sheets would be marked "1 of 56" through "56 of 56"). The estimator must account for every sheet, since if one is lost, chances are that the costs on that sheet will never be included in the bid price. Few people can write so legibly that others may easily understand what they have written; it is therefore suggested that the work should be printed. Never alter or destroy calculations; if you wish to change them, simply draw a line through them and rewrite. Numbers that are written down must be clear beyond a shadow of a doubt. Too often a "4" can be confused with a "9," a "2" with a "7," and so on.

All work done in compiling the estimate must be totally clear and self-explanatory. It should be clear enough to allow another person to come in and follow all work completed and all computations made each step of the way.

Figure 4–4 is intended to show the correct method for listing items on the sheet.

Building	Salem Fire Station							Estimate no.	735
Location	New Salem Rd - Salem Conn							Sheet no.	3 of
Architect	HY Assoc.							Date	Nov. 15 —
Subject	Conc. Work - Ftg & Block							Estimator	3D CH. T.D.

| Description of work | Dimensions | | | Forms S.F. | Conc. C.Y. | | | |
	L	W	H					
Conc. Footings: 3000 psi								
Front	~~79~~ 72	2	1	158	6			
LEFT SIDE	68	2	1	136	5			
Rear	109	2	1	218	8			
Right Side	62	2	1	124	5			
Interior	22	2	1	44	2			
				580	26 c.y			26 c.y.
			X	2				
Forms 2×12				1160 B.F.				1160 B.F.
Conc. Pier Footings 3000 psi No.	L	W	H	FORMS S.F.	CONC C.Y.			
Ext. 2	5	3	1	32	1.5			
Int. 4	7½	7	1	112	5			
Ext. 2	4	4	1	32	1.5			
				176	8 c.y			8 c.y.
Forms 2×12 - 4 sides			X	2				
2×12				352 B.F.				352 B.F.

Figure 4-4 Workup Sheet

The measurements are taken from the plans, appropriate specification notes are included, the quantities are determined, and material costs are introduced and entered into the right-hand column, which is reserved for totals.

When taking off the quantities, make it a point to break each item down into different sizes, types, and materials. This involves checking the specifications for each item you are listing. For example, in listing concrete blocks, you must consider the different sizes required, the bond pattern, the color of the unit, and the color of the mortar joint. If any of these items varies, it should be listed separately. It is important that the take off be complete in all details; do not simply write "wire mesh," but "wire mesh 6 × 6 10/10"—the size and type are very important. If the mesh is galvanized, it will increase your material cost by about 20 percent, so note it on your sheet.

4-10 SUMMARY SHEET

All of the costs contained on the workup sheets are condensed, totaled, and included on the summary sheet. All items of labor, equipment, material, plant, overhead, and profit must be likewise included. Only the most pertinent information is mentioned: the project name, location, architect, type of building, date, initials of the estimator and checker, number of stories, an estimate number, and sheet number are placed on each sheet. Figure 4-5 gives a typical summary sheet breakdown.

RECAPITULATION

Project: GUILD HALL — Estimate no. 754
Location: WEST SALEM — Sheet No. 2 of
Architect Engineer: MAC ASSOC. — Date MARCH 12 —
Summary by Ø   Priced by Ø   Checked by D

No.	Description	Quantity	Unit	Unit Price	Total Est. Materials	Unit Price	Total Est. Labor Cost	Total
0310	Concrete Formwork	10,300	M.F. B.M.	380		390-		
		12,500	S.F.	1.40		170		
0320	Concrete Reinforcing							
	#6 bars	5,250	lbs	.26		.09		
	#3 bars	1,450	lbs	.29		.11		
	Mesh 6x6 ¹⁰/₁₀ plain	42	rolls	38⁵⁰		13 -		
0330	Cast-in-place conc.							
	Footings - 3500 psi	112	c.y.	67-		52⁵⁰		
	Slab-on-grade-2500psi	185	c.y.	62⁵⁰		47-		
0335	FINISHES							
	Slab-machine trowel	15,800	S.F.			.09		
0340	Precast Concrete							
0343	Prestressed tees	31,600	S.F.					94800-

Figure 4-5 Summary Sheet

It should be noted that the work is broken down into classifications that coincide with the company's cost account numbers, which are used in connection with its accounting systems.

The summary sheet should list all of the information required, but none of the calculations and sketches that were used on the workup sheets. It should list only the essentials, yet still provide information complete enough for the person pricing the job not to have to continually look up required sizes, thicknesses, strengths, and similar types of information.

4-11 ERRORS AND OMISSIONS

No matter how much care is taken in the preparation of the contract documents, it is inevitable that certain errors will occur. Errors in the specifications were discussed in Sec. 3-12, and the same note-taking procedure is used for all other discrepancies, errors, and omissions. The Instructions to Bidders or Supplementary General Conditions generally state that, if there are discrepancies, the specifications take precedence over drawings and dimensioned figures, and detailed drawings take precedence over scaled measurements from drawings. All important discrepancies (those that affect the estimate) should be checked with the architect-engineer's office.

REVIEW QUESTIONS

1. Keep a daily log of your activities for the next week, including the amount of time spent on each item.

2. What factors may cause contractors not to bid a particular project even though they are trying to get work?

3. Why is it important to check the contract documents to be certain you have everything?

4. What makes the site investigation important?

5. What are subcontractors and what is their relationship to the owner?

6. A subcontractor installs a piece of equipment incorrectly and it does not work; when the owner notifies the prime contractor the owner is told there is nothing the prime contractor can do about it. According to the contract documents, whose responsibility would it be to remedy the situation and why?

7. Why should material quotations be in writing? What items must be checked for in these proposals?

8. According to the AIA General Conditions, does the architect-engineer have the right to object to the use of a particular subcontractor? If so, under what circumstances?

9. Why should all subcontractors and suppliers be notified early in the bidding period if the estimator wants to request a bid from them?

10. Go to a local plan room and review a set of contract documents. Stop at the desk, introduce yourself, and explain what you are doing; there is no doubt they will allow you to use the facilities.

5
OVERHEAD
AND
CONTINGENCIES

5-1 OVERHEAD

Overhead costs are generally divided into *general overhead costs* and *job overhead costs*. The general costs include items that cannot be readily charged to any one project, while the job overhead costs include all items that can be readily charged to any one project and yet are not for materials or labor.

Overhead constitutes a large percentage of costs on the job. Failure to allow sufficient amounts for overhead has forced many firms to go out of business. Consider overhead carefully, make a complete list of all required items, and the cost of each item.

Depending on where the various costs are included, the cost of overhead can vary from 12 to 30 percent of the sum of materials, labor, and equipment. No one figure is correct, since the costs of running each office must be analyzed carefully to determine its overhead. Don't be tempted to use "average" figures for this item.

5-2 GENERAL OVERHEAD (INDIRECT)

General overhead costs are not readily chargeable to one particular project—they represent the cost of doing business and fixed expenses that must be paid by the contractor. These expenses must be shared proportionally among the projects undertaken; usually the items are estimated based on a year and reduced to a percentage of the total annual business. The following are items that should be included:

Office. Rent (if owned—the cost plus a return on investment), electricity, heat, office supplies, postage, insurances (fire, theft, liability), taxes (property, water), telephone, and office machines and furnishings.

Salaries. Office employees such as executives, accountants, estimators, purchasing agent, bookkeeper, stenographer, and anyone else working in office.

Miscellaneous. Advertising, literature (magazines, books for library), legal fees, (not applicable to one particular project), professional services (architects, engineers, CPA), donations, travel (including company vehicles), and club and association dues.

Depreciation. A separate account should be kept for expenditures on office equipment, calculators, typewriters, and any other equipment. A certain percentage of the cost is written off as depreciation each year and is part of the general overhead expense of running a business.

The list of expenses shown in Fig. 5-1 is a typical general overhead list. Obviously, for smaller contractors the list would contain considerably fewer items and

---

General Overhead Expenses For One Year

Office Rental	$ 6,000.00
Heat and Electricity	1,800.00
Telephone	850.00
Office Supplies (inc. stationery)	1,150.00
Postage	320.00
Insurance (fire and office liability)	660.00
Bookkeeper-Secretary	17,500.00
Estimator-Purchaser	32,500.00
Executive Salaries	52,000.00
Legal Services	3,000.00
Travel	4,500.00
Advertising, Donations	1,200.00
Literature	750.00
Club and Association Dues	1,200.00
Depreciation (office furniture and equipment)	2,500.00
Miscellaneous Expenses	2,750.00
	$128,680.00

Annual business approximately $1,500,000
General Overhead expense = 8.6% of total annual business
Assuming the project being bid had an estimated cost of about $450,000 the amount of general overhead chargeable to that project would be

$$\frac{450,000}{1,500,000} \times 128,680 = \$38,604 \text{ (general overhead)}$$

Figure 5-1 General Overhead Expenses for One Year

general overhead (indirect)

for large contractors it could fill pages; but the idea is the same. From this, it should be obvious that the more work that can be handled by a given setup, the smaller the amount that must be allowed for general overhead and the better the chance of being low bidder. As a contractor's operation grows and develops, particular attention should be paid to office rental expense versus that involved in purchasing a building or moving to larger quarters and the addition of staff. The original staff may be able to handle the extra work if the load is laid out carefully.

Some contractors do not allow for "General Overhead Expense" separately in their estimates; instead they figure a larger percentage for profit, or group overhead and profit together. This, in effect, "buries" or hides part of the expenses. From the estimator's point of view, it is desirable that all expenses be listed separately so they may be analyzed periodically. In this manner, the amount allowed for profit is actually figured as profit, the amount left after *all* expenses are figured.

5-3 JOB OVERHEAD (GENERAL CONDITIONS, DIRECT OVERHEAD)

Also referred to as *general conditions expense, job overhead* comprises all costs that can be readily charged to a specific project. It includes the items of job expense that cannot be charged directly to a particular branch of work, but are required to construct the project. The list of job overhead items is placed on the first of the general estimate sheets under the heading of "Job Overhead" or "General Conditions."

Itemizing each cost gives the estimator a basis for determining the amount of that expense and also provides for comparison of projects. A percentage should not be added to the cost simply to cover this item. It is important that each portion of the estimate can be analyzed for accuracy to determine whether the estimator should bid an item higher or lower next time.

Salaries. This will include the salaries paid to the project superintendent, assistant superintendent, timekeeper, material clerk, all foremen required, and security, if needed. The cost must include all traveling expenses of these people.

The salaries of the various workers required are estimated per week or per month, and that amount is multiplied by the estimated time it is expected that each will be required on the project. The estimator must be neither overly optimistic nor pessimistic in regard to the time each person will be required to spend on the job.

Temporary office. The cost of providing a temporary job office for use by the contractor and architect during the construction of the project should also include office expenses such as light, heat, water, telephone, and office equipment. Check the specification for special requirements pertaining to the office. A particular size may be required; the architect may require a temporary office, or other requirements may be included.

If the contractor owns the temporary office, a charge is still made against the project for depreciation and return on investment. If the temporary office is rented, the rental cost is charged to the project. Since the rental charges are generally based on a monthly fee, estimate the number of months required carefully. At $250.00 per month, three extra months amount to $750.00 out of the profit of the project. Check whether or not the monthly fee includes setup and return of the office. If not, this must also be included.

Temporary buildings, barricades, enclosures. The cost of temporary buildings includes all toolsheds, workshops, and material storage spaces. The cost of building and maintaining barricades and providing signal lights in conjunction with the barricades must also be included. Necessary enclosures include fences, temporary doors and windows, ramps, platforms, and protection over equipment.

Temporary utilities. The costs of temporary water, light, power, and heat must also be included. For each of these items the specifications must be carefully read to determine which contractor must arrange for the installation of the temporary utilities and who will pay for the actual amounts of each item used (power, fuel, water).

Water may have to be supplied to all contractors by the general contractor. This information is included in the specifications and must be checked for each project. The contractor may be able to tie into existing water mains. In this case a plumber will have to be hired to make the connection. Sometimes the people who own the adjoining property will allow use of their water supply, usually for a fee. Water may be drawn from a nearby creek with a pump, or perhaps a water truck will have to be used. No matter where the water comes from, temporary water will be required on many projects, and the contractor must include the cost. This is also one of the items the estimator should investigate at the site. Remember that many water departments charge for the water used and the estimator will have to estimate the volume of water that will be required.

Electrical requirements may include light and power for the project. This item is sometimes covered in the electrical specifications, and the estimator must review it to be certain that all of the requirements of the project will be met. In small projects it may be sufficient to tap existing power lines run to a meter and string out extension lines through the project, from which lights will be added and power taken off. On large complexes it may be necessary to install poles, transformers, and extensive wiring so that all power equipment being used to construct the project will be supplied. If you, the estimator, find that the temporary electric required will not meet the needs, first call the architect-engineer to discuss it. The architect may decide to issue an addendum revising the specifications. If not, you will have to include the cost in the general construction estimate. All power consumed must be paid for, and how this is handled should be in the specifications; often the cost is split on a percentage basis among the contractors, and it is necessary to make an allowance for this item also.

Heat is required if the project will be under construction in the winter. Much of the construction process requires maintenance of temperatures at a certain point. The required heat may be supplied by a portable heater using kerosene and LP or natural gas for small projects—the total cost includes the costs of equipment, fuel, and required labor (remember the kerosene type must be filled twice a day). On large projects the heating system is sometimes put into use before the rest of the project is complete; however, the labor costs run quite high since it may require around the clock tending. Since most of the heaters use electricity for ignition and to run the fan, the cost of power must also be included and a place to "plug in" provided.

Sanitary facilities. All projects must provide toilets for the workers. The most common type in use are the portable toilets, which are most often rented. Large projects will require several portable toilets throughout the job.

Drinking water. The cost of providing drinking water in the temporary office and throughout the project must be included. It is customary to provide cold or ice water throughout the summer. Keep in mind that the estimated cost must include the containers, cups, and someone to service them.

Photographs. Many project specifications require photographs at various stages of the construction and, even if they are not required, it is strongly suggested a camera, film, and flashbulbs be kept at each job site. The superintendent should make use of them at all important phases of the project to record the progress. In addition to the cost of the above item, the cost of developing and any required enlarging of the film should be considered.

Surveys. If a specification requests a survey of the project location on the property, the estimator must include the cost for the survey in the estimate. Check the specifications to see if a licensed surveyor is required, and then ask several local surveyors to submit a proposal.

Cleanup. Throughout the construction's progress, rubbish will have to be removed from the project site. Estimate how many trips will be required and a cost per trip; this means the estimator will have to determine where the rubbish can be dumped.

Before acceptance of the project by the owner, the contractor will have to clean the floors, clean up the job, and in some cases even wash the windows.

Winter construction. When construction will run into or through the winter, several items of extra cost must be considered, including the cost of temporary enclosures, that of heating the enclosure, that of heated concrete and materials, and the cost of protecting equipment from the elements.

Miscellaneous items that should be contemplated include the possibility of damaging adjacent buildings, such as breaking windows, and the possible undermining of foundations or damages by workmen or equipment. All sidewalks and paved areas that are torn up or damaged during construction must be repaired. Many items of new construction require protection in order that they will not be damaged during construction; among these are wood floors, carpeting, finished hardware, and wall finishes in heavily traveled areas. New work that is damaged will have to be repaired or replaced. Often, no one will admit to damaging an item, so the contractor must absorb the cost of replacing the item.

The supplementary general conditions should be carefully checked for other requirements that will add to the job overhead expense. Examples of these may be job signs, billboards, building permits, testing of soil and concrete, and written progress reports.

5-4 SCHEDULING

One of the basic underlying effects of the cost of a project is how long it will take to complete it once the contracts have been awarded to a contractor. This length of time especially affects the estimator in regard to overhead items, wages paid to supervisory and home office personnel, rental on trailers and toilet facilities, and any guards, traffic directors, and barricading required. It also affects the estimate in terms of how long equipment will be required on the project. Traditionally, the estimator has estimated an approximate length of time and that was the basis used in the estimate.

Using computers enables us to calculate the approximate length of time required for a project, to plan what the sequence of the work trades would be on the project and when materials should be ordered and delivered to the project, and to do it in such a way that the dates and lengths of time were tabulated electronically. When carefully planned, this procedure provides an accurate guide to the estimator of how long it will take to build the project.

But use of the computer goes far beyond just the estimator. To the owner it means that the project is actually being planned out and that the estimated completion date means more than the approach of the contractor stating, "Well, it should take about six months."

On many large projects, the owner requires each contractor submitting bids to also provide a schedule of time for the work required to build the building. Commonly referred to as CPM, ProMIS, CPM precedence method, CPM sequence method, PERT, these computer programs provide a valuable service from bidding time to the end of construction.

The basics of scheduling the project break down into 5 areas:

1. List all activities required for the completion of the project.

   a. Work—each item or portion of work required must be listed. For example, it is not sufficient to list "concrete"; instead each portion of the work must be listed, such as: footings, interior slabs, sidewalks, driveways, etc.

   b. Ordering materials—while some materials are readily available and can be ordered the day before they are required, many materials require special ordering months in advance of when they are actually required. For example, special precast concrete may require months to be delivered to the project by the time shop drawings are approved, forms for the precast ordered and delivered, precast made and finished if required.

   c. Delivery of materials—while it is sometimes possible that material ordered in the morning will be delivered in the afternoon, it is not always so simple. For materials from out of state the time required for truck or train delivery may run several days. In addition, the firms may be so busy that they have their delivery schedule worked out two and three weeks ahead. In this portion of the activities it is necessary to determine about how long it will take for delivery and order the material in ample time to receive it when needed.

2. Assign the time required for each of the activities listed in step 1. It is most important that *all* of the times be accurate; inaccurate times (or lengths of time) will result in inaccurate results.

3. For each activity it is then necessary to list the succeeding activity which can be completed once it is done. For example, the concrete can be poured after the reinforcing is installed. Similarly, the reinforcing can be installed after the forms are built and the reinforcing has been delivered.

4. For easy reference (and for entry into the computer) each activity is then given a code (number, letters, or a combination of numbers and letters).

5. The information may now be fed into a properly programmed computer and the time required for the project can be obtained. The computer can also:

   a. Determine when materials must be ordered.

   b. Determine when delivery must be arranged.

c. Determine how much variation in time is available, or what is the latest date an activity can be performed and still keep the project on schedule for completion.

Continuous use of these computerized printouts may be used for:

1. Updating analysis of actual cost for each portion of work as compared with the estimate.
2. Calculating how many workers and how much equipment will be required for a portion of work.
3. Determining the totals of workers and equipment required at any given time. (Keep in mind that one front-end loader cannot work in two places at once. This helps coordinate how many of what type equipment and number of workers will be required at any given time.)
4. Preparing monthly job status cost reports, broken down into labor, materials, equipment, and overhead if desired.
5. Figuring unit cost of installed work.
6. Establishing what equipment is inactive and what is available for use.
7. Determining where each piece of equipment is being used.
8. Processing payroll and labor costs.

It is beyond the scope of this text to show the complete workings of computerized scheduling. However, by the use of a small example the basics can be explained. The small office shown in Fig. 5-2 will be used.

Now, in the general order given for breaking up scheduling:

1. List work activities (Fig. 5-3).
2. List ordering of materials required (Fig. 5-4).
3. List delivery of materials required (Fig. 5-5).
4. Assign time required for each activity.
5. Give each activity a code.
6. List succeeding activity required for each activity (Many contractors list both preceding and succeeding activities.) (Fig. 5-6).

Now the work activities may be organized into the required flow. If a computer were being used, this information would be fed into it and the scheduling would be done. But a computer isn't necessary, and the principles can be shown by following these steps:

1. Cut out pieces of paper, each with one activity on it.
2. Lay out the pieces of paper in the sequence in which the work must occur. For example, the very first item is getting the permits, which leads to the surveyor coming and laying out the project, on to removing the topsoil, then to digging the footings, and so on.
3. Once satisfied with the layout of activities, make a sketch of the chart of activities (Fig. 5-7).
4. Next, write the duration of time for each activity (Fig. 5-8).

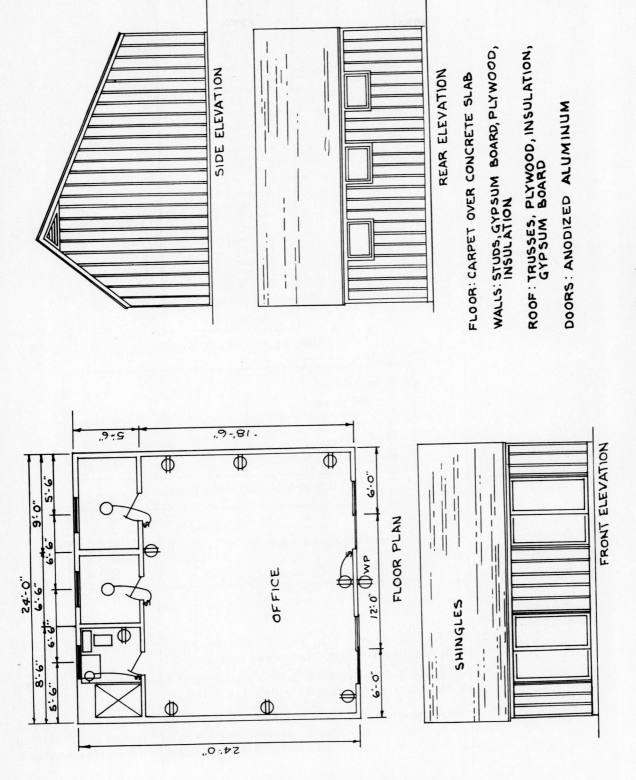

SIDE ELEVATION

REAR ELEVATION

FLOOR: CARPET OVER CONCRETE SLAB

WALLS: STUDS, GYPSUM BOARD, PLYWOOD,
INSULATION

ROOF: TRUSSES, PLYWOOD, INSULATION,
GYPSUM BOARD

DOORS: ANODIZED ALUMINUM

OFFICE

FLOOR PLAN

SHINGLES

FRONT ELEVATION

Figure 5-2   Plan and Three Elevations

59

```
Work Activities

Clear site
Scrap topsoil
Gravel fill
Plumbing, rough in
Forms, concrete slab
Pour slab and finish
Rough carpentry
Electrical, rough in
Insulation
Roofing
Plumbing, set shower
Drywall
Interior trim
Exterior trim
Telephone, rough
Plumbing, finish
Heating, rough in
Heating, finish
Painting
Stain, exterior
Carpet
Windows
Glass doors
Wood doors
Final grade
Seed
Electrical, finish
```

Figure 5-3   Work Activities

Ordering Activities	Duration	Event
Plumbing fixtures	15	15403A
8' anodizied aluminum doors	8	08100A
Windows	10	08600A
Heating unit	12	15802A
Carpet	18	09680A
Electrical fixtures	3	16012A

Figure 5-4   Materials Required

Delivery Activities	Duration	Event
8' anodizied aluminum doors	2	08100B
Windows	2	08600B
Gravel	1	03301B

Figure 5-5   Deliveries Required

5. By checking the lengths of time through the "circuits" of the chart, the time required for each circuit or path can be determined (Fig. 5-9).

6. The circuit with the most time is the "critical path" which will govern the length of time it takes to complete the project (Fig. 5-10).

Such a chart or computerized report is only as good, or as accurate, as the information used. For example, if two days are allowed for ordering and delivery of windows and it will actually take four to five weeks, it will be difficult to be accurate.

overhead and contingencies

Work Activities	Duration	Event	Succeeding Event
Clear site	1/2	02100	02200
Scrap topsoil	1/2	02200	03100
Gravel fill	1/2	03301	15401
Plumbing, rough in	2	15401	03300
Forms, concrete slab	1/2	03100	15401
Pour slab and finish	2	03300	06100
Rough carpentry	3	06100	15402, 08100, 16700, 16011, 15801, 07300
Electrical, rough in	1	16011	07200
Insulation	1-1/2	07200	09250
Roofing	1	07300	07200, 06201
Plumbing, set shower	1/2	15402	09250
Drywall	4	09250	08200
Interior trim	2	06200	09900
Exterior trim	1-1/2	06201	09901
Telephone, rough	1/2	16700	09250
Plumbing, finish	1	15403	02201
Heating, rough in	2-1/2	15801	09250
Heating, finish	1-1/2	15802	
Painting	3-1/2	09900	15802, 16012, 09680
Stain, exterior	2	09901	02201
Carpet	1/2	09680	15403
Windows	1/2	08600	
Glass doors	1/2	08100	06200
Wood doors	1/2	08200	06200
Final grade	1	02201	02800
Seed	1/2	02800	
Electrical, finish	1	16012	09900

Figure 5-6   Activity Layout

Therefore, a careful, accurate list of all information is the most important requirement of scheduling.

The procedure of planning a scheduling chart clearly shows the flow of materials and activities on a project and emphasizes the need for orderly management of a project.

5-5 CONTINGENCIES

On virtually every construction job there are some items that are left out or not foreseen when the estimates are prepared. In some cases the items left out could not have been anticipated at the time of estimating.

Should a *contingency* amount be included; that is, should a sum of money (or percentage) be added to the bid for items overlooked or left out, this money would provide a fund from which the items can be purchased.

If the money is allowed, then it is not necessary to be quite so careful in the preparation of an estimate. But, if an accurate estimate is not made, you never know how much to allow for these forgotten items.

Contingencies are an excuse for using poor estimating practices. When used, you have not truly estimated the project. Instead of adding this amount, the proper approach is to be as careful as possible in listing all items from the plans and specifications. This listing should include everything you will be required to furnish, should estimate labor costs carefully and accurately, and should include the job overhead expense. To these items you can then add the profit you honestly hope to get from the job.

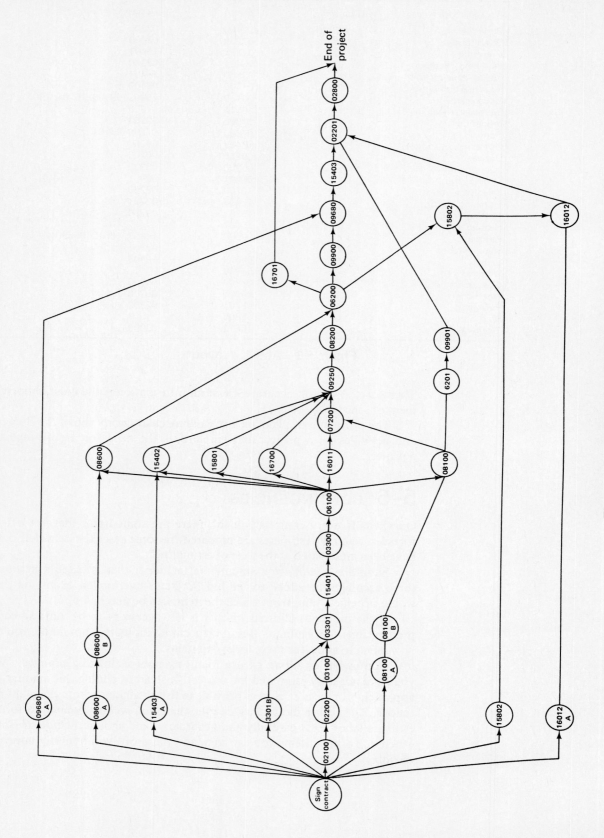

Figure 5-7 Sketch the Chart of Activities

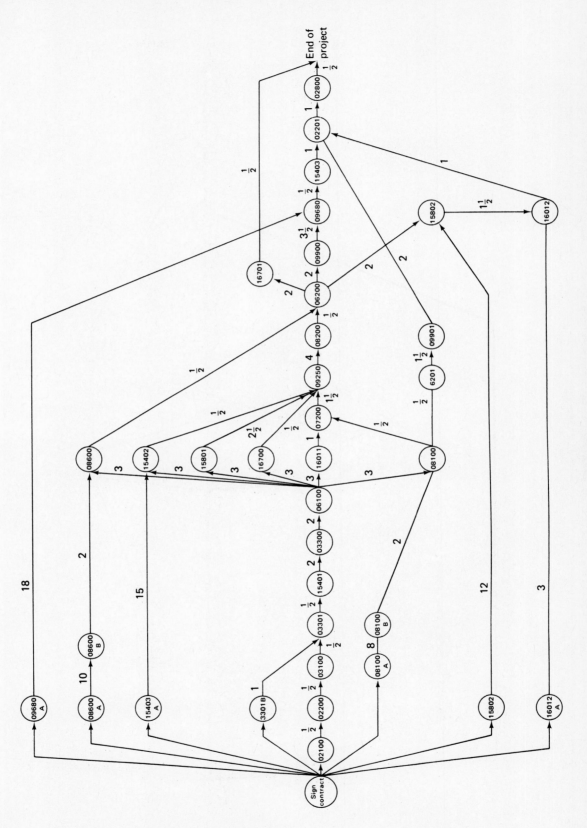

Figure 5-8   Add Time Duration

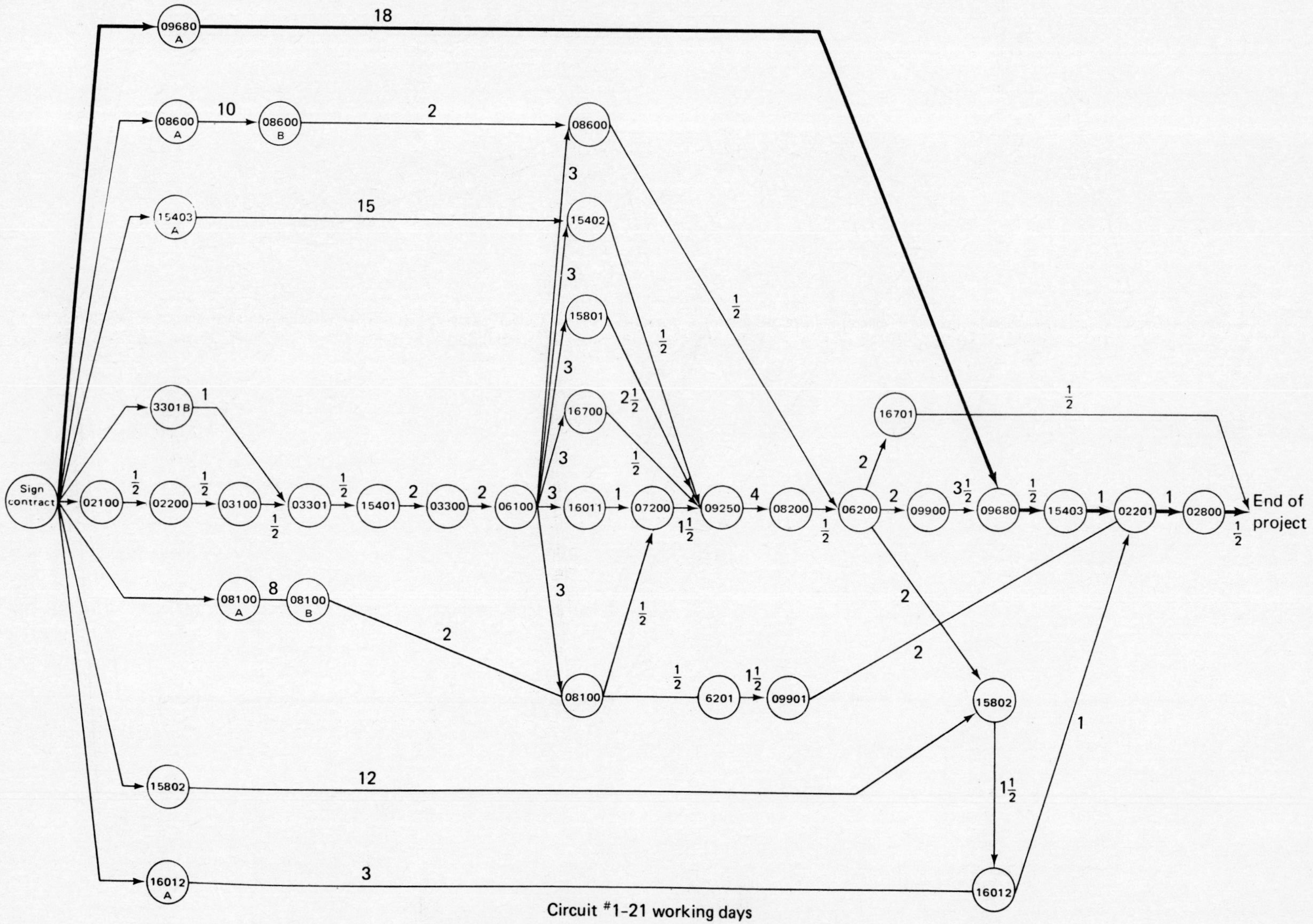

Figure 5-9 Determine the Time Required for Each Circuit

Circuit #1–21 working days

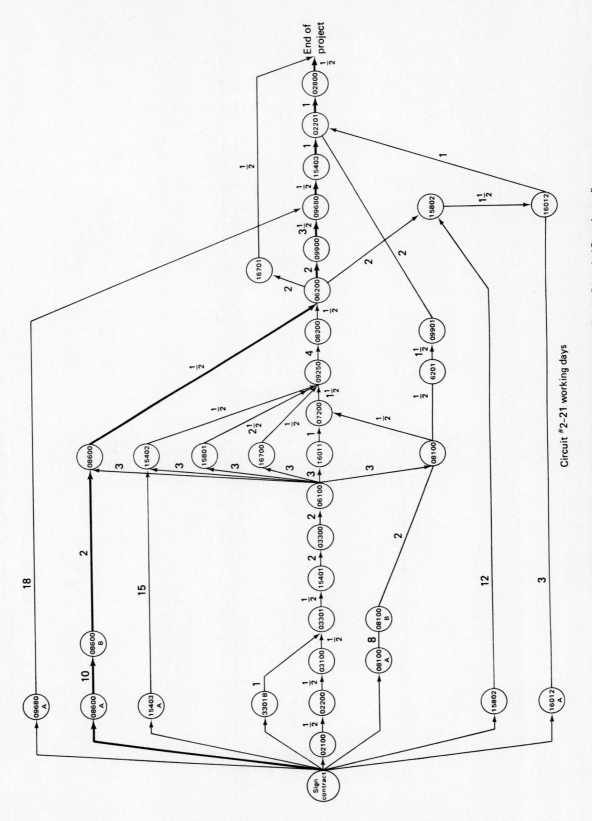

Circuit #2–21 working days

Figure 59 Determine the Time Required for Each Circuit (*Continued*)

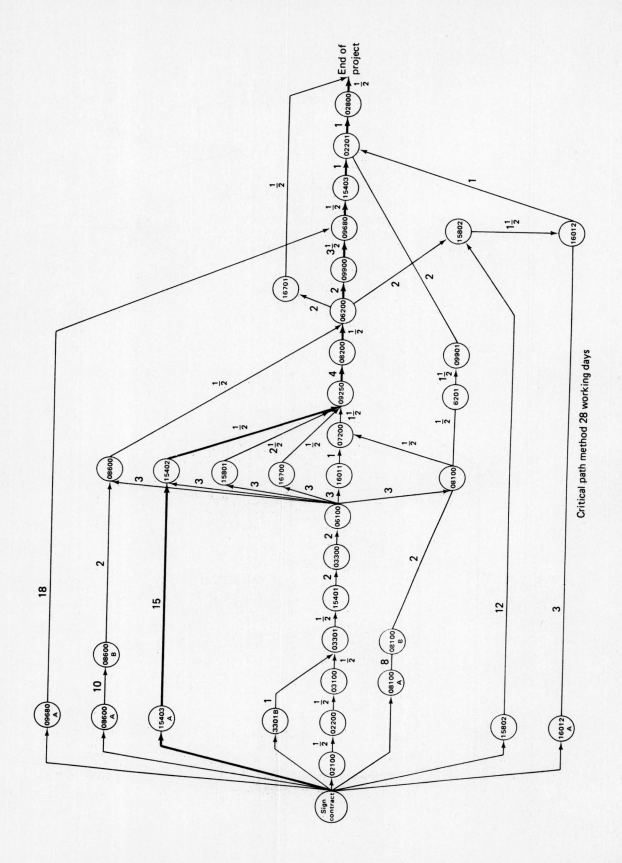

Figure 5-10 Critical Path Chart

66

5-6 CHECKLIST

A checklist is included at the end of many chapters so that items required will not be missed—*USE IT*.

General Overhead (Indirect)

Salaries:
    executives
    secretaries
    estimators
    purchasing agents
    cost-keeping and bookkeeping
    drafters
    office personnel

Office:
    rent (or cost of owning)
    electricity
    heat
    water
    office supplies
    postage
    telephone
    furniture and fixture depreciation
    office machine depreciation
    advertising
    literature
    miscellaneous
    club and association dues

Professional Services:
    lawyers
    accountants
    architects
    engineers

Vehicle:
    depreciation on company cars
    depreciation on general use vehicles

Insurance: (aside from those required on a particular project)
    fire and theft
    public liability
    automobile and vehicle
    property damage
    workmen's compensation
    social security
    unemployment

Job Overhead (Direct, General Conditions)

Salaries:
    superintendent
    assistant superintendent

timekeepers
material clerks
foremen
security
traveling expenses

Temporary Office:
number of offices required
rent (or cost of owning)
setup and removal
light
heat
water
telephone
office supplies
office equipment depreciation

Bonds:
bid (security)
performance
lien
labor and material

Insurance:
fire
property damage
windstorm
automobile and vehicle liability
public liability
workmen's compensation
unemployment
social security
boiler
flood
theft
elevator

Utilities:
temporary heat
temporary light
temporary power
water
sanitary facilities

Miscellaneous:
temporary buildings and enclosures
barricades and signal lights
photographs
engineering services
clean-up
repair of streets and pavements
damage to adjoining buildings
cutting and patching for other trades

removal and replacement of public utilities
permits
licenses
tools and equipment
job signs
pumping and shoring
dust control

REVIEW QUESTIONS

1. What are the types of overhead and how do they differ?
2. Give several examples of each type of overhead.
3. How are the items to be included under each type of overhead arrived at?
4. How does the length of time to complete the project affect overhead?
5. How does the climate and seasons during which the construction will take place affect the overhead?
6. What type of overhead expenses are commonly found in the supplementary general conditions?
7. Define contingency amounts. Why is the use of contingency amounts undesirable?
8. If a full set of drawings and specifications is available, make a list of overhead expenses required for the project. If a set is not available, make another trip to the plan room and use the documents on file there.
9. Using the plans and outline specifications in Appendix A, prepare a schedule for the construction of the project.

6
LABOR

6-1 GENERAL

Estimating the labor required is not the most difficult portion of the estimate to the experienced estimator, who will use work experience and job work studies. But the inexperienced estimator must tread carefully in this area.

The unit of time for estimating labor is the hour. This will eliminate the effects of varying lengths of work days. The estimator must consider that the skill and mental attitude of the worker will affect the length of time required to do a certain piece of work.

Allowances must also be made for variations in wages during the construction of the job, working conditions, availability of skilled and unskilled workers, climatic conditions, and supervision. When there are a lot of jobs for workers, there is a tendency for them to "slow down," and the time required for completion of a piece of work tends to increase. When jobs are scarce, workers tend to perform their tasks at a greater rate of speed. Further, the estimator must realize that a worker seldom works 60 minutes during the hour. Studies of the amount of actual working time ranges from about 30 to 50 minutes per hour. Keep in mind the time taken to "start up" in the morning, coffee breaks, trips to the toilet and for a drink of water, lunches that start a little early and end a little late, and clean-up time all tend to shorten the work day. It is this long list of variables, as well as other possible work interruptions such as waiting for materials, that makes estimating labor the most difficult portion of the estimate.

The basic principles of estimating labor costs are discussed in this chapter to form a basis of understanding for the labor costs which will be illustrated in each chapter that covers quantity takeoff.

6-2 UNIONS—WAGES AND RULES

The local labor situation must be surveyed carefully in advance of making the estimate. Local unions and their work rules should be given particular attention since they may affect the contractor in a given community. There are localities in which the union rules and regulations limit the hourly output of the worker; if these rules are not followed, chances are there will be trouble with the unions—perhaps a strike. There are labor rules that can cause the increase of construction costs by as much as 35 to 80 percent. The estimator will have to determine whether the local union is cooperative or not, whether the union mechanics would tend to be militant in their approach to strike or would prefer to talk first and strike as a last resort. The estimator will also have to determine if the unions can supply the skilled workers required for the construction, if workers will have to be brought in, or if it will be necessary to use a large percentage of unskilled workers. All of these items must be considered in determining how much work will be accomplished on any particular job in one hour.

While surveying the unions, the estimator will have to get information on the prevailing hourly wages, fringe benefits, and holidays. Also find out the date raises have been negotiated for, when the present contract expires, and what the results and attitudes during past negotiations have been. If the contract will run through the expiration date, the estimator will have to include enough in the prices to cover all work done after the expiration of the contract. Often this takes a bit of research into price trends throughout the country; at best it is risky, and many contractors refuse to bid just before the expiration of union contracts unless the terms of the contract provide for adjustment of the contract amount for such increases.

6-3 FIELD REPORTS

The estimator should receive reports from the construction site. These *field reports* (Fig. 6-1) should include an accurate record of the number of units of work completed; the number of workers used (broken down into trades); the number of hours spent on each portion of the work (broken down into trades); and a description of job conditions, climatic conditions, and any other circumstances that might affect the production of the labor force. Equipment utilized is also important and should be noted. These reports are made out daily.

Field Report				
Proj. No. Supt.			Date Weather	
Work code	Total work hours	Skilled	Laborers	Work
06100	32	24	8	Erected roof trusses, sheathed walls and roof

Figure 6-1 Field Report

Most field personnel dislike paper work—it represents tedious labor to them, so they avoid it as much as possible. When they do fill out a report, there is a tendency to "get it done" whether it is accurate or not. If they know that a particular portion of the work is running high, they often "hide" or "bury" the true cost by charging off some of the labor to another area. The estimator must take the time to explain why the reports are so important, show the field personnel how the information is used, and provide a full explanation of how each portion is utilized for future planning and estimating in the office.

The accumulation of accurate labor costs is the estimator's fund of knowledge and is one of the most important and valuable items in the contractor's office.

6-4 PRICING LABOR

To price labor, first estimate the work hours required to do a unit of work. Be sure to include all trades required for the work—masons, carpenters, laborers, etc. Remember that all will probably not work exactly the same number of hours on the work, since they may only be performing one portion of the job. Once the work hours for each trade have been estimated, they are multiplied by the wages per hour and totaled. This will give the total labor cost per unit of work. The hourly scale should include all fringe benefits required in the agreement.

The estimator is responsible for deciding the number of work hours required to perform a unit of work. This can be done only through experience. If you have some doubt about how to handle a particular labor cost, check past project records and then consult with the general superintendent.

EXAMPLE 6-1 — Labor costs per square foot of surface area for brick are as follows: For the first 4 feet in height on the wall, it will require 15 mason-hours per 100 square feet and 18 helper-(laborer-) hours to mix the mortar, and bring mortar and brick to the masons. The mason will lay the brick and square it, "tool" the joints, and clean up at the end of the day.

The wage per hour is multiplied by the appropriate number of work hours, and a cost per 100 square feet is calculated. Assuming that the mason earns $28.50 per hour and the laborer's wage scale is $13.50 per hour, we can compute the total cost of labor for 100 square feet to be $28.50 (15) + $13.50 (18) = $670.50. Therefore, the cost per square foot is $6.71 for labor.

Remember, as the building rises in the air, the labor costs also go up. The higher brickwork will cost more, since the laborers will have to set up scaffolds and move the brick and mortar farther. The time required for the masons to actually place the brick in the wall will remain about the same providing they are working on a solid scaffold and do not have to wait for materials.

An obvious caution should be added to the estimator at this point: It is imperative that wages be kept up-to-date and that the level of productivity upon which the estimate is based be accurate and confirmed by experience.

6-5 LABOR DIAGRAMS

Tables and diagrams are used by estimators to simplify the work, but often shortcuts may be taken to make the work conform to the diagrams. Never revise work hours just to be able to use a table or diagram; it is better simply to work them out as shown in the example.

Generally, the wage scale per hour is listed on the horizontal scale and the unit costs are listed on the vertical scale. The lines representing the varying work hours required for labor for a given unit of work are drawn on the diagram.

Figure 6-2 shows a typical labor diagram for finishing concrete. To arrive at a cost per square foot, the estimator must first decide the number of work hours required per 100 square feet and must know the hourly wage scale. The number of work hours will vary, depending on the type of finish required. For this example assume that a trowel finish for slab on grade is desired and that two work hours are required for 100 square feet. With an hourly wage scale of $28.50 the cost per square foot is $28.50 times 2 divided by 100 or $0.57 per square foot. This result may also be read from the diagram. The estimator plots the lines on the diagram by designating three of the points and connecting them with a line. In this manner the estimator may make as many diagrams as needed. Many firms use computer programs to calculate this type of information and to track labor costs.

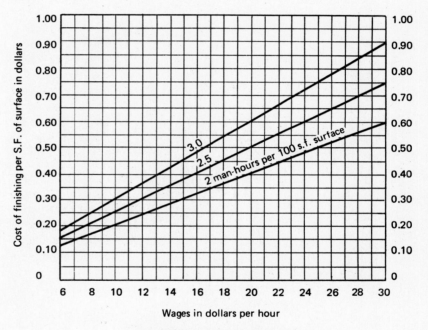

Figure 6-2   Labor Diagram

REVIEW QUESTIONS

1. What unit of time is used in estimating labor? Why?

2. In estimating the amount of labor required there are many variables which must be considered; what are five of the variables?

3. What factors influence the amount of time that a person works in an hour?

4. How may union work rules affect the pricing of labor?

5. Why do many contractors hesitate to bid just before the expiration of union contracts?

6. What are field reports? Who makes them out?

7. Why are field reports important to the estimator?

8. What can be done to try to ensure that the people making out field reports are as accurate as possible?

7
EQUIPMENT

7-1 GENERAL

One of the problems an estimator faces is the selection of suitable equipment to use on a given project. The equipment must pay for itself and unless a piece of equipment will earn money for the contractor, it should not be used.

Since it is impossible for contractors to own all types and sizes of equipment, the selection of equipment will be primarily from that which they own. However, new equipment is purchased if the cost can be justified. If the cost of the equipment can be charged off to the one project or written off in combination with other proposed uses of the equipment, the equipment will pay for itself and should be purchased. For example, if a piece of equipment costing $15,000 will save $20,000 on a project, it should be purchased regardless of whether it will be used on future projects or whether it can be sold at the end of the current one.

Figuring the cost of equipment required for a project presents the same type of problems to estimators as figuring labor. It is necessary for the estimator to decide what equipment is required for each phase of the work and for what length of time it will have to be used.

If the equipment is to be used for a while and then not needed again for a few weeks, the estimator should ask: What will be done with it? Will it be returned to the main yard? Is there room to store it on the project? If rented, will it be returned so that the rental charge will be saved?

Equipment that is required throughout the project is included under equipment expenses, since it cannot be charged to any particular item of work. This equipment often includes the hoist towers and material-handling equipment such as lift trucks.

As the estimator does a takeoff of each item, all equipment required should be listed so that the cost can be totaled in the appropriate column.

Equipment required for one project only or equipment that might be used only infrequently is often purchased for the one project and sold when it is no longer needed. The difference between the purchase price and the selling price would then be charged to the project.

7–2 OPERATION COSTS

The costs of operating the construction equipment should be calculated on the basis of the working hour since the ownership or rental cost is also a cost per hour. Included are items such as fuel, grease, oil, electricity, miscellaneous supplies, and repairs. Wages of the operator and mobilization costs are not included in equipment-operation costs.

Costs for power equipment are usually based on the horsepower of the equipment. Generally, a gasoline engine will use between 0.06 to 0.07 gallon of gasoline per horsepower per hour when operating at full capacity. However, the equipment will probably operate at 55 to 80 percent of full capacity per working hour and will not operate for the full hour, but only for a portion of it.

EXAMPLE 7–1 — What are the estimated fuel costs of a 120-horsepower payloader? A job-condition analysis indicates that the unit will operate about 45 minutes per hour at about 70 percent of its rated horsepower.

Assuming for this project a fuel cost of $1.10 per gallon including all taxes and fuel consumption rate of 0.06 gallon, we can calculate the probable cost of fuel consumption to be

$$120 \times 0.75 \times 0.70 \times 1.10 \times 0.06 = \$4.16 \text{ per hour}$$

The diesel engine requires about 0.04 to 0.05 gallon of fuel per horsepower per hour when operating at full capacity. Since the equipment is usually operated at 55 to 80 percent of capacity and will not operate continuously each hour, the amount of fuel actually used will be less than the full per hour requirement. The full capacity at which the equipment works, the portion of each hour it will be operated, the horsepower, and the cost of fuel must all be determined from a job-condition analysis.

EXAMPLE 7–2 — The job-condition analysis indicates that an 85 horsepower shovel will be used on a job. It further indicates that the unit will operate an average of 50 minutes per hour and at 80 percent of rated capacity, and that the fuel will cost 1.10 per gallon.

The probable fuel consumption based on a fuel consumption rate of 0.04 gallon will be

$$85 \times 0.833 \times 0.80 \times 1.10 \times 0.04 = \$2.49 \text{ per hour}$$

Lubrication.   The amount of oil and grease required by any given piece of equipment varies with the oil capacity, type of equipment, and job conditions. A piece of equipment usually has its oil changed and is greased every 100 to 150 hours. Under severe conditions the equipment may need much more frequent servicing. Any oil consumed between oil changes must also be included in the cost.

equipment

EXAMPLE 7-3 — A piece of equipment has its oil changed and is greased every 120 hours. It requires 6 quarts of oil for the change. The time required for the oil change and greasing is estimated at 2.5 hours.

Allowing $1.30 per quart of oil and $17.50 per hour for labor, you can calculate the lubrication cost to be

$$\frac{6\ (\$1.30)\ +\ \$17.50\ (2.5)}{120} = \frac{7.80\ +\ 43.75}{120} = \$0.43 \text{ per hour}$$

Tires.   The cost of tires can be quite high on an hourly basis. Since the cost of tires is part of the original cost, it is left in when figuring the cost of interest, but taken out for the cost of repairs and salvage values. The cost of tires, replacement, repair, and depreciation should be figured separately. The cost of the tires is depreciated over the useful life of the tires and the cost of repairs taken as a percentage of the depreciation, based on past experience.

EXAMPLE 7-4 — Tires for a piece of equipment cost $5,000 and have a useful life of about 3,500 hours; the average cost for repairs to the tires is 15 percent of depreciation. What is the average cost of tires per hour?

Tires, depreciation $\dfrac{\$5,000}{3,500} = \$1.43$

Tire repairs 15% of $ 1.43 = $\underline{\quad 0.22 \quad}$
Tire cost, per hour $1.65

7-3 DEPRECIATION

As soon as a piece of equipment is purchased, it begins to decrease (*depreciate*) in value. As the equipment is used on the projects, it begins to wear out and in a given amount of time it will have become completely worn out or obsolete. If an allowance for depreciation is not included in the estimate, then when the equipment is worn out, there will be no money set aside to purchase new equipment. This is not profit, and the money for equipment should not be taken from profit.

On a yearly basis, for tax purposes, depreciation can be figured a number of ways, but for practical purposes, the total depreciation for any piece of equipment will be 100 percent of the capital investment minus the scrap or salvage value, divided by the number of years it will be used. For estimating the depreciation costs, you should assign the equipment a useful life expressed in years, hours, or units of production—whichever is the most appropriate for a given piece of equipment.

EXAMPLE 7-5 — If a unit of equipment had an original cost (first cost less salvage value) of $57,500 and an estimated life of five years, the depreciation would be $11,500 per year. If the unit had a useful life of 10,000 hours, the depreciation cost would be $5.75 per hour.

If the equipment had executed an estimated 150,000 units of work during its life, the cost of depreciation would be $0.39 per work unit. The estimator may use whichever method is most appropriate.

7-4 INTEREST

Interest rates must be checked by the estimator. The interest should be charged against the entire cost of the equipment, even though the contractor paid part of the cost in cash. Contractors should figure that the least they should get for the use of their money is the current rate of interest. Interest is paid on the unpaid balance. On this basis the balance due begins at the cost price and decreases to virtually nothing when the last payment is made. Since the balance on which interest is being charged ranges from 100 to 0 percent, the average amount that the interest is paid on is 50 percent of the cost. (This is an approximate method.)

$$\frac{C}{2} \times I \times L = \text{approximate interest cost}$$

$$\frac{C \times I \times L}{2} = \text{approximate interest cost}$$

where

C = cost of equipment (purchase price — salvage value)

I = interest rate (expressed as a decimal)

L = life of loan (in years)

Since the estimator will want to have the interest costs in terms of cost per hour, the projected useful life in terms of working hours must be assumed. The formula used to figure the interest cost per hour would be

$$\frac{C \times I \times L}{2 \times H} = \text{approximate interest cost per working hour}$$

where

H = useful life of equipment (working hours)

The last formula is the one used to determine interest costs toward the total fixed cost per hour for a piece of equipment, as shown in the discussion of ownership costs (Sec. 7-5). Remember that 8 percent, when written as a decimal, is 0.08 since the percent is divided by 100.

Other fixed costs are figured in a similar manner to that used for figuring interest. The costs to be considered include insurance, taxes, storage, and repairs. Depreciation (Sec. 7-3) must also be considered. These items are taken as percentages of the cost of equipment minus the cost of the tires and are expressed as decimals in the formula. Where the expenses are expressed in terms of percent per year, they must be multiplied by the number of years of useful life to determine accurate costs.

7-5 OWNERSHIP COSTS

To estimate the cost of using a piece of equipment owned by the contractor, you must consider depreciation, major repairs, and overhaul as well as interest, insurance, taxes, and storage. These items are most often taken as a percentage of the initial cost to the owner. Also to be added later is the cost for fuel, oil, and tires. The cost to the owner should include all freight costs, sales taxes, and preparation charges.

EXAMPLE 7–6 — Estimate the cost of owning and operating a piece of equipment on a project with the costs listed below:

Actual cost f.o.b. delivered = $47,600

Cost of tires = $4,500

Salvage or scrap value = 3 percent

Depreciation = 100—3 = 97 percent

Useful life = 7 years and 14,000 hours

Total interest = 8 percent per year

Total insurance, taxes, and storage = 6 percent per year

Fuel consumed = 150 horsepower, 0.06 gallons/horsepower, 62 percent operating factor, $1.10 per gallon

Lubrication = 4 quarts oil at $1.25, 2 hours' labor at $16.50,every 150 hours

Life of tires = 4,000 hours

Repair to tires = 12 percent of depreciation

Repairs to equipment = 65 percent over useful life

Fixed Cost (per hour)

$$\text{Interest} = \frac{\$47,600(7)(0.08)}{2(14,000)} = \qquad \$0.95$$

$$\text{Other fixed costs} = \frac{(\$47,600\text{-}4,500)[0.97 + 0.65 + (7 \times 0.06)]}{14,000} = \$6.28$$

Operating Costs (per hour)

$$\text{Tire depreciation} = \frac{\$4,500}{4,000} = (1.125) = \qquad 1.13$$

$$\text{Tire repairs} = 12 \text{ percent of } \$1.13 = (0.1356) = \qquad 0.14$$

$$\text{Fuel} = 150 \times 0.06 \times 0.62 \times 1.10 = (6.138) = \qquad 6.14$$

$$\text{Lubrication} = \frac{[(4 \times 1.25) + (16.50)] \times 2}{150} = (0.2866) = \qquad 0.29$$

TOTAL COST per hour, excluding labor $13.98

7–6 RENTAL COSTS

If a project is a long distance from the contractor's home base or if the construction involves the use of equipment that the contractor does not own and will not likely use often after the completion of this one project, the estimator should seriously consider renting equipment. In considering the rental of equipment, the estimator must investigate the available rental agencies for the type and condition of equipment available, the costs, and the services the rental firm provides. The estimator must be certain that all terms of rental are understood, especially those concerning repair of the equipment.

There is a tendency for contractors to buy equipment even when it is more reasonable to rent. Many rental firms have newer equipment than a contractor might purchase and they also may have a better maintenance program. Check the rental firms carefully, especially the first time you are doing work in a given locale. Price of rental is important, but the emphasis should be on equipment condition and service. If no reputable rental agency is available, the contractor may be forced to purchase the required equipment.

Equipment is generally rented for short periods of time, and lease agreements are arranged when the period of time extends to one year or more.

Rental rates are usually quoted by the month, week, or day. These costs must be broken down into costs per hour or per unit of work so that they may be accurately included in the estimate and checked during construction. The rental charge will be based on a day of 8 hours (or less). If the equipment is to be used more than 8 hours per day, a proportional charge will be added.

EXAMPLE 7-7 — What is the cost per hour of renting a payloader if the monthly rental charge is $1,600? Assume the equipment will actually be used an average of 7 hours per day for 18 days during the month.

$$7 \text{ hours per day} \times 18 \text{ days} = 126 \text{ total hours}$$
$$\frac{\$1,600}{126} = \$12.70 \text{ per hour}$$

To this must be added the other costs of operating the equipment; usually these include mobilization, repairs (except ordinary wear and tear), and day-to-day maintenance, as well as the costs of fuel, insurance, taxes, and cleaning.

7-7 MISCELLANEOUS TOOLS

Examples of miscellaneous tools are wheelbarrows, shovels, picks, crowbars, hammers, hoses, buckets, ropes, and the like. The mechanics that work on the projects have their own small tools, but the contractor will still need a supply of miscellaneous tools and equipment. The estimator should list the equipment required and estimate its cost. The life of this type of equipment and tools is generally taken as an average of one year. Loss of miscellaneous tools and equipment due to disappearance (theft) is common, and all attempts must be made to keep it under control.

7-8 COST ACCOUNTING

The costs for equipment cannot come from thin air; the estimator must rely heavily on equipment expense data for future bids. Especially in heavy construction, the cost accounting is important since the contractor has a great deal of money invested, and the equipment costs become a large percentage of the costs of the project. It is important that equipment costs be constantly analyzed and kept under control.

Small, miscellaneous equipment and tools are not subjected to this cost control analysis and are generally charged to each project on a flat-rate basis.

The procedure for determining equipment expenses varies from contractor to contractor, but the important point is that *the expenses must be determined*. Generally the equipment expense is broken down into a charge per hour or a charge per unit of work. Field reports of equipment time must include only the time during which the equipment is in use. If there is excessive idle time, it must be checked to see whether it can be attributed to bad weather, poor working conditions, or management problems on the project. Management problems sometimes include poor field supervision, poor equipment maintenance, poor equipment selection, and an excessive amount of equipment on the project.

A report on quantities of work performed is required if a cost per unit of work is desired. Generally, the work is measured on a weekly basis; sometimes the work completed is estimated as a percentage of the total work to be performed. This type of report must be stated in work units that are compatible with the estimate.

7-9 MOBILIZATION

The estimate must also include the cost of transporting all equipment required for the project to the job and then back again when the work is completed. Obviously, this cost will vary with the distance, type and amount of equipment, method of transportation used, and the amount of dismantling required for the various equipment. Mobilization costs must also be considered for rental equipment since it must be brought to the job site. The cost of erecting some types of equipment, such as hoists, scaffolding, cranes, etc., must also be included, as must the costs of loading the equipment at the contractor's yard and unloading it at the job site.

7-10 CHECKLIST

Equipment listings are made in each chapter and are considered in relation to the work required on the project. Equipment that may be required throughout the project includes:

Lifting cranes

Hoisting towers

Lift trucks

Hoisting engines

Heaters

Scaffolding

REVIEW QUESTIONS

1. What are advantages of a small contractor's renting equipment instead of owning?
2. What is depreciation on equipment?
3. What operating costs must be considered?
4. Why should interest be included in the equipment costs if the contractor paid cash for the equipment?
5. What is the approximate interest cost if the cost of equipment is $75,000, the interest rate 9.5%, and the life of the loan is 6 years?
6. Why must mobilization be included in the cost of equipment?

7. Why is it important that reports from the field, pertaining to equipment, be kept?

8. If there is excessive idle time for equipment on the job, what factors may this be attributed to?

8
EXCAVATION

8-1 GENERAL

Figuring the quantities of work that must be excavated is considered to be another of the more difficult portions of the estimator's task—it is loaded with problem areas. Under the title of "Excavation" in the specifications there is often a great deal of work included. The number of cubic yards to excavate is sometimes easy enough to compute, but figuring the cost for this portion of the work is difficult because of the various hidden items that may affect the cost. These include such variables as type of soil, required slope of bank in the excavated area, whether bracing or sheet-piling will be required, whether water will be encountered and pumping required, etc.

8-2 SPECIFICATIONS

Carefully check the specifications to see exactly what is included under the title "Excavation." Several questions demand answers: What is the extent of work covered? What happens to the excess excavated material? Can it be left on the site or must it be removed? If the excess must be removed, how far must it be hauled? All of these things and more must be considered. Who does the clearing and grubbing?; removes trees? *Read your specs.* Must the topsoil be stockpiled for future use? Where? Who is responsible for any trenching required for the electrical and mechanical trades? *Read your specs.*

It is important that the estimator understand exactly what work each contractor is responsible for. If the specifications are not clear, call the architect-engineer and explain what problems have been encountered. A typical excavation specification is shown in Fig. 8–32 at the end of the chapter.

8–3 SOIL

One of the first items the estimator must consider is the type of soil that will be encountered at the site. The estimator may begin by investigating the soil borings shown on the drawings or included in the specifications. When soil borings are provided, the contract documents often absolve the architect-engineer and the owner of any responsibility for their correctness. Be certain to check any notes on the drawings and the specifications in this regard. Because of such notes and because the specifications for some projects provide no soil information, it is common practice for the estimator to investigate the soil conditions when visiting the site. Bring a long-handled shovel or a post hole digger and check the soil for yourself. Be certain that you record all that you see in your notebook.

8–4 EXCAVATION—UNIT OF MEASURE

Excavation is measured by the cubic yard for a quantity takeoff (27 cubic feet = 1 cubic yard). Before excavation and in an undisturbed condition, earth weighs about 100 pounds per cubic foot; rock weighs about 150 pounds per cubic foot.

Once excavation begins, the earth and rocks are disturbed and they begin to swell. This means they expand to assume a larger volume; this expansion represents the amount of *swell* and is generally expressed as a percentage gained above the original volume. When loose material is placed and compacted (as fill) on a project, it will be compressed into a smaller volume than when it was loose; this reduction in volume is referred to as *shrinkage*. Refer to Fig. 8–1 for typical percentages of swell.

In surveying the dimensions are given in feet and decimals of a foot. There is usually no reason to change to units of feet and inches; however, there are times when inches must be changed to decimals. Remember that when estimating quantities, the computations need not be worked out to an exact calculation. The following example illustrates the degree of accuracy required.

EXAMPLE 8–1 — In the calculation of an excavation for an area of 52.83 × 75.75 feet with an average depth of 6.33 feet, we could give an acceptable answer of 25,330 cubic feet. There is no reason to give the exact answer of 25,331.852 cubic feet.

Conversion of inches to decimals, and decimals to inches is shown in Appendix D. Estimators will have to learn the approximate values so that they will not have to continuously check for them.

8–5 EQUIPMENT

Selecting and using suitable equipment is of prime importance. The methods available vary considerably depending on the size project and the equipment owned by the contractor. Hand digging and shoveling should be kept to a bare minimum, but almost every job requires some hand work. Equipment used includes trenching

excavation

Percentages of Swell	
Material	Swell Percentages
Sand and gravel	10 to 18%
Loam	15 to 25%
Dense clay	20 to 35%
Solid rock	40 to 60%

Example: If 180 cubic yards of dense clay were excavated with an average swell of 30% how many cubic yards would have to be hauled away on trucks?
180 × 1.30 = 234 cubic yards to be hauled away.

Figure 8-1 Percentages of Swell

machines, bulldozers, power shovels, scrapers, front-end loaders, back-hoes, and clamshells. Each piece of equipment has its use and as the estimator does the takeoff, the appropriate equipment for each phase of the excavation must be selected.

If material must be hauled some distance, either as excavated material hauled out or fill material hauled in, equipment such as trucks or tractor-pulled wagons may be required.

The front-end loader is frequently used for excavating basements and can load directly into the trucks to haul excavated material away. The bulldozer and front-end loader are often used in shallow excavations provided the soil excavated is spread out on a lot near the excavation area. If the equipment must travel over 100 feet in one direction, it will probably be more economical to select other types of equipment.

The backhoe is used for digging trenches for strip footings and utilities, and for excavating individual pier footings, manholes, catch basins, and septic tanks. The excavated material is placed alongside the excavation. For large projects a trenching machine may be economically used for footing and utility trenches.

A power shovel is used in large excavations as an economical method of excavating and loading the trucks quickly and efficiently. On large-sized grading projects, tractor-hauled and self-propelled scrapers are used for the cutting and filling requirements.

8-6 EARTHWORK—NEW SITE GRADES; ROUGH GRADING

Virtually every project requires a certain amount of earthwork. It generally requires cutting and filling to reshape the grade. *Cutting* consists of bringing the ground to a lower level by removing earth. *Filling* is bringing soil in to build the land to a higher level than it is.

The estimator with little or no surveying or engineering knowledge can still handle smaller, uncomplicated projects, but should obtain help from someone more experienced if a complex project is being bid.

Some firms use a special cut-and-fill estimate sheet like the one in Fig. 8-2. Special sheet or not, the main point is to work each grid area separately, and to always keep depth of cut and fill separate until they are totalled up. Then the estimator can see if it will be possible to borrow fill or if there is excess cut to be taken away. The totals at the bottom should be multiplied by the swelling factor for loosened soil.

TEDMAR CONSTRUCTION CO. INC.

Cut and Fill Estimate Sheet

Project __736__

Due date __3/28/7-__
By __FD__
Date __3/13/7-__

Grid	Cut			Fill		
	Area (S.F.)	Ave. cut (ft.)	Volume (C.Y.)	Area (S.F.)	Ave. fill (ft.)	Volume (C.Y.)
1	2500	0.4	37.0			
2	2500	0.7	64.8			
3	2200	0.6	48.9	300	0.2	22.2
4	1600	0.3	17.8	900	0.4	13.3
5	2500	1.1	101.8			
6	2100	0.85	66.1	400	0.7	10.4
7	1200	0.75	33.4	1300	0.55	26.5
8	2500	1.85	171.0			
9	2500	1.15	106.5			
10	600	0.3	6.7	1900	0.9	63.3
11				2500	1.1	102.0
12				2500	1.9	176.0
13				2500	1.3	120.0
14	1400	0.35	18.1	1100	0.45	18.3
15	2500	1.3	120.0			
16	1000	0.65	24.1	1500	0.8	44.5
			816.2			596.5

(This is a typical cut and fill sheet and not the cut and fill for the example which begins in Fig. 8-3A).

Figure 8-2 Cut-and-Fill Estimate Sheet

EXAMPLE 8-2 — Cross-Section Method: Estimate the amount of cut and fill required for the parking area shown on Fig. 8-3.
The desired grade is 103.0 feet. As long as the slope of grade is reasonably

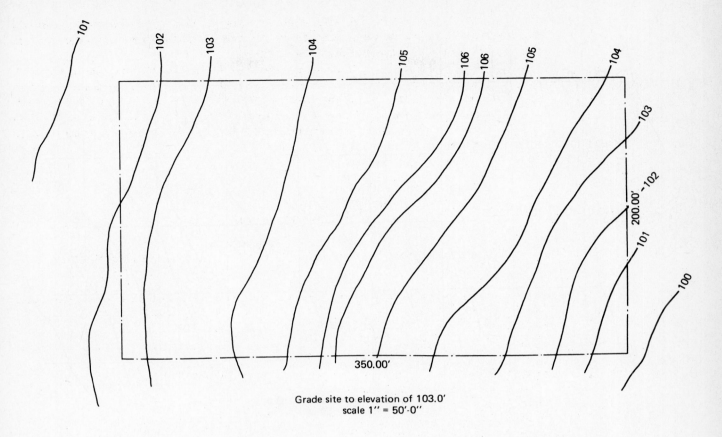

101 102 103 104 105 106 106 105 104 103 200.00' 102 101 100

350.00'

Grade site to elevation of 103.0'
scale 1'' = 50'-0''

Figure 8-3A Typical Plot Plan

regular, as indicated by the fairly even spacing of the contour lines, a grid is placed over the site; in this case, the grid spacing is 50'-0" in both directions. Then each grid line is given a number or letter. If the horizontal lines are indicated numerically, the vertical lines would be indicated alphabetically; the opposite is also true. In this manner you can make reference to any given point on the plan. The area within each grid is also given a number so that it may be referred to readily.

The next step is to calculate the approximate elevation for each grid line intersection. This is done by interpolating between the contour lines shown on the drawing. The results are marked on the drawings in the upper right quadrant of the intersection. The proposed grade elevation is marked in the upper left quadrant, and the two elevations can be compared. If fill is required, it is noted in the lower right quadrant. If cut is required, it is noted in the lower left quadrant (Fig. 8–4).

Once the entire grid has been laid out with existing and proposed elevations, cut, or fill, examine to see which grids show a change of grade from cut to fill. These grids must be calculated separately. For example, to calculate the volume of cut required for area 3, all grid intersections show that a cut is required. In order to determine the amount of cut, the amounts of cut required at the corners are added together and divided by the number of corners. This

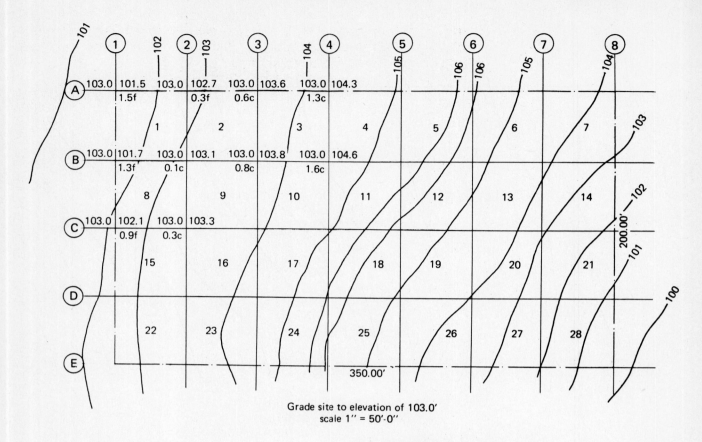

Grade site to elevation of 103.0'
scale 1'' = 50'-0''

Figure 8-3B Grid Layout

Future grade	Present grade
cut	fill

Figure 8-4 Typical Quadrant Layout

average cut is multiplied by the area of the space to give the cubic feet of cut. Divide by 27 and you have the average in cubic yards.

$$\frac{0.6' + 1.3' + 1.6' + 0.8'}{4} \times 50 \times 50 =$$

$$1.075 \times 50 \times 50 = 2688 \text{ c.f.} = 99.5 \text{ c.y. cut}$$

The cut or fill for each grid is calculated, and the cuts and fills are kept separate.

Those grids that show a change from cut to fill within them must be considered separately. Grid 8 is a good example of this. It is necessary in this case to determine the approximate amounts of each grid area that is cut and the amount that is fill.

excavation

Using the approximate scale at which the plan was drawn, scale off the distance from the corner of the grid to the point that separates the cut and fill, in this case the elevation of 103.0 feet. Grid 8 is shown in Fig. 8-5. The area of cut is shaped like a trapezoid and the area is equal to

$$50'\text{-}0'' \ \times \ \frac{3.5' + 23'}{2} = 50 \times 13.25 = 662.5 \text{ s.f.}$$

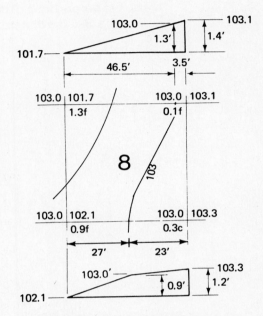

Figure 8-5   Grid 8 of Parking Lot

The heights are added together and divided by two in order to arrive at the average height. The average depth of cut is

$$\frac{0 + 0.1 + 0.3 + 0}{4} = 0.1 \text{ foot average depth of cut}$$

Total cut = 662.5 s.f. × 0.1 = 66.25 c.f. = 2.5 c.y.

The number of cubic yards of fill is calculated after finding the area of cut. The total area, minus the area of cut, gives the area of fill.

$$(50 \times 50) - 662.5 = 2500 - 662.5 = 1837.5 \text{ s.f.}$$

The area of fill is multiplied by the average depth of fill.

$$\frac{1.3 + 0 + 0 + 0.9}{4} = \frac{2.2}{4} = 0.55 \text{ foot average depth of fill}$$

The cubic yards of fill required equals the area of fill times the average depth of fill divided by 27.

$$\frac{1837.5 \times 0.55}{27} = 37.4 \text{ c.y. of fill required}$$

This process must be repeated for all grids requiring both cut and fill.

Another approach to determining the cut and fill areas required for Grid 8 is to separate the grid into cut and fill areas.

EXAMPLE 8-3 — To determine the dimensions to be used for the cut and fill areas, the proportions of cut and fill to grid size must be used.
For Fig. 8-6 the dimensions are:

$$x = \frac{a}{c} \times y$$

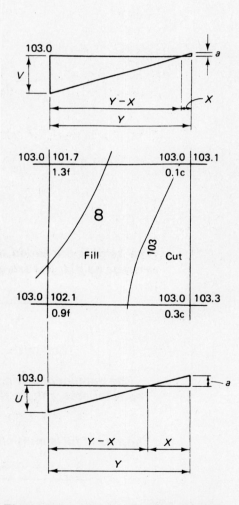

Figure 8-6   Cut and Fill Dimensions

Along bottom of grid:

$$x = \frac{0.3}{1.2} \times 50 = 12.5 \text{ ft.}$$

$$y - x = 50 - 12.5 = 37.5 \text{ ft.}$$

Along top of grid:

$$x = \frac{0.1}{1.4} \times 50 = 3.5 \text{ ft.}$$

$$y - x = 50 - 3.5 = 46.5 \text{ ft.}$$

Cut area:

$$50 \times \frac{12.5 + 3.5}{2} = 50 \times 8 = 400 \text{ s.f.}$$

Fill area:

$$50 \times \frac{37.5 + 46.5}{2} = 50 \times 42 = 2{,}100 \text{ s.f.}$$

Note that the answer is different than when the excavated area is scaled off in the more accurate approach used in Fig. 8–5.

8-7 PERIMETERS AND AREAS

Throughout the estimate there is some basic information that is used over and over again. The *perimeter* of a building is one such basic dimension that must be calculated. The perimeter is the distance around the building; it is the total length around the building expressed in lineal feet.

The *area* of a plan is the surface included within specified limits—in this case the building, roadway, parking lot, or plot. It is expressed in square feet.

EXAMPLE 8-4 — What are the perimeter and the area of the building shown in Fig. 8-7? (Refer also to Appendix A for building drawings and specifications.)

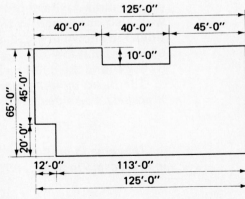

Figure 8-7   Plan—Building Lines

Perimeter

Basic perimeter: 65′ + 125′ + 65′ + 125′ = 380 l.f.
Recessed walls: 10′ + 10′ = 20 l.f.
                                                 400 l.f. (lineal feet)

*This could also have been figured by adding up the footage around the build-
ing, starting in the upper left corner of the building and proceeding clockwise:*

40 + 10 + 40 + 10 + 45 + 65 + 113 + 20 + 12 + 45 = 400 l.f.

Area
*Find the basic area of the plan and deduct the areas which are recessed and
offset.*

Basic area: 125′ - 0″ × 65′ - 0″ = 8125 s.f. (square feet)
Recesses and offsets:
   Southwest offset 12 × 20 = 240 s.f.
   North recess   10 × 40 = 400 s.f.

      Total deductions   − 640 s.f.
      Net area of building   7485 s.f.

8-8 TOPSOIL REMOVAL

The removal of topsoil to a designated area where it is to be stockpiled for finished
grading and future use is included in many specifications. Thus, the estimator must
determine the depth of topsoil, where it will be stockpiled, and what equipment
should be used to strip the topsoil and move it to the stockpile area. Topsoil is
generally removed from all building, walk, roadway, and parking areas. Volume of
topsoil is figured in cubic yards. A clearance around the entire basic plan must also
be left to allow for the slope required for the general excavation—usually about
5′-0″ (5 feet, 0 inches) is allowed on each side of a building and 1′-0″ to 2′-0″ for
walks, roadways, and parking areas.

**EXAMPLE 8-5—Estimate the amount of topsoil to be excavated and
stockpiled on the site for the building shown in Fig. 8-7. From the drawings
(in Appendix A) the depth of the topsoil is 9 inches. Assume that this depth
was also confirmed in the site investigation (Sec.4-6). The topsoil quantities are shown in Fig. 8-8. Also assume that the depth of topsoil averages
9 inches, that the lot is generally level, and that the topsoil will be stripped
5′-0″ beyond the building lines.**

*The basic building size is 65′-0″ × 125′-0″, and if a 5′-0″ clearance
is added on all sides, the dimensions become:*

(5 + 65 + 5) × (5 + 125 + 5) = 75 × 135 = 10,125 s.f.

*This area is multiplied by the average depth of topsoil to be excavated—
in this case, 9 inches or 0.75 feet—to arrive at the cubic feet of excavation.*

10, 125 × 0.75 = 7,594 c.f.

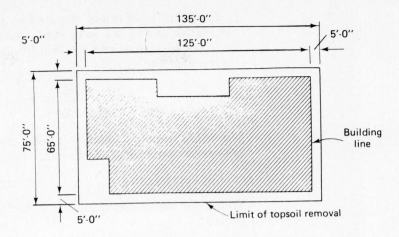

Figure 8-8 Topsoil Quantities

The unit of measure used for excavation is the cubic yard (1 cubic yard = 27 cubic feet). Therefore:

$$\frac{7{,}594}{27} = 281.25 = 281 \text{ c.y.}$$

Equipment selection will probably be limited to either a bulldozer or a front-end loader. Assume that a front-end loader with a 1-cubic yard capacity bucket was selected and that it is estimated it can scrap and stockpile an average of 24 cubic yards per hour. Mobilization time is estimated at 2.5 hours, the operating cost per hour for the equipment is estimated at $11.35, and the cost for an operator is $17.75 per hour. Estimate the number of hours required for the project and the cost for this item.

Soil	Dozer				Tractor shovel		Front end loader		Backhoe	
	50' haul		100' haul		No haul		50' haul	100' haul	No haul	
	50 hp	120 hp	50 hp	120 hp	1 c.y.	$2\frac{1}{4}$ c.y.	1 c.y.	$2\frac{1}{4}$ c.y.	$\frac{1}{2}$ c.y.	1 c.y.
Medium	40	100	30	75	40	70	24	30	25	55
Soft, sand	45	110	35	85	45	90	30	40	25	60
Heavy soil or stiff clay	15–20	40	10–15	30–35	15–20	35	10	12	10	15

Load and haul		
Truck size	Haul	c.y.
6 c.y.	1 mile	12–16
	2 miles	8–12
12 c.y.	1 mile	18–22
	2 miles	12–14

Figure 8-9 Equipment Capacity

topsoil removal

First, the total hours required to complete this phase of work must be calculated. Divide the total cubic yards to be excavated by the rate of work done per hour, and add the mobilization time; the answer is the total hours for the phase of work.

$$\frac{281}{24} + 2.5 = 14.2 \text{ hours}$$

The total number of hours is then multiplied by the cost of operating the equipment per hour plus the cost of crew for the period of time.

$$14.2 \times (\$11.35 + \$17.75) = 14.2 \text{ hours} \times \$29.10 \text{ per hour} = \$413.22$$

This is the estimated cost to strip and stockpile the topsoil on this particular project.

8-9 GENERAL EXCAVATION

Included under *general (mass) excavation* is the removal of all types of soil that can be handled in fairly large quantities, such as excavations required for a basement, mat footing, or a cut for a highway or parking area. Power equipment such as power shovels, front-end loaders, bulldozers, and graders are typically used in this type of project.

When calculating the amount of excavation to be done for a project, be certain that the dimensions used are the measurements of the outside face of the footings and not those of the outside buildings. The footings usually project at least 6 inches (and sometimes much more) beyond the wall. Also, add about 6" to 1'-0" on all sides of the footing to allow the worker to install and remove forms. The estimator must also allow for the sloping of the banks so that they will not cave in. The amount of slope required must be determined by the estimator, who considers the depth of excavation, type of soil, and possible water conditions. Some commonly used slopes (referred to as "angles of repose") are given in Fig. 8-10. If job conditions will not allow the sloping of soil, the estimator will have to consider using sheet-piling (Fig. 8-11) or some type of bracing to shore up the bank. Any building or column footing projection is a separate calculation, and the result is added to the amount of excavation for the main portion of the building. Allow about 1'-0" all around for working space.

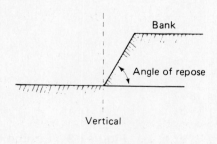

Material	Angle		
	Wet	Moist	Dry
Gravel	15–25	20–30	25–40
Clay	15–25	25–40	40–60
Sand	20–35	35–50	25–40

Figure 8-10   Earthwork Slopes

excavation

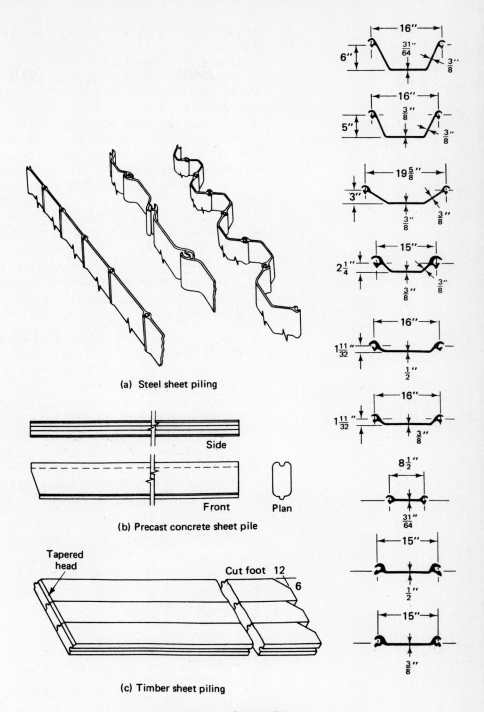

(a) Steel sheet piling

Side

Front

Plan

(b) Precast concrete sheet pile

Tapered head

Cut foot 12

6

(c) Timber sheet piling

Figure 8-11 Sheet Piles

The actual depth of cut is the distance from the top of grade to the bottom of the fill material used under the concrete floor slab. If topsoil has been stripped, the average depth of topsoil is deducted from the depth of cut. If fill material such as gravel were not used under the concrete floor, the depth would be measured to the bottom of the floor slab. Since the footings usually extend below the fill material, a certain amount of excavation will be required to bring the excavation down to the proper elevation before footings can be placed. This would also be included under the heading of "general excavation," but kept separate from topsoil.

Before estimators can select equipment, they will have to determine what must be done with the excess excavation—whether it can be placed elsewhere on the site or whether it must be hauled away. If it must be hauled away, they should decide how far. The answers to these questions will help determine the types and amount of equipment required for the most economical completion of this phase of the work.

To determine the amount of general excavation it is necessary to determine:

1. Size of building (building dimensions).
2. The distance that the footing projects out beyond the wall (the distance labeled "A" in Fig. 8–12).
3. The amount of working space required between the edge of the footing and the beginning of excavation (the distance labeled "B" in Fig. 8–12).
4. The elevation of the *existing* land, by checking the existing contour lines on the plot (site) plan.
5. The type of soil which will be encountered. This is determined by first checking the soil borings (either on the drawings), but this *must* be checked during the site investigation (Sec. 4–6). Almost every specification clearly states that the soil borings are for the contractor's information *but* they are not guaranteed.
6. Whether the excavation will be sloped as shown in Fig. 8–12 or shored. Slope angles (angles of repose) are given in Fig. 8–10.
7. The depth that the excavation will have to be. This is done by determining the bottom elevation of the cut to be made. Then take the existing elevation, deduct any topsoil removed, and subtract the bottom elevation of the cut. This will determine the depth of the general excavation.

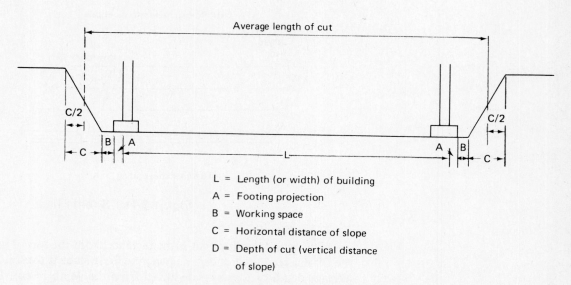

L = Length (or width) of building

A = Footing projection

B = Working space

C = Horizontal distance of slope

D = Depth of cut (vertical distance of slope)

Figure 8-12   Typical Excavation

excavation

To practice estimating general excavation, examples are included which will show:

1. Rectangular building with basement (Appendix F).
2. Rectangular building with column footing projection.
3. Irregular building with slab on grade (Appendix A).
4. Residence with crawl space (Appendix G).

EXAMPLE 8-6 — Determine the amount of general excavation required for the building shown in Appendix F.

1. From the floor plan, the building size is 70'-0" × 120'-0".
2. From the wall section, the footing projects out 6 in. from the wall.
3. The work space between the edge of the footing and the beginning of the excavation will be 1'-0" in this example.
4. The elevation of the existing land, by checking the existing contour lines on the plot (site) plan, is about 113'-0".

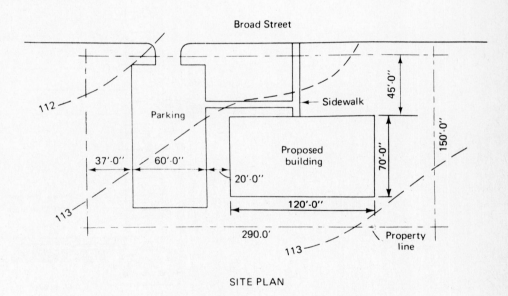

SITE PLAN

Figure 8-13 Plot Plan

5. A check of the soil borings (assuming that this has been confirmed in the site investigation) indicates that:
    a. the soil is a gravel-clay mixture.
    b. the topsoil averages about 8 in. deep.
6. The gravel-clay soil mixture could (from experience and the table in Fig. 8–10) have an angle of repose of about 60°, or a 1.5:1 slope (which means for every 1.5 foot of vertical depth an additional 1.0 feet of

general excavation   97 horizontal width is needed. Since the alternative is shoring or sheet-piling on this project, the sloped excavation will be used.

7. The bottom elevation of the general excavation cut will be at the bottom of the gravel. Since this elevation is rarely given it may have to be calculated. Generally, the drawings will give the elevation of the basement slab or bottom of the footing and the depth of cut is calculated from these. In Fig. 8-14 the depth of cut is shown on a wall section when the basement slab elevation is given. In Fig. 8-15 the depth of cut is shown when the bottom of footing is given. In this excavation the total depth is 9'-8", and once the 8 in. of topsoil is deducted, a 9'-0" cut is required.

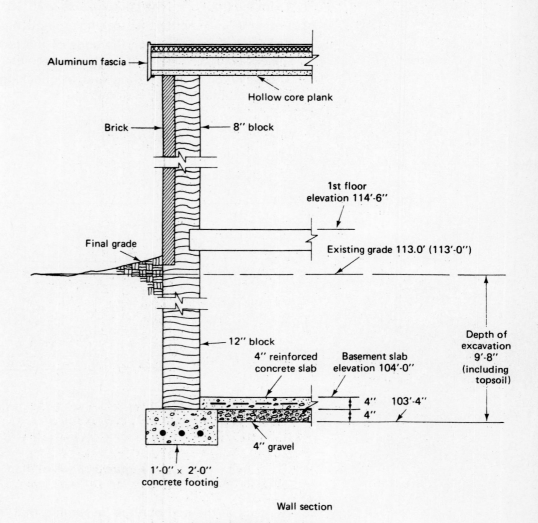

Wall section

Figure 8-14 Depth of Cut

To calculate the total area to be excavated, add the areas of the column footings and working space. The working space and the allowance for the slope of the soil are added only to the length of the column footing. The area of

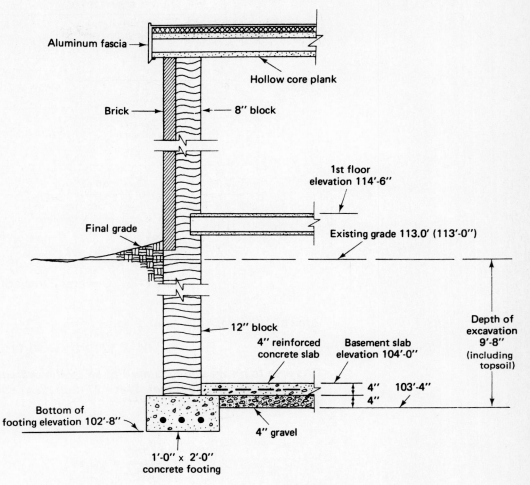

Figure 8-15   Depth of Cut

excavation for column projections equals the number of columns times the projection from the footing times the sum of the length of projection plus working space plus the amount required to slope the soil.

All of the information is put on a workup sheet as it is gathered. The sketch of this excavation problem is shown in Fig. 8–16.

The area of the major portion of the excavation is the length times the width of the cut.

Average length of cut (left to right):

3'-0" + 1'-0" + 0'-6" + 70'-0" + 0'-6" + 1'-0" + 3'-0" = 79'-0" ft.

Average width of cut:

3'-0" + 1'-0" + 0'-6" + 120'-0" + 0'-6" + 1'-0" + 3'-0" = 129'0" ft.

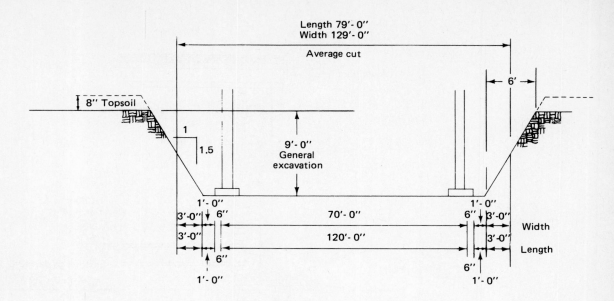

Figure 8-16 General Excavation

Area of building:

$$79'\text{-}0'' \times 129'\text{-}0'' = 10,191 \text{ s.f.}$$

The area to be excavated is multiplied by the average depth of the cut (in feet) to determine the cubic feet. Since 27 cubic feet equal 1 cubic yard, the conversion is easily done.

$$10,191 \times 9'\text{-}0'' = 91,719 \text{ c.f.}$$

$$\frac{91,719}{27} = 3,397 \text{ c.y. to be excavated}$$

EXAMPLE 8-7 — For the same building as estimated in the example above, this shows the calculations required when column footings extend out from the building footings, as illustrated in Fig. 8–17.

Area of building (from Example 8–6): 10,191 s.f.
Column projections: Assume 8 projections, each 3'-0" × 7'-0"

$$\text{Column projections} = 8 \times 3'\text{-}0''^* \times 15'\text{-}0'' = 360 \text{ s.f.}$$
$$\text{Total area} = 10,191 + 360 = 10,551 \text{ s.f.}$$
$$\text{Cubic feet} = 10,551 \times 9'\text{-}0'' = 94,959$$
$$\text{Cubic yards} = \frac{94,959}{27} = 3,517 \text{ c.y. to be excavated}$$

*Slope and working space dimensions are not added to the 3'-0" dimension since they were previously accounted for. See Fig. 8–17.

EXAMPLE 8-8 — For the small commercial building in Appendix A, the general excavation sketch is illustrated in Fig. 8–18.

excavation

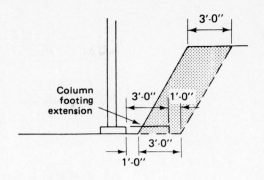

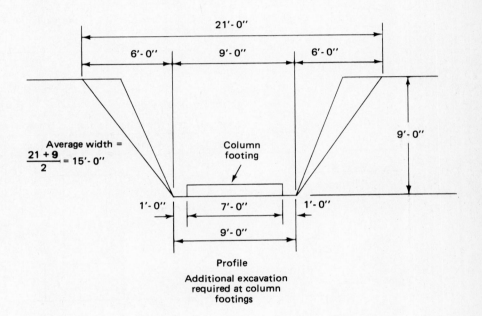

Figure 8-17 General Excavation

Soil: Sand
Slope: Use 45° (1:1)
Depth of excavation: 2'-9"
Footing widths: 1'-8" and 2'-0" (use 2'-0" throughout)
Excavation, average width: 7'-6" (11'-0" to 4'-0")
Length of excavation: 400 l.f. (Example 8-4)

NOTE: The exact length is along the center line of the footing; however, the perimeter is usually used for convenience. If the centerline dimension were used in this problem, it would be 397 l.f.

$$\text{Excavation} = \frac{400 \times 7.5 \times 2.75}{27} = 305.5 \text{ (use 306 c.y.)}$$

general excavation

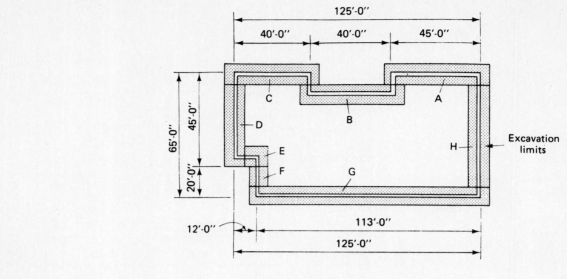

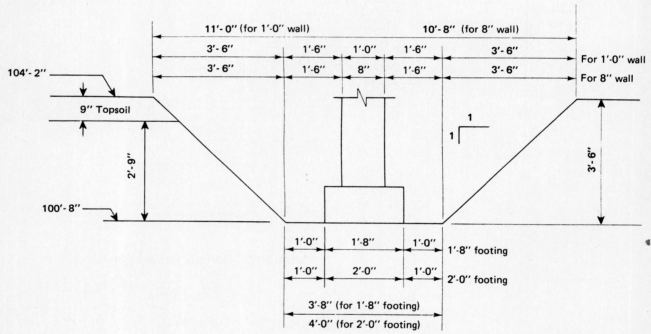

Figure 8-18 General Excavation

EXAMPLE 8–9 — For the small residence with a crawl space in Appendix G, the general excavation sketch is illustrated in Fig. 8–19.

Soil: Gravel and clay
Slope: About 65°
Depth of excavation: 2′ –8″
Footing width: 1′ –4″

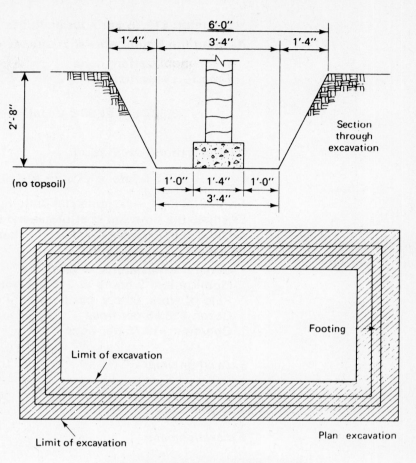

Figure 8-19 General Excavation

Excavation, average width: 4′-8″
Length of excavation: 148 l.f.

$$\text{Excavation} = \frac{148 \times 4.67 \times 2.67}{27} = 68.4 \text{ c.y.}$$

Labor.   In these examples, the labor costs for the general excavation will be determined by using the time charts in Fig. 8-9 and the labor costs as shown in the examples. Labor examples are given for each of the general excavation examples worked above.

EXAMPLE 8-10 — Commercial Building (Appendix F): Estimate the costs of general excavation for the building in Appendix F, with general excavation quantities taken off in Example 8-6.

$$\text{General excavation} = 3{,}397 \text{ c.y.}$$

Equipment:
  Tractor shovel
  Front-end loader: \$15.75 per hour

general excavation

103

Operator: $19.75 per hour (includes all fringe benefits, etc.)

Assume the tractor shovel is already on the site.

Truck mobilization: none

Time to excavate:

$$3{,}397 \text{ c.y. at } 80 \text{ c.y. per hour} = \frac{3{,}397}{80} = 42.5 \text{ hours}$$

Cost of general excavation:

$$42.5 \text{ hours} \times (\$15.75 + \$19.75 \text{ per hour}) = \$1508.75$$

EXAMPLE 8-11 — Commercial Building (Appendix F, with column footings): Estimate the time and cost of general excavation for the general excavation calculated in Example 8-7. Assume that a 120 h.p. dozer with a 100′ haul will be used.

General excavations = 3,517 c.y.
Mobilization: 3 hours at $48 per hour
Rate of work: 85 c.y. per hour
Dozer: $16.85 per hour
Operator: $19.75 per hour

Excavation time:

$$\frac{3{,}517}{85} = 41.4 \text{ hours}$$

Excavation cost:

Mobilization (3 hours at $48 per hour) = $ 144.00
Excavation (41.4 hours at $36.60 per hour) =   1,515.24
                                                         1,659.24

EXAMPLE 8-12 — Commercial Building (Appendix A): Estimate the time and cost of general excavation for the general excavation calculated in Example 8-8. Assume that a backhoe with 1-c.y. capacity will be used.

General excavation = 362 c.y.
Mobilization: 1.5 hours
Rate of work: 55 c.y. per hour
Backhoe: $13.45 per hour
Operator: $16.75 per hour

Excavation time:

$$\frac{362}{55} + 1.5 = 6.6 + 1.5 = 8.1 \text{ hours}$$

Excavation cost:

$$8.1 \text{ hours} \times \$30.20 \text{ per hour} = \$244.62$$

EXAMPLE 8-13 — Residential Building (Appendix G): Estimate the time and cost of general excavation for the general excavation calculated in Fig. 8-19. Assume a backhoe with a 1/2 = c.y. capacity is used.

General excavation = 68.4 c.y.
Mobilization: 1.5 hours
Rate of work: 15 c.y. per hour
Backhoe: $9.75 per hour
Operator: $11.35 per hour

Excavation time:

$$\frac{68.4}{15} + 1.5 = 4.6 + 1.5 = 6.1 \text{ hours}$$

Excavation cost:

$$6.1 \text{ hours} \times \$21.10 \text{ per hour} = \$128.71$$

NOTE: Many subcontractors have a minimum charge for work done—perhaps a $100, $200, or $300 minimum to come and do any work at all.

8-10 SPECIAL EXCAVATION

Usually the *special excavations* are the portions of the work that require hand excavation, but also included may be any excavation that requires special equipment; that is, equipment used for a particular excavating portion other than general (mass) excavation.

Portions of work most often included under this heading are footing holes, small trenches, and the trench-out below the general excavation for wall and column footings, if required. On a large project a backhoe may be brought in to perform this work, but a certain amount of hand labor is required on almost every project.

The various types of excavation must be kept separately on the estimate and, if there is more than one type of special or general excavation involved, each should be considered separately and then grouped together under the headings "special excavation" or "general excavation."

In calculating the special excavation, the estimator must calculate the cubic yards, select the method of excavation, and determine the cost.

Examples of special excavation are shown for two previous examples (the commercial building in Appendix F, with and without column footings) which require special excavation to dig out for the footing. This special excavation is not required in the previous examples of the commercial building in Appendix A and the small residence in Appendix G, since all of the excavation was done under general excavation (Sec. 8-9).

EXAMPLE 8-14 — Commercial Building (Appendix F): Special excavation areas (Figs. 8-20 and 8-21):

2'-0" wide footing, 2'-6" wide excavation
l.f. of excavation: 120 + 70 + 120 + 70 = 380 l.f.
Depth of excavation: 8"

special excavation

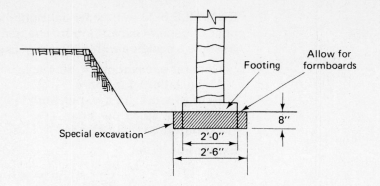

Figure 8-20 Special Excavation

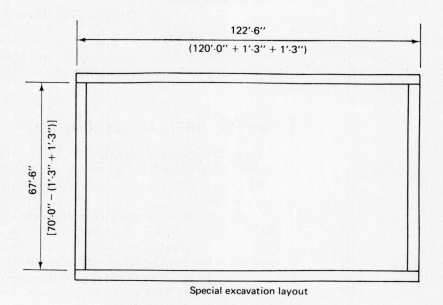

Special excavation layout

Figure 8-21 Special Excavation

$$\text{Volume} = \frac{2'\text{-}6'' \times 8'' \times 380'\text{-}0''}{27} = \frac{2.5 \times 0.67 \times 380}{27} = 23.6 \text{ c.y.}$$

$$\text{Special excavation} = 23.6 \text{ c.y.}$$

EXAMPLE 8-15 — Commercial Building (Appendix F, with column footings): Projections: 8 required, 8″ deep (Fig. 8-22)

$$\text{Volume} = \frac{8 \times 8'' \times 3'\text{-}0'' \times 7'\text{-}0''}{27} = \frac{8 \times 0.67 \times 3 \times 7}{27} = 4.2 \text{ c.y.}$$

$$
\begin{aligned}
\text{Wall footing} &= 23.6 \text{ c.y.} \\
\text{Column footing} &= \underline{\ 4.2 \text{ c.y.}} \\
\text{Special excavation} &= 27.8 \text{ c.y.}
\end{aligned}
$$

excavation

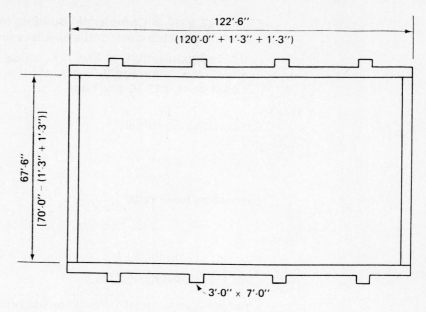

Special excavation layout

Figure 8-22 Special Excavation

Labor. Labor is estimated using the work chart shown in Fig. 8–23 and the hourly wages in the locale the work is being done. The labor for the special excavation was calculated in Sec. 8–9.

Soil	Depth of excavation		
	3'	6'	12'
	Output (c.y. per hour)		
Light	0.90–1.80	0.80–1.40	0.60–1.10
Medium	0.70–1.30	0.60–1.10	0.50–0.90
Heavy	0.60–1.10	0.50–0.90	0.35–0.70

Figure 8-23 Work Chart

EXAMPLE 8–16 — Commercial Building (Appendix F): Calculate the labor cost for the special excavation required in Example 8–14.

  Special excavation = 23.6 c.y.
  Hand labor: Average 0.75 c.y. per hour
  Labor cost: $12.50 per hour

Excavation labor time:

$$\frac{23.6}{0.75} = 31.5 \text{ hours}$$

Excavation labor cost:

31.5 hours × $12.50 per hour = $393.75

special excavation

EXAMPLE 8-17 — Commercial Building (Appendix F, with column footings):
Calculate the labor cost for the special excavation required in Example 8-15.

    Special excavation = 27.8 c.y.
    Hand labor: Average 0.75 c.y. per hour
    Labor cost: $12.50 per hour

Excavation labor time:

$$\frac{27.8}{0.75} = 37.1 \text{ hours}$$

Excavation labor cost:

$$37.1 \text{ hours} \times \$12.50 \text{ per hour} = \$463.75$$

8-11 BACKFILLING

Once the foundation of the building has been constructed, one of the next steps in construction is the backfilling required around the building. *Backfilling* is the putting back of excess soil that was removed from around the building during the general excavation. After the topsoil, and general and special excavation have been estimated, it is customary to calculate the amount of backfill.

The material may be transported by wheelbarrows, scrapers, front-end loaders with scoops or buckets, scrapers, bulldozers, and perhaps trucks if the soil must be transported a long distance. The selection of equipment will depend on type of soil, weather conditions, and distance the material must be moved. If tamping or compaction is required, special equipment will be needed, and the rate of work per hour will be considerably lower than if no tamping or compaction were required.

One method for calculating the amount of backfill to be moved is to determine the total volume of the building that will be built within the area of the excavation. This would be the total volume of the basement area, figured from the underside of fill material, and would include the volume of all footings, piers, and foundation walls. This volume is deducted from the volume of excavation that had been previously calculated. The volume of backfill required is the result of this subtraction. The figures should not include the data for topsoil, which should be calculated separately.

A second method for calculating backfill is to compute the actual volume of backfill required. The estimator usually makes a sketch of the actual backfill dimensions and finds the required amount of backfill.

The following examples illustrate how to calculate backfill quantities.

EXAMPLE 8-18 — Commercial Building, (Appendix F):
Building volume (Fig. 8-24):

$$\frac{70' \times 120' \times 9'}{27} = \frac{75,600}{27} = 2,800 \text{ c.y.}$$

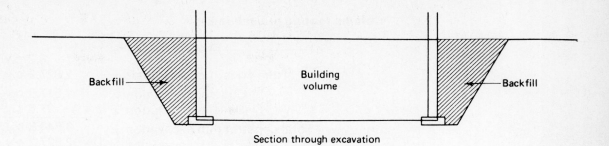

Section through excavation

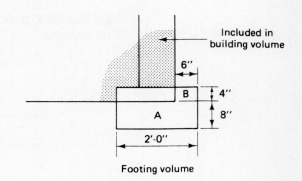

Footing volume

Figure 8-24   Backfilling

Footing volume (Fig. 8–24):

A. $\dfrac{378'\text{-}0'' \times 2'\text{-}0'' \times 8''}{27} = \dfrac{378 \times 2 \times 0.67}{27} = 18.7$ c.y.

B. $\dfrac{378'\text{-}0'' \times 4'' \times 6''}{27} = \dfrac{378 \times 0.33 \times 0.5}{27} = 2.3$ c.y.

Total construction volume = 2,821.1 c.y.

General excavation = 3,397.0 c.y.
Special excavation = + 23.6 c.y.

Total excavation = 3,420.6 c.y.
Total construction
volume = − 2,821.1 c.y.

Backfill = 599.5 c.y.

EXAMPLE 8-19 — Commercial Building (Appendix F, with column footings):
Building volume:

$$\frac{70 \times 120 \times 9}{27} = 2,800 \text{ c.y.}$$

Footing volume (same as in Example 8-18):

$$18.7 + 2.3 = 21.0 \text{ c.y.}$$

backfilling

Column footing projections:

$$\frac{8 \times 3'-0'' \times 7'-0'' \times 1'-0''}{27} =$$

	6.2 c.y.
Total construction volume =	2,827.2 c.y.
General excavation =	3,517.0 c.y.
Special excavation =	(+) 27.8 c.y.
Total construction excavation =	3,544.8 c.y.
Total volume =	(−)2,827.2 c.y.
Backfill =	717.6 c.y.

EXAMPLE 8-20 — Commercial Building (Appendix A) (see Fig. 8-25 for backfill sketch):

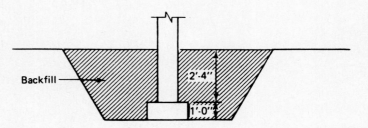

Figure 8-25   Backfilling

12' wall volume (1'-0" thick, 145 l.f., 2'-4" high):

$$\frac{1 \times 145 \times 2.33}{27} = 12.5 \text{ c.y.}$$

2'-0" footing volume (1'-0" thick, 145 l.f., 2'-0" wide):

$$\frac{1 \times 145 \times 2}{27} = 10.7 \text{ c.y.}$$

8" wall volume (8" thick, 255 l.f., 2'-4" high):

$$\frac{0.67 \times 255 \times 2.33}{27} = 14.7 \text{ c.y.}$$

1'-8" footing volume (1'-0" thick, 255 l.f., 1'-8" wide):

$$\frac{1 \times 255 \times 1.67}{27} = 15.8 \text{ c.y.}$$

Total construction volume = 53.7 c.y.

General excavation = 306.0 c.y.   (From Example 8-8)
Total volume   = 53.7 c.y.
Backfill   = 252.3 c.y.

Backfill volume [1′-0″ (12″) wall. A second approach to calculating backfill is to calculate the areas of backfill required (Figs. 8–26 and 8–27):

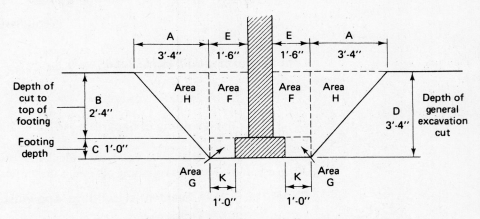

Figure 8-26   Backfill Volume (12″ wall)

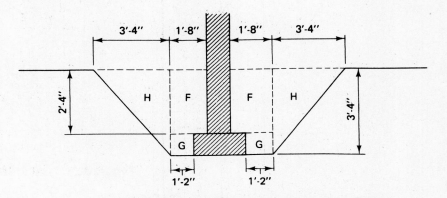

Figure 8-27   Backfill Volume (8″ wall)

Area H = $\dfrac{A \times D}{2} = \dfrac{3'\text{-}4'' \times 3'\text{-}4''}{2} = \dfrac{3.33 \times 3.33}{2} = 5.55 \times 2$ areas = 11.0 c.f.

Area F = $E \times B = 1'\text{-}6'' \times 2'\text{-}4'' = 1.5 \times 2.33 = 3.50 \times 2$ areas = 7.0 c.f.

Area G = $K \times C = 1'\text{-}0'' \times 1'\text{-}0'' = \qquad\qquad 1.0 \times 2$ areas = 2.0 c.f.

Area of backfill per lineal foot of wall = 20.0 c.f.

Length of excavation: 145 l.f. (of 12″ wall) 20.0 c.f. per lineal foot:

Volume of backfill = $\dfrac{145 \times 20}{27} = 107.4$ c.y.

backfilling

Backfill volume of 8″ wall:

Area H   = 5.5 × 2 areas = 11.0 c.f.
Area F = 1′-8″ × 2′-4″ = 1.67 × 2.33 = 3.9 × 2 areas = 7.8 c.f.
Area G = 1′-0″ × 1′-2″ = 1 × 1.17 = 1.2 × 2 areas = 2.4 c.f.

Total area of backfill = 21.2 c.f.

Length of excavation: 255 l.f. (of 8″ wall)

$$\text{Backfill volume} = \frac{255 \times 21.2}{27} = 200.2 \text{ c.y.}$$

Total volume of backfill = 107.4 + 200.2 = 307.6 c.y.

NOTE: The difference between the calculated backfill volumes of 308.3 and 307.6 c.y. is the result of rounding off and is insignificant.

EXAMPLE 8–21 — Residential Building (Appendix G):

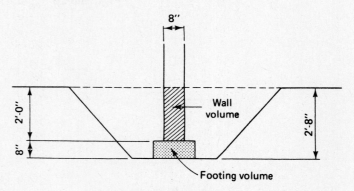

Figure 8-28 Backfilling

Wall volume (8″ thick, 148 l.f., 2′-0″ high):

$$\frac{0.67 \times 148 \times 2}{27} = 7.3 \text{ c.y.}$$

Footing volume (8″ high, 148 l.f., 1′-4″ wide):

$$\frac{0.67 \times 148 \times 1.33}{27} = 4.9 \text{ c.y.}$$

Wall and footing volume = 12.2 c.y.
General excavation = 68.4 c.y.
Wall and footing volume = -12.2 c.y.
Backfill = 56.2 c.y.

Labor. The time required to do the backfilling is taken from the charts in Fig. 8–9. The hourly wages used are based on local labor and union conditions.

EXAMPLE 8-22 — Commercial Building (Appendix F):

Backfill = 599.5 c.y. (Example 8-18)
Tractor shovel (2-1/4 c.y. capacity): $18.50 per hour
Operator: $18.35 per hour
Mobilization: 2.5 hours
Rate of work: 70 c.y. per hour
Laborer: $11.70 per hour (to work along with operator)

Backfill time:

$$\frac{599.5}{70} + 2.5 = 8.6 + 2.5 \text{ hours} = 11.1 \text{ hours}$$

Backfill cost:

11.1 hours at $11.70 per hour = $129.87
11.1 hours at $36.85 per hour = $409.04
$538.91

EXAMPLE 8-23 — Commercial Building (Appendix F, with column projections):

Backfill = 717.6 c.y. (Example 8-19)
Dozer (120 H.P.): $16.85 per hour
Rate of work: 120 c.y. per hour
Operator: $19.75 per hour
Mobilization: 1.5 hours
Laborer: $11.47 per hour

Backfill time:

$$\frac{717.6}{120} + 1.5 = 6.0 + 1.5 = 7.5 \text{ hours}$$

Backfill cost:

7.5 hours at $36.60 per hour = $274.50
7.5 hours at $11.47 per hour = $ 86.03
$360.53

EXAMPLE 8-24 — Commercial Building (Appendix A):

Backfill = 252.3 c.y. (Example 8-20)
Tractor shovel (2-1/4 c.y. capacity): $21.50 per hour
Operator: $22.18 per hour
Laborer: $12.45 per hour
Rate of work: 50 c.y. per hour
Mobilization: 2.5 hours

Backfill time:

$$\frac{252.3}{50} + 2.5 = 5.1 + 2.5 = 7.6 \text{ hours}$$

Backfill cost:

$$7.6 \text{ hours at } \$43.68 \text{ per hour } = \$331.97$$
$$7.6 \text{ hours at } \$12.45 \text{ per hour } = \underline{87.15}$$
$$419.12$$

EXAMPLE 8–25 — Residential Building (Appendix G):

Backfill = 56.2 c.y. (Example 8–21)
Tractor shovel (1 c.y. capacity): $9.50 per hour
Operator: $12.35 per hour
Laborer: $5.25 per hour
Rate of work: 15 c.y. per hour
Mobilization: 1 hour

Backfill time:

$$\frac{56.2}{15} \text{ [allow 3.7 hours] } + 1 = 3.7 + 1 = 4.7 \text{ hours}$$

Backfill cost:

$$4.7 \text{ hours at } \$21.85 \text{ per hour } = \$102.70$$
$$4.7 \text{ hours at } \$5.25 \text{ per hour } = \underline{24.68}$$
$$127.38$$

8–12 EXCESS AND BORROW

Once the excavation has been calculated in terms of excavation, backfill, and grading, the estimator must compare the total amounts of cut and fill required and determine if there will be an *excess* of materials that must be disposed of, or if there is a shortage of materials that must be brought in *(borrow)*. Topsoil is not included in the comparison at this time; topsoil must be compared separately since it is much more expensive than the other fill that might be required, and excess topsoil is easily sold.

The specifications must be checked for what must be done with the excess material. Some specifications state that it may be placed in a particular location on the site, but many times they direct the removal of excess materials from the site. If the material is to be hauled away, the first thing the estimator must know is how many cubic yards are required and then he must find a place to haul them to. Remember that soil swells; if you calculated a haul-away of 100 cubic yards and the swell is estimated at 15 percent, then you will actually have to haul away 115 cubic yards. Finding a place that the excess material can be hauled to is not always a simple matter, since it is desirable to keep the distance as short as possible from the site. The estimator should check into this when visiting the site. If material must be brought in, the estimator must first calculate the amount required and then set out to find a supply of material as close to the job site as is practical. Check the specifications for any special requirements pertaining to the type of soil that may be used. Keep in mind that the material being brought in is loose and will be compacted on the job. If it is calculated that 100 cubic yards are required, the contractor will have to haul in at least 106 to 114 cubic yards of soil, and even more if it is clay or loam.

The next step is to select equipment for the work to be done; this will depend on the amount of material, type of soil, and the distance it must be hauled.

8-13 SPREADING TOPSOIL, FINISH GRADE

Many specifications call for topsoil to be placed over the rough grade to make it a *finish grade*. The topsoil that was stockpiled may be used for this purpose, but if there is no topsoil on the site, it will have to be purchased and hauled in—a rather costly proposition. Be certain to check the specification requirements and the soil at the site to see if the existing soil can be used as topsoil. On most projects the equipment will consist of a bulldozer or scraper. The quantity of topsoil is usually calculated by multiplying the area used for rough grading by the depth of topsoil required.

After the topsoil has been spread throughout the site in certain areas, such as in the courtyards, around the buildings, along the parking areas, etc., it is necessary to hand rake the topsoil to a finish grade. The volume of soil to be moved must be calculated and entered on the estimate.

8-14 LANDSCAPING

Most specifications require at least some landscaping work. If the work is seeding and fertilizing, the general contractor may do it or subcontract it out. The estimator should check the specifications for the type of fertilizer required. The specifications may also state the amount of pounds per square that must be spread. The seed type will be included in the specifications, which will also state the number of pounds to be spread per square (1 square = 100 square feet). Who is responsible for the growth of grass? If the contractor is, be certain that it will receive adequate water and that the soil will not erode before the grass begins to grow. Often the seeded area is covered with straw or special cloth, which helps keep in moisture and cut down erosion. The estimator will have to calculate the area to be fertilized and seeded, determine what equipment will be used, arrange covering and water, and then arrange the removal of the covering if it is straw.

Sodding is often required when the owner wants an "instant lawn." The specifications will state the type of sod required, and the estimator must determine the area to be filled. The general contractor or a subcontractor may handle this phase of work. The estimator, in calculating labor costs, must consider how close to the actual area being sodded the truck bringing the sod can get, or whether the sod will have to be transferred from the truck to wheelbarrows or lift trucks and brought in.

Trees and shrubs required for the project will generally be taken care of by a subcontractor who specializes in landscaping. The estimator should note the number, size, and species required. The landscape contractor may submit a proposal to perform all the landscape work; however, it is up to the estimator to make a preliminary decision about how it will be handled.

In the estimate there must also be an allowance for the maintenance of the landscaping for whatever period is required in the specifications. Also many specifications require a guarantee period during which the contractor must replace landscaping that does not grow or dies.

8-15 PUMPING (DEWATERING)

Almost all specifications state that the contractor provide all pumping and dewatering required. The estimator should examine any soil investigation reports made on the property for the possibility of a high water table. Also to be considered is the time of year that the soil investigation was made, since the water table varies throughout the year depending on the particular area or locale in which the project is located.

Water can present a problem to the contractor in almost any location. For even if the ground water table presents no problem, there is always the possibility of ground water, after a rainfall, affecting the construction and requiring a couple of pumps for its removal.

Some projects require constant pumping to keep the excavation dry, while some require a small pump simply to remove excess ground water. The variation in costs is extreme, and the estimator must rely on past experience in a given area as a guide. If the area is unfamiliar, then the problem should be discussed with people who are familiar with the locale. Who should be asked? If the contacts in a given area are limited, consider the following sources of information: local building department, *all* retired construction engineers, superintendents, surveyors (who are the most readily available source of information)—generally the people listed above will be most honest in their appraisal of a situation and will be delighted to be of help. Other sources might be local sales representatives, the architect-engineer, and any local subcontractor, such as an excavator.

The problem of water will vary depending on season of year, location, type of work, general topography, and weather. The superintendent should be certain that during construction, the general slope of the land is away from the excavation so, in case of a sudden rainfall, it is not "washed out."

8-16 ROCK EXCAVATION

The excavation of rock is different from the excavation of ordinary soils. Rock is generally classed as soft, medium, or hard—depending on the difficulty of drilling it. Almost all rock excavation requires drilling holes into the rock and then blasting the rock into smaller pieces so it may be handled on the job. Types of drills include rotary and core jackhammers. The bits employed may be detachable, carbide, or diamond-cutting edges, solid or hollow. Among the types of explosives used are blasting powder and dynamite. Where a contractor has had no experience in blasting, a specialist is called in to perform the work required. The estimator working for a contractor who has had blasting experience will make use of the cost information from past projects.

Factors that affect rock excavation include the type of rock encountered, amount of rock to be excavated, whether bracing is required, manner of loosening the rock, equipment required, length of haul, delays, and special safety requirements. Also, the estimator must decide whether it is to be an open cut or a tunnel in which water might be encountered.

Once loosened, a cubic yard of rock will require as much as 50 percent more space than it initially occupied, depending on the type of rock encountered.

The cost of blasting will be affected by the type of rock, equipment, explosive, and depth of drilling. Only the most experienced personnel should be used and all precautions taken. Mats are often used to control the possibility of flying debris.

8-17 SUBCONTRACTORS

Contractors who specialize in excavation are available in most areas. There are certain advantages the specialized subcontractor has over many general contractors. For instance, they own a large variety of equipment, are familiar with the soil encountered throughout a given area, and know where fill can be obtained and excess cut can be hauled to. The subcontractor may bid the project as a lump sum or by the cubic yard. Either way the estimator must still prepare a complete estimate—in the first place to check the subcontractor's price since it must be checked to be certain it is neither too high or too low, and in the second place since the estimator will need the quantities to arrive at a bid price and must check all subcontractors' bids anyway.

Always discuss with subcontractors exactly what will be included in their proposals. Put it in writing so that both parties agree on what is being bid. It is customary that the general contractor perform all hand excavation and sometimes the trenching. Selection of a subcontractor is most often based on cost, but also to be considered are the equipment owned by the subcontractor and a reputation for speed and dependability.

If the subcontractor doesn't meet the construction schedule, it will probably cost far more than the couple of hundred dollars that may have been saved on initial cost.

8-18 OVERHEAD AND PROFIT

Some estimators include overhead and profit at each stage of the estimate. However, the preferable method is to consider all overhead separately and enter it as one figure, instead of in bits and pieces throughout the project. The same general approach should be used with profit. After pricing the entire project as closely as possible, the estimator should thoroughly review the problems, difficulties, etc., and put a figure in for profit.

8-19 EXCAVATION CHECKLIST

Clearing Site:
    removing trees and stumps
    clearing underbrush
    removing old materials from premises
    removing fences and rails
    removing boulders
    wrecking old buildings, foundations

underpinning existing buildings
disconnecting existing utilities
clearing shrubbery

Excavation (including backfilling):
basement
footings
foundation walls
sheet-piling
pumping
manholes
catchbasins
backfilling
tamping
blasting
grading (rough and fine)
utility trenches
grading and seeding lawns
trees
shrubbery
topsoil removal
topsoil—brought in

8-20 PILES

Piles used to support loads are referred to as *bearing piles*. They may be wood, steel "H" sections, poured-in-place concrete with metal casing removed, poured-in-place with metal casing left in place, wood and concrete, or precast concrete.

The wide variations in design, conditions of use, types of soil, and depth to be driven make accurate cost details difficult to arrive at unless the details of each particular job are known. The various types and shapes available in piles also add to the difficulty, particularly in concrete piles. It is suggested that the estimator approach subcontractors who specialize in this type of work until experience has been gained. Figure 8–29 gives approximate requirements for labor when placing bearing piles.

Sheet-piling may be wood, steel, or concrete. It is used where the excavation adjoins a property line or where the soil is not self-supporting. Sheet-piling is taken off by the surface area to be braced or sheet-piled.

Wood sheet-piling is purchased by board feet, steel piling by 100 pounds (cwt), and concrete piling by the piece. Depending on the type used, the quantity of materials required must also be determined.

The sheet-piling may be placed in the excavation and braced to hold it erect or it may be driven into the soil. No one way is least expensive. Often, after the work is done, the piles are removed *(pulled)*. Pulling may require the employment of a pile extractor for difficult jobs, or hand tools for the simple jobs. Figure 8–30 gives approximate requirements for labor when placing sheet-piling.

Estimating—The first step is to determine exactly what type and shape pile is required; next, determine the number of piles required; and third, decide the length

Type and size of pile	Approximate no. of feet driven per hour
Wood	80–130
Steel	
30–100	50–100
100–200	30–70
Concrete	
100–200	70–100
200–400	10–40

Figure 8-29 Driving Bearing Piles

Type of sheet pile	Place, drive — by hand	Place, drive — by power hammer	Brace	Pulling
Wood	4.0–9.0	3.3–8.0	2.0–3.0	3.0–5.0
Steel	3.0–9.0	3.0–7.5	2.0–3.0	3.0–5.0
Concrete	——	4.0–10.0	2.0–4.5	2.0–8.0

*Work hours required per 100 s.f. of surface area of the wall.

Figure 8-30 Sheet Piles, Approximate Work Hours Required

that they will have to be driven. With this information the material quantities and their cost can be determined.

If the contractor's crew will drive the piles, the next determination should be the type of equipment required, how many hours the work will take, and cost per hour. Then, the crew size, along with the hours and cost per hour required, are calculated. To arrive at this, an estimate of the lineal footage placed per hour (including all cutting off, moving around, etc.) must be made so that the total hours required for the completion of the pile driving is determined. In dealing with sheet-piling, if the sheets are to be pulled at the end of construction, the estimator must also include a figure for this. Costs must also comprise mobilization of workers, equipment, and material for pile driving, and demobilization when the former has been completed.

Check the soil borings for the kind of soil recorded and the approximate depth the piles must be driven. The accuracy of this information is very important.

Specifications. Check the type of piles required and under what conditions sheet-piling is required. Any requirements regarding soil conditions, equipment, or other special items should be noted. Many specifications require that only firms experienced in driving piles will be allowed to do so on the job.

8-21 PILE CHECKLIST

Type:
    wood
    wood and concrete
    "H" piles
    casing and poured-in-place
    poured-in-place
    precast
    wood sheets
    steel sheets
    precast sheets
    bracing

Equipment:
    pile driver
    compressors
    derricks
    cranes

8-22 ASPHALT PAVING

The asphalt paving required on the project is generally subcontracted out to someone specializing in paving. The general contractor's estimator will make an estimate to check the subcontractor's price.

Asphalt paving will, most commonly, be hot-mix and generally classified by traffic (heavy, medium, or light) and use (walks, driveways, courts, streets, driveways, etc.).

The estimator will be concerned with subgrade preparation, subdrains, soil sterilization, insulation course, subbase course, base courses, prime coats, and the asphalt paving required. Not all items are required on any given project so the estimator should determine which items will be required, material and equipment necessary for each portion of the work, and the requisite thickness and amount of compaction.

Specifications. Check the requirements for compaction, thickness of layers, total thicknesses, and materials required for each portion of the work. The drawings will also have to be checked for some of these items. The drawings will show the location of most of the work to be completed, but the specifications should also be checked. The specifications and drawings will list different requirements for the various uses. (These are called traffic requirements.)

Estimate. The number of square feet (or square yards) of surface area to be covered is determined, and the thickness (compacted) of each course and type of materials required is noted. Base courses and the asphalt paving are often taken off by the ton since this is the unit in which these materials must be bought. The type of asphalt and aggregate size required must also be noted. Two layers of asphalt paving are required on some projects: a coarse base mix may be used with a fine topping mix.

Equipment required may include a steel-wheel roller, trailers to transport equipment, dump trucks, paving machines, and various small tools.

To estimate the tons of material required per 1,000 s.f. of surface area, refer to Fig. 8-31. Different requirements will be listed for the various uses (walk, driveway, etc.), and the different spaces must be kept separately.

Compacted thickness (inches)	Asphalt[1] paving	Granular[2] material	Subgrade[3] material
1″	6.5	5.25	4.6
2″	13.0	10.5	9.2
3″	19.5	15.75	13.8
4″	26.0	21.0	18.4
5″	32.5	26.25	23.0
6″	39.0	31.5	27.6
8″	52.0	42.0	36.8
10″	65.0	52.5	46.0
12″	78.0	63.0	55.2

* Per 1,000 S.F. of surface area, figures include 10% waste.

1. Asphalt paving, 140–150 lbs per cu. ft.

2. Granular material, 110–120 lbs per cu. ft.

3. Subgrade material, 95–105 lbs per cu. ft.

Figure 8-31 Approximate Asphalt Paving Tonnage

In many climates the asphalt paving has a cutoff date in cold weather and the paving that is not placed when the mixing plants shut down will not be laid until the start-up time in the spring. The plants may be shut down for as long as four months or more, depending on the locale.

EXAMPLE 8–26 — Calculate the asphalt paving required for the commercial building in Appendix A.

Driveway and parking—4″ coarse asphalt over 6″ compacted crushed gravel:

$$30' \times 20' = 600 \text{ s.f.}$$
$$40' \times 85.5' = 3,420 \text{ s.f.}$$
$$20' \times 22' = 440 \text{ s.f.}$$
$$= 4,460 \text{ s.f.}$$

Sidewalk—2" fine asphalt over 4" compacted sand:

$$10.5' \times 80' = 840 \text{ s.f.}$$

Labor. Paving is done by specialty subcontractors who bid on a square-foot or lump-sum basis.

REVIEW QUESTIONS

1. Calculate all required excavation and asphalt paving required for the building shown in Appendix E. Use workup sheets and sketches.

2. How does the type of soil to be excavated affect the estimate?

3. What is soil swell? How does it affect excavation takeoffs?

4. Convert the following feet and inches to decimals:

   a. $8'3''$ d. $103'4\frac{1}{2}''$
   b. $142'8\frac{1}{2}''$ e. $5'1''$
   c. $36'7''$ f. $18'-10\frac{3}{4}''$

5. What type of equipment would you use for the excavation required in problem 1? Give your reasons for each piece of equipment chosen.

6. What are the types (classes) of excavation encountered? Give examples of each.

7. Define *excess* and *borrow*.

8. What are the prime sources of information when an estimator is unfamiliar with a given area?

9. How will the possibility of a high water table or underground stream affect the bid?

10. If a portion of the excavated material must be hauled away, what variables are involved and how will they affect the estimate?

GENERAL

All provisions of the "Sections 1A and 1B" form a part of this section.

SCOPE

The Contractor shall furnish all labor, materials, and equipment to do excavation, backfilling, grading, etc., in connection with the new building and site as may be shown or as may be found to be necessary in the fields. It shall be the responsibility of the Contractor to take whatever measures as may be necessary in the field. It shall be the responsibility of the Contractor to take whatever measures as may be necessary to keep all excavation safe and free from water.

The Contractor shall continuously maintain adequate and lawful protection of all excavations and job site. He shall protect adjacent property as provided by law. He shall provide and maintain all guard rails, fences, lights, and other facilities as required for the protection of the public and workmen for the duration of all work or until no longer required by the Owner.

SUBSURFACE SOIL DATA:

Subsurface soil investigations have been made at the project site. The location of each exploration and the information obtained are indicated on the drawings. The information was obtained for use in preparing the foundation design; however, it is also made available for the general information of bidders. Bidders are expected to examine the site and the record of investigations and then decide for themselves the character of materials to be encountered.

BENCH MARKS AND MONUMENTS:

Maintain carefully all bench marks, monuments, and other reference points; if disturbed or destroyed, replace as directed.

EXCAVATION

Excavation shall include the removal of all materials of every description existing in the areas to be excavated. Do all excavation necessary to attain the required depths for all work noted on the Drawings and/or specified.

Excavation in areas to receive earth-borne slabs shall be carried to sufficient depths to allow for the installation of slab and layer of underfill.

Extend excavation work sufficiently on each side of walls, footings, piers, etc. to permit installing forms, bracing of banks, and inspection of finished work.

Excavate all topsoil and stockpile on site for finished grade and future use.

FILL UNDER FLOOR SLABS ON GRADE:

Where fill is required to raise the sub-grade for concrete floor or terrace slabs to the elevations indicated on drawings, such fills shall be of earth or bank-run gravel, placed and compacted as specified. Either earth or bank-run gravel shall be used for fills not exceeding 12 inches deep; only bank-run gravel or other approved granular material shall be used for fills greater than 12 inches. The type and quality of material for fills shall be approved by Architects; cinders or other similar material that will corrode piping or other metal shall not be used for fills. The placing and compaction of fill under slabs after foundation walls are in place shall be coordinated with the backfilling against the outside of the walls, or walls shall be adequately braced to prevent damage.

Before depositing fill, remove all loam, vegetation, and other unsuitable material from areas to receive fill. Do not deposit fill until the subgrade has been checked and approved by Architect. In no case shall fill be placed on a subgrade that is muddy, frozen, or that contains frost.

Figure 8-32   Typical Excavation Specification

excavation

BACKFILL

Backfill material shall be suitable clean material and shall be properly compacted to insure no settlement. Material shall be subject to Architect's approval.

BACKFILLING FOR BUILDING AND STRUCTURES:

Backfill against foundation walls only after approval of the Architect has been obtained. Place and compact backfill so as to minimize settlement and avoid damage to the walls and other work in place.

Before placing fill, remove all debris subject to termite attack, rot, or corrosion, and all other deleterious materials from areas to be backfilled. Deposit backfill in layers not more than 8 inches thick. All fill material shall be reasonably free from roots, plaster, bats, and unsuitable material. Stones larger than 4 inches, maximum dimension, shall not be permitted in the upper 6 inches of fill. Place the fill material in successive horizontal layers, in loose depth as specified, for the full width of the cross section. Thoroughly compact each layer by rolling or pneumatic tamping after a light sprinkling with water. The finished subgrade shall be brought to elevations indicated and sloped to drain water away from the building walls. Fill to required elevations any areas where settlement occurs.

SITE GRADING

GRADES:

Do all cutting, filling, compacting of fills, and rough grading required to bring the entire project area, outside of buildings, to subgrades as follows:

For surfaced areas, roadways, parking areas, service courts, steps, and walks to the underside of the respective surfacing or base course, as fixed by the finished grades therefor.

For lawn and planted areas, to 4 inches below finished grade.

FINISH GRADE:

Finish grade all areas within the project. Use topsoil to bring rough grading to finished grade. Place no topsoil until subgrade has been approved. Topsoil shall be placed so that after final settlement there will be good drainage and conforming to elevations shown on the drawings. Maintain surfaces and place any additional topsoil necessary to replace that eroded before acceptance.

Figure 8-32   Typical Excavation Specification *(Continued)*

9
CONCRETE

9-1 CONCRETE WORK

The concrete for a project may be either ready-mixed or mixed on the job. Most of the concrete used on commercial and residential work is ready-mixed and delivered to the job by the ready-mix company. Quality control, proper gradation, water, and design mixes are easily obtained by the ready-mix producers, which have, in many cases, fully automated and computerized their operations. When ready-mix is used, the estimator must determine the amount of concrete required, the type of cement, aggregates, and admixtures, which are discussed with the supplier, who will then give a proposal for supplying the specific material.

Concrete for large projects, which will be mixed at the job site, will require a field batching plant. It is necessary for the estimator to have a thorough understanding of the design of concrete mixes. The basic materials required for concrete are cement, aggregates (fine and coarse), and water. Various admixtures may also be required in the mix. The estimator will have to compute the amounts of each material required, considering the materials available or the materials that will be used (perhaps shipped in on railroad cars from 1,000 miles away in special cases). No attempt will be made here to show the design of concrete mixes. It is suggested that the estimator who is unfamiliar with mix design take a course in concrete mix design and control. Courses are offered at many college campuses (night school) and in conjunction with local building construction organizations. Contact your local organization for information concerning such classes. If no classes are available, perhaps the local builders' organization would be interested in sponsoring such a course at a local high school, technical school, or college. Publications are available from the Portland Cement Association, 5420 Old Orchard Road, Skokie, Illinois 60077, which explain the design and should be a part of your library.

Concrete is one of the oldest materials used and still one of the least understood by those who use it. Subscribe to the various technical magazines to keep up with the developments.

9-2 ESTIMATING CONCRETE

Concrete is estimated by the cubic yard, or by the cubic foot and converted into cubic yards. The cubic yard is used since it is the pricing unit of the ready-mix companies, and most tables and charts available relate to the cubic yard.

Roof and floor slabs, slabs on grade, pavements, and sidewalks are most commonly measured and taken off in length, width, and thickness and converted to cubic feet and cubic yards (27 cubic feet = 1 cubic yard). Often, irregularly shaped projects are broken down into smaller areas for more accurate and convenient manipulation.

Footings, columns, beams, and girders are estimated by taking the lineal footage of each item times its width and depth. The cubic footage of the various items may then be tabulated and converted to cubic yards.

In estimating the footings for buildings with irregular shapes and many jogs, include the corners only once. If necessary, mark lightly on the plans which portions of the footing have been figured as you proceed. When taking measurements, keep in mind that the footing projects out from the wall, and therefore the footing length is greater than the wall length.

In completing quantities, make no deductions for holes smaller than 2 square feet or for the space that reinforcing bars or other miscellaneous accessories take up. Waste ranges from 5 percent on footings, columns, and beams to 8 percent for slabs.

The procedure which should be used to estimate the concrete on a project would be:

1. Review the specifications to determine the requirements for each area in which concrete is used separately (such as footings, floor slabs, and walkways) and list the following:
   a. type of concrete
   b. strength of concrete
   c. color of concrete
   d. any special curing or testing.

2. Review the drawings to be certain that all concrete items shown on the drawings are covered in the specification. If not, a call will have to be made to the architect-engineer so an addendum can be issued. (Refer also to Secs. 1-4 and 3-10.)

3. List each of the concrete items required on the project. (Precast concrete is covered in Sec. 12-1.)

4. Determine the quantities required for the working drawings. Footing sizes are checked on the wall sections and foundation plans. Watch for different size footings under different walls.

Concrete slab information will most commonly be found on wall sections and floor plans. Exterior walks and driveways will most likely be identified on the plot (site) plan and in sections and details.

An estimate of the concrete required for the commercial building in Appendix A (see examples in Chapter 8, "Excavation") and for the residential building in Appendix C (see examples in Chapter 8, "Excavation") will be prepared.

EXAMPLE 9-1 — Commercial Building (Appendix A): Find the amount of concrete required for footings.

First, determine the lengths of the various size footings. Under the 1'-0" wide wall a 2'-0" footing is required, while under the 8" wide wall a 1'-8" footing is required, 1'-0" deep.

$$2'-0'' \text{ wide footings}—146 \text{ l.f.} = \quad 2 \times 146 \times 1 = 292 \text{ c.f.}$$
$$1'-8'' \text{ wide footings}—250 \text{ l.f.} = 1.67 \times 250 \times 1 = \underline{417 \text{ c.f.}}$$
$$709 \text{ c.f.}$$

Concrete for footings:

$$\frac{709}{27} = 26.3 \text{ c.y.}$$

$$+ 5\% \text{ waste} = \underline{1.3 \text{ c.y.}}$$
$$27.6 \text{ c.y.} = 28 \text{ c.y. required}$$

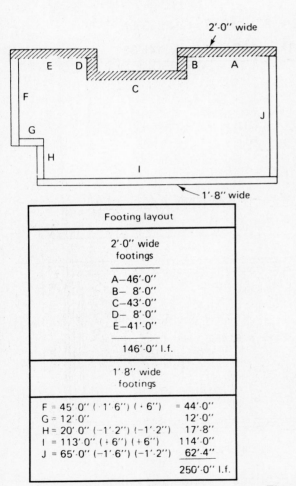

Footing layout
2'-0'' wide footings
A—46'-0''
B— 8'-0''
C—43'-0''
D— 8'-0''
E—41'-0''
146'-0'' l.f.
1'-8'' wide footings

F = 45'-0'' (-1'-6'') (+6'')	= 44'-0''
G = 12'-0''	12'-0''
H = 20'-0'' (-1'-2'') (-1'-2'')	17'-8''
I = 113'-0'' (+6'') (+6'')	114'-0''
J = 65'-0'' (-1'-6'') (-1'-2'')	62'-4''
	250'-0'' l.f.

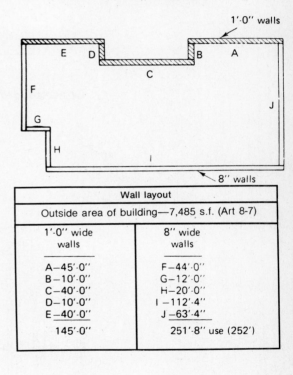

Wall layout	
Outside area of building—7,485 s.f. (Art 8-7)	
1'-0'' wide walls	8'' wide walls
A—45'-0''	F—44'-0''
B—10'-0''	G—12'-0''
C—40'-0''	H—20'-0''
D—10'-0''	I —112'-4''
E—40'-0''	J —63'-4''
145'-0''	251'-8'' use (252')

Figure 9-1 Concrete Footings—Commercial Building

The building area (Sec. 8-7) is 7,485 square feet. To determine the square feet of concrete slab required, it is necessary to subtract the area of the exterior walls. 252 lineal feet of exterior wall is 8-inch block, 145 lineal feet is 12-inch block, and the slab is 4 inches thick. (Building drawings and specifications in Appendix A).

$$\begin{aligned}
&\text{Area of building} && = 7,485 \text{ s.f.}\\
\text{Area of 8-inch block } (252 \times 0.67) &= 168 \text{ s.f.}\\
\text{Area of 12-inch block } (145 \times 1) &= \underline{145} \text{ s.f.}\\
& \quad\quad\quad\quad\quad 313 \text{ s.f.} && \underline{\quad 313} \text{ s.f.}\\
& && 7,172 \text{ net s.f. of conc. slab}
\end{aligned}$$

$$7,172 \text{ s.f.} \times 0.33 = 2,367 \text{ c.f.} = 87.7 \text{ c.y.}$$
$$+ \ 8\% \text{ waste} = \underline{\ \ 7.0} \text{ c.y.}$$
$$94.7 \text{ c.y.} = 95 \text{ c.y. required}$$

EXAMPLE 9-2 — Residential Building (Appendix G):
Footings:

$$50.67 + 22 + 22 + 50.67 = 145.33 \text{ l.f.}$$

$$\frac{8'' \times 1'\text{-}4'' \times 145.33'}{27} = \frac{0.67 \times 1.33 \times 145.33}{27} = 4.8 \text{ c.y.}$$

Concrete for footings:

$$\begin{aligned}
&\quad\quad\quad\quad\quad 4.80 \text{ c.y.}\\
+ \text{ waste} &= \underline{\ \ .20} \text{ c.y.}\\
&\quad\quad\quad 5 \quad \text{c.y. required}
\end{aligned}$$

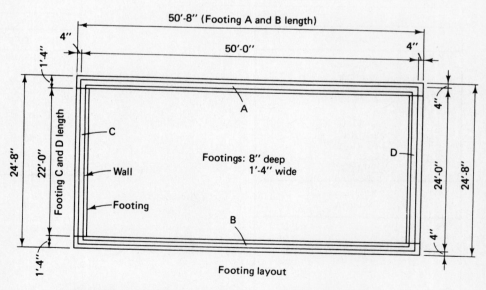

Figure 9-2 Concrete Footings—Residential Buildings

concrete

128

Labor. The time required for the concrete work has been estimated from the chart in Fig. 9-3. The hourly wages are based on local labor and union conditions.

Type of Placement	Work hours per c.y.
Columns	1–4
Footings (up to 8 c.y.)	1–1.5
Footings (over 8 c.y.)	2–4
Slabs (basement, 1st floor)	
up to 1,000 s.f.	1–2
1,000 to 5,000 s.f.	2–3
over 5,000 s.f.	3–4
Stairs	1.5–4
Structural slabs	1.5–3
Thin walls (6 in. or less)	1.5–4
Thick walls	2–5

Figure 9-3 Work Hours for Concrete Work

EXAMPLE 9–3 — Commercial Building:

Footings = 28 c.y.
Rate of work: 2.5 c.y. per work hour
Labor: $11.75 per hour

$$Time = \frac{28}{2.5} = 11.2 \ work \ hours$$

Cost = 11.2 hours × $11.75 per hour = $131.60

NOTE: Forms, reinforcing, finishing, etc., are covered later in this chapter.

Slab = 95 c.y.
Rate of work: 4 c.y. per work hour
Labor: $11.75 per hour

$$Time = \frac{95}{4} = 23.75 \ work \ hours$$

Cost = 23.75 hours × $11.75 per hour = $279.06

EXAMPLE 9-4 — Residential Building

Footings = 5 c.y.
Rate of work: 2.5 c.y. per work hour
Labor: $7.50 per hour

$$Time = \frac{5}{2.5} = 2 \ work \ hours$$

Cost = 2 hours × $7.50 per hour = $15.00

For example, when you are taking off the wall footings, list all of the work to be considered.

Wall Footings:
    forms—
    accessories—
    reinforcing—

concrete—(3,000 psi)

curing—

If the portion of work being considered is a concrete floor, it may include the following items:

Concrete Floors:
vapor barrier—
reinforcing—
expansion joints—
concrete—(2,500 psi)
finishing—

9-3 REINFORCING

The reinforcing used in concrete may be either reinforcing bars, welded wire mesh, or a combination of the two. Reinforcing bars are listed (noted) by the bar number, which corresponds to the bar diameter in eighths of an inch. For example, a #7 bar (deformed) contains as much metal as a $\frac{7}{8}$-inch diameter bar. The #2 bar is a plain round bar, but all the rest are deformed round bars. The bar numbers, diameters, areas, and weights are given in Fig. 9–4.

Bar no.	Nominal size — inches	Area sq. in.	Weight lbs/ft
2	$\frac{1}{4}''$ ϕ	0.05	0.167
3	$\frac{3}{8}''$ ϕ	0.11	0.376
4	$\frac{1}{2}''$ ϕ	0.20	0.668
5	$\frac{5}{8}''$ ϕ	0.31	1.043
6	$\frac{3}{4}''$ ϕ	0.44	1.502
7	$\frac{7}{8}''$ ϕ	0.60	2.044
8	1.0'' ϕ	0.79	2.670
9	1.13'' ϕ	1.00	3.400
10	1.27'' ϕ	1.27	4.303
11	1.41'' ϕ	1.56	5.313
14	1.	2.41	8.18
18		3.98	13.52

No. 2 bar is a plain round bar. All other bars are deformed.

Figure 9-4 Weights and Areas of Reinforcing Bars

Reinforcing bars are taken off by lineal feet. The takeoff (workup) sheet should be set up to include the numbers of the bars, number of pieces, lengths, and bends. Since reinforcing bars are usually priced by the hundredweight (100 pounds, cwt), the weight of reinforcing required must be calculated. Usually, the steel can be bought from the mill or main warehouses, and the required bars will be cut, bundled, and tied. Bars will also be bent to job requirements at these central points. Bars purchased at smaller local warehouses are generally bought in 20-foot lengths, and cut and bends are made by the contractor. This process is usually more expensive and there is more waste. Whenever time permits, order the reinforcing bars from the mill or main warehouse. Often they will provide shop drawings for the bars; also, this facilitates the handling and placing of the bars on the job. The types of steel commonly used include billet steel (A15) and rail steel (A16). The type of steel required should be noted on the takeoff and summary sheet. Refer to Fig. 9-6 for work hours required for required for reinforcing bar bending and placing. Check the specifications to determine the type steel required and whether it is plain or must be zinc-coated or galvanized. Zinc coating and galvanizing can increase material cost by as much as 150 percent and often delay delivery by two weeks or more. An allowance for splicing (lapping) the bars (Fig. 9-5) must be included also (lap splicing costs may range from 5 to 15 percent, depending on the size of the bar and yield strength of steel used). Waste may range from less than 1 percent for precut and preformed bars to 10 percent when the bars are cut and bent on the job site.

Mesh reinforcing may be welded wire mesh or expanded metal. The former is

Bar size	Splice required when specified as number of bar diameter	
	24 d	30 d
3	N.R.	N.R.
4	1'-0''	1'-3''
5	1'-3''	1'-7''
6	1'-6''	1'-11''
7	1'-9''	2'-3''
8	2'-0''	2'-6''
9	2'-4''	2'-10''
10	2'-7''	3'-3''
11	2'10''	3'-7''

The recommended minimum length of splice is 1'-0'' or 24 bar diameters whichever is greater.

N.R. — No recommended, use 1'-0''.

Figure 9-5 Splice Requirements

Size of bar inches	Hooks Hrs. per 100 hooks		Bends Hrs. per 100 bends	
	By hand	By machine	By hand	By machine
$\frac{1}{2}$ and less	2–5	1–3	1.5 – 3	0.6 – 2
$\frac{5}{8}$ and $\frac{3}{4}$	3–7	1.4–4	3 – 6	0.8 – 2.5
$\frac{7}{8}$ and up	—	2–6	—	1.5 – 4

*Approximate

Size of bar inches	Place 1000 L.F. inc. chairs, stirrups, ties and wiring	Place 1000 L.F. tie wire only	Place 100 pieces short lengths and tie wire
$\frac{1}{2}$ and less	2 – 4	1.5 – 3	1.5 – 4
$\frac{5}{8}$ and $\frac{3}{4}$	2.5 – 5	2.0 – 4	1.8 – 4.5
$\frac{7}{8}$, 1 and $1\frac{1}{8}$	3.0 – 6	2.5 – 5	2.3 – 5
$1\frac{1}{4}$ and $1\frac{1}{2}$	3.5 – 7	3.0 – 6	3 – 6

*Approximate

Figure 9-6 Work Hours Required for Hooks and Bends (top) and for Placing Reinforcing Bars (bottom)

an economical reinforcing for floor and driveways, and is commonly used as temperature reinforcing and beam and column wrapping. It is usually furnished in a square or rectangular mesh arrangement of the wires and usually sold in a roll or in flat sheets (of a variety of sizes, usually specially ordered). The rolls are 5 feet wide and 150 feet long. Wire mesh is designated on the drawings by wire spacing and wire gauge in the following manner: 6 × 6 10/10. This designation shows that the longitudinal and transverse wires are spaced 6 inches on center, while both wires are No. 10 gauge. Another example, 4 × 8 8/12, means that the longitudinal wires are spaced 4 inches on center while the transverse wires are spaced 8 inches on center; the longitudinal wire is No. 8 gauge while the transverse is No. 12 gauge. The takeoff must be broken up into the various sizes required and the number of square feet required of each type. It is commonly specified that wire mesh have a lap of one square and this allowance must be included; for example, a 6-inch lap requires 10 percent extra mesh while a 4-inch lap requires almost 6.7 percent extra. The mesh may be either plain or galvanized, and this information is included in the specifications. Galvanized mesh may require special ordering and delivery times of two and three weeks.

The ¾-inch rib lath is designed primarily as a reinforcement for concrete floor and roof construction. The ¾-inch ribs are spaced 6 inches on centers, and this reinforcement is available only in coated copper alloy steel, 0.60 and 0.75 pounds per square foot. The width available is 24 inches with lengths of 96, 120, and 144 inches, packed six sheets to a bundle (96, 120, and 144 square feet, respectively). Allow for any required lapping and 3 percent for waste (for rectangular spaces).

Let us consider at this time how the steel will be held in the forms. It may be supported by concrete bricks, bar chairs, spacers, or bolsters, or it may be suspended down by wires. The supports may be plastic, steel galvanized, or zinc-coated steel, steel with plastic coated legs, and other materials. If the finished concrete will be exposed to view and the supports are touching the portion to be exposed, consideration should be given to using noncorrosive supports. Steel (and even zinc or galvanized steel) has a tendency to rust and this rust may show through on the exposed finish. There is a wide selection of accessories and supports available. Complete brochures should be included in your files.

When the reinforcing must be fabricated into a special shape, perhaps round or rectangular, it is usually cheaper to have this done at a fabricating plant, which has the equipment for forming the shapes required. There will thus be a charge in addition to the base cost of the reinforcing, but the process provides speed and economy in most cases.

Also used for reinforcing concrete are corrugated steel subfloor systems. When corrugated steel floor deck material is used as reinforcing for the concrete, it acts also as a form for the concrete that is to be poured on top of it. The system may simply be corrugated deck with concrete or may be as elaborate as supplying in-floor distribution of electric, hot air, and telephone requirements. The more elaborate the system, the more coordination between trades that is required.

Steel deck subfloors are taken off by the square footage required (also discussed in Chapter 12). Available in a variety of heights and widths, the type used will depend on the span and loading requirements of the particular job. Finishes include galvanized, galvanized with primer on the underside, and phosphate-treated on upper surface with primer on the underside.

Wire mesh is sometimes specified, for use as a temperature steel, in conjunction with the steel decking. Don't forget to include it in your takeoff when it is required.

Estimating reinforcing bars. The lineal footage of re-bars can most often be worked up from the concrete calculations. The sections and details must be checked to determine the reinforcing requirements of the various footings. The various footing sizes can generally be taken from the concrete calculations and adapted to the reinforcing takeoff.

Estimating wire mesh. The square footage of floor area to be covered may be taken from the slab concrete calculations. Check the sections and details for the size of the mesh required. To determine the number of rolls required, add the lap required to the area to be covered and divide by 750 (the square footage in a roll). Waste averages about 5 percent unless a good deal of cutting is required; only full rolls may be purchased in most cases.

EXAMPLES 9-5 — Commercial Building (Appendix A): The 2'-0" wide footing requires four #4 re-bars, while the 1'-8" wide footing requires three #4 re-bars (#4 bar weighs 0.668 lb./foot). See footing layout sketch (Fig. 9-1).

$$2'\text{-}0'' \text{ wide footing}-146 \text{ l.f. (4 per foot)} = 584 \text{ l.f.}$$
$$1'\text{-}8'' \text{ wide footing}-250 \text{ l.f. (3 per foot)} = 750 \text{ l.f.}$$
$$\underline{}$$
$$1,334 \text{ l.f.}$$
$$+ \ 10\% \text{ lap and waste} = 134 \text{ l.f.}$$
$$\underline{}$$
$$1,468 \text{ l.f.}$$

$$\text{Weight} = 1,468 \text{ l.f.} \times 0.668 \text{ lb./ft.} = 980 \text{ lb.}$$

The square footage of the floor area is taken from Sec. 9-2. 6 × 6 10/10 mesh is required (10% lap).

$$\text{Floor area} = 7,172 \text{ s.f.}$$
$$+ \ 10\% \text{ lap} = 718 \text{ s.f.}$$
$$\underline{}$$
$$7,890 \text{ s.f.}$$

$$\frac{7,890}{750} = 10.5 \text{ rolls (order 11 rolls}-8,250 \text{ s.f.)}$$

EXAMPLE 9-6 — Residential Building (Appendix G):

Footing: 1'-4" wide
Re-bars: Two #4 bars
Footing length: 146 l.f. (Fig. 9-2)

$$146 \text{ l.f.} \times 2 = 292 \text{ l.f.}$$
$$+ \ 10\% \text{ lap and waste} = 28 \text{ l.f.}$$
$$\underline{}$$
$$320 \text{ l.f. required}$$

$$\text{Weight} = 320 \text{ l.f.} \times 0.668 \text{ lb./ft.} = 214 \text{ lb.}$$

Labor. The time required for the reinforcing has been estimated from the charts in Fig. 9-3. The hourly wages are based on local labor and union conditions.

EXAMPLE 9-7—Commercial Building (Appendix A):

Footing re-bars: 1,468 l.f. #4 bars (1/2" diam.)
Rate of work: 2.5 work hours per 1,000 l.f.
Labor: $11.75 per hour

$$\text{Time} = 1.468 \times 2.5 = 3.67 \text{ work hours}$$
$$\text{Cost} = 3.67 \text{ hours} \times \$11.75 \text{ per hour} = \$49.12$$

Wire Mesh: 11 rolls, 8,250 s.f. 6 × 6 10/10
Rate of work: 3 work hours per 1,000 s.f.
Labor: $11.75 per hour

$$\text{Time} = 8.25 \times 3 = 24.75 \text{ work hours}$$
$$\text{Cost} = 24.75 \text{ hours} \times \$11.75 \text{ per hour} = \$290.81$$

EXAMPLE 9-8 — Residential Building (Appendix G):

Footing re-bars: 320 l.f. #4 bars
Rate of work: 4.0 work hours per 1,000 l.f.
Labor: $6.75 per hour

$$\text{Time} = 0.32 \times 4.0 = 1.28 \text{ work hours}$$
$$\text{Cost} = 1.28 \text{ hours} \times \$6.75 \text{ per hour} = \$8.64$$

9-4 VAPOR BARRIER

The *vapor barrier* placed between the gravel and the slab poured on it is usually included in the concrete portion of the takeoff. This vapor barrier most commonly consists of polyethylene films or kraft papers. The polyethylene films are designated by the required mil thickness (usually 4 or 6 mil). The material used should be lapped about 6 inches, so an allowance for this must be made depending on the widths available. Polyethylene rolls are available in widths of 3, 4, 6, 8, 10, 12, 16, 18, and 20 feet, and are 100 feet long. Careful planning can significantly cut down on waste, which should average 5 percent plus lapping. Two workers can place 1,000 square feet of vapor barrier on the gravel in about one hour, including the time required to get the material from storage, place, and secure it. Large areas can be covered in proportionately less time.

Estimating vapor barrier. The most accurate method of determining the vapor barrier is to sketch a layout of how it might be on the job. Often, several sketches must be made, trying rolls of various sizes, before the most economical arrangement is arrived at.

EXAMPLE 9-9 — Commercial Building: Vapor barrier (from sketch—Fig. 9-7):

3 rolls 20 feet wide	=	6,000 s.f.
1 roll 16 feet wide	=	1,600 s.f.
		7,600 s.f. of vapor barrier (4 mil)

Note that only 7,172 square feet of slab is required, and yet 7,600 square feet of vapor barrier must be purchased.

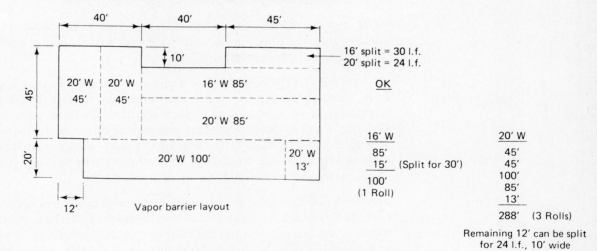

Figure 9-7 Vapor Barrier—Commercial Building

(A careful check of the specifications and drawings shows that there is no vapor barrier required on this residence).

Labor. The time required for the installation of the vapor barrier has been estimated from the chart in Fig. 9–8. The hourly wages are based on local labor and union conditions.

Sq. ft. of vapor barrier	Hours per 1,000 s.f.
Up to 1,000	1-1.5
1,000 to 5,000	0.75-1.5
Over 5,000	0.5-1.5

Figure 9-8 Work Hours Required for Installing Vapor Barriers

EXAMPLE 9–10 — Commercial Building:

Vapor barrier: 7,600 s.f.
Rate of work: 1.1 work hours per 1,000 s.f.
Labor: $11.75 per hour

$$\text{Time} = 7,600 \times 1.1 = 8.4 \text{ work hours}$$
$$\text{Cost} = 8.4 \times \$11.75 = \$98.70$$

9–5 ACCESSORIES

Any item cast into the concrete should be included in the concrete take-off. The list of items that might be included is extensive; the materials vary depending on the item and intended usage. The accessory items may include:

EXPANSION JOINT COVERS — Made of aluminum and bronze, expansion joint covers are available in the usual wide variety of shapes for various uses. Take-off by the lineal footage required.

EXPANSION JOINT FILLER — Materials commonly used as fillers are asphalt, fiber, sponge rubber, cork, and asphalt impregnated fiber. These are available in thicknesses of ¼, ⅜, ½, ¾, and 1 inch; widths of 2 to 8 inches are most common. However, sheets of filler are available and may be cut to the desired width on the job. Whenever possible, filler of the width to be used should be ordered to save labor costs and reduce waste. Lengths of filler strips may be up to 10 feet, and the filler should be taken off by lineal feet plus 5 percent waste.

WATERSTOPS — Used to seal construction and expansion joints in poured concrete structures against leakage caused by hydrostatic pressure, waterstops are commonly composed of polyvinyl chloride, rubber, and neoprene in a variety of widths and shapes. Do your takeoff in lineal feet, check the size roll in which the specified waterstop is available, and add 5 percent for waste. No partial rolls may be purchased, only full rolls (usually 50 feet long).

MANHOLE FRAMES AND COVERS — Manhole frames and covers are available in round or square shapes. The frame is cast in the concrete, and the cover put in the frame later. The materials used are aluminum (lighter duty) and cast iron (light and heavy duty). The covers may be recessed to receive tile, plain, abrasive, or have holes put in it, depending on the intended usage and desired appearance. The size, material, type, and installation appear in the specifications and on the drawings. They are taken off by the actual number of each type required. The various size frames and types of covers all cost varying amounts, so keep all different items separate.

TRENCH FRAMES AND COVERS — The trench frame is cast in the concrete. The lineal footage of frames required must be determined and the number of inside and outside corners noted as well as the frame type. Cast-iron frames are available in 3-foot lengths, while aluminum frames are available up to 20 feet long. Covers may also be cast iron or aluminum in a variety of finishes (perforated, abrasive, plain, recessed).

The takeoff is made in lineal feet with the widths, material, and finish all noted. Cast-iron covers are available, most commonly, in 2-foot lengths while aluminum frames are available in 10-, 12-, and 20-foot lengths, depending on the finish required. Many cast-iron trench manufacturers will supply fractional sizes of covers and frames to fit whatever size trench length is required. Unless the aluminum frames and covers can be purchased in such a manner, sizable waste may be incurred on small jobs.

Miscellaneous accessories such as anchor bolts, bar supports, screed chairs, screw anchors, screw anchor bolts, plugs, inserts of all types (to receive screws and bolts), anchors, and splices for reinforcing bars must be included in the takeoff. These accessories are taken off by the number required; they may be priced individually or in 10's, 25's, 50's, or 100's, depending on the type and particular manufacturer. Carefully note the size required since so many sizes are available. The material (usually steel, cast iron, or plastic) should be listed on the drawings or in the specifications. Some inserts are also available in bronze and stainless steel.

Accessories such as reglets, dovetail anchor slots, and slotted inserts are taken off by the lineal footage required. Available in various widths, thicknesses, and lengths, they may be made out of cold rolled steel, galvanized, or zinc-coated steel, bronze, stainless steel, copper, and aluminum.

Estimate expansion joint filler. Determine from the plans, details and sections exactly how many lineal feet are required. A ½-inch thick expansion joint filler is required where any concrete slab abuts a vertical surface. This would mean that filler would be required around the outside of the slab, between it and the wall.

EXAMPLE 9-11 — Accessories, Commercial Building:

Expansion joint filler (from sketch—Fig. 9-9):

Expansion joint filler = 400 l.f.

Labor. The time required for the installation of the expansion joint filler has been estimated from the chart in Fig. 9-10. The hourly wages are based on local labor and union conditions.

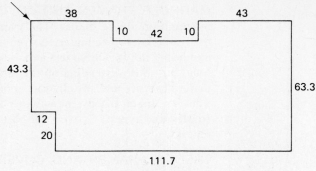

I.f. = 38 + 10 + 42 + 10 + 43 + 63.3 + 111.7 + 20 + 12 + 43.3 = 393.3 l.f.
Order 400 l.f. of $\frac{1}{2}''$ × 4'' expansion joint filler

Figure 9-9 Expansion Joint Filler

I.f. of expansion joint	Hours per 100 l.f.
Up to 500	1- 1.5
500 to 1,000	0.75 1.5
Over 1,000	0.5 1.5

Figure 9-10 Work Hours Required for Installing Expansion Joint Filler

EXAMPLE 9-12 — Commercial Building:

Rate of work: 1.0 work hour per 100 l.f. (Fig. 9-10)
Labor: $11.75 per hour

Time = 4.0 × 1.0 = 4 work hours

Cost = 4 × $11.75 = $47.00

9-6 CONCRETE FINISHING

All exposed concrete surfaces require some type of *finishing*. Basically finishing consists of the patch-up work after the removal of forms and the dressing up of the surface by troweling, sandblasting, and other methods.

Patch-up work may include patching voids and stone packets, and removing fins and patching chips. Except for some floor slabs (on grade), there is always a certain amount of this type of work on exposed surfaces. It varies considerably from job to job and can be kept to a minimum with good quality concrete, with use of forms that are tight and in good repair, and with careful workmanship, especially in stripping the forms. This may be included with the form stripping costs, or it may be a separate item. As a separate item, it is much easier to get cost figures and keep a cost control on the particular item rather than "bury" it in with stripping costs.

Small patches are usually made with a cement-sand grout mix of 1:2; be certain the type of cement (even the brand name) is the same used in the pour, since different cements are varying shades of gray. The work hours required will depend on the type of surface, number of blemishes, and the quality of the patch job required. Scaffolding will be required for work above 6 feet.

The finishes required on the concrete surfaces will vary throughout the project. The finishes are included in the specifications and finish schedules; sections and details should also be checked. Finishes commonly required for floors include hand or machine troweled, carborundum rubbing (machine or hand), wood float, broom, floor hardeners, and sealers. Walls and ceilings may also be troweled, but they often receive decorative surfaces such as bushhammered, exposed aggregate, rubbed, sandblasted, ground, and lightly sanded. Finishes such as troweled, ground, sanded, wood float, broom finishes, and bushhammered require no materials to get the desired finish, only labor and equipment. Exposed aggregate finishes may be of two types. In the first, a retarder is used on the form liner, and then the retarder is sprayed off and surface cleaned. This requires the purchase of a retarding agent, spray equipment to coat the liner, and a hose with water and brushes to clean the surface. (These must be added to material costs.) The second method is to spray or trowel an exposed aggregate finish on the concrete; it may be a two- or three-coat process, and both materials and equipment are required. For best results it is recommended only experienced technicians should place this finish. Subcontractors will price this application by the square foot.

Rubbed finishes, either with burlap and grout or with float, require both materials and work hours plus a few hand tools. The burlap and grout rubbed finish requires less material and more labor than the float finish. A mixer may be required to mix the grout.

Sandblasting requires equipment, labor, and the grit to sandblast the surface. It may be a light, medium, or heavy sandblasting job with best results usually occurring with green to partially cured concrete.

Bushhammering, surface finish technique, is done to expose portions of the aggregates and concrete. It may be done by hand, with chisel and hammer, or with pneumatic hammers. The hammers are commonly used, but hand chiseling is not uncommon. Obviously hand chiseling will raise the cost of finishing considerably.

Other surface finishes may also be encountered. Analyze thoroughly for each finish the operations involved, material and equipment required, and work hours needed to do the work.

The finishing of concrete surfaces is estimated by the square foot, except bases, curbs, and sills, which are estimated by the lineal foot. Since various finishes will be required throughout, keep the takeoff for each one separately. The work hours required (approximate) for the various types of finishes are shown in Fig. 9–8. Charts of this type should never take the place of experience and common sense, and are included only as a guide.

Materials for most operations (except exposed aggregate or other coating) will cost only 10 to 20 percent of labor. The equipment required will depend on the type of finishing work being done. Trowels (hand and machine), floats, burlap, sandblasting equipment, sprayers, small mixers, scaffolding, and small hand tools must be included with the costs of their respective items of finishing.

Estimate concrete finishing.   Areas to be finished may be taken from other concrete calculations, either for the actual concrete required or the square footage of forms required.

Roof and floor slabs, and slabs on grade, pavements, and sidewalk areas can most easily be taken from actual concrete required. Be careful to separate each area requiring a different finish. Footing, column, walls, beam and girder areas are most commonly found in the form calculations.

EXAMPLE 9–13 — Commercial Building:

Concrete finishing: slab 7,172 s.f. (from Fig. 9–1)
Machine trowel

The residence requires no concrete finishing; leveling the top of the footing is considered by most estimators to be part of the job of placing the concrete.

Labor. The time required for the finishing of the concrete has been estimated from the chart in Fig. 9-11. The hourly wages are based on local labor and union conditions.

Type of finishing		Work hours
Troweling, machine	— slabs on ground (100 S.F.)	1—2
hand	— slabs on ground (100 S.F.)	2—4
Broom finish (brush)	— slabs (100 S.F.)	$\frac{1}{2}$—2
	— stairs (100 S.F.)	3—6
Float finish	— slabs on grade (100 S.F.)	1—3
	— slabs suspended (+ 10%)	
	— walls (100 S.F.)	2—5
	— stairs (100 S.F.)	2—4
	— curbs (100 L.F.)	5—12
Bushhammer, machine	— green concrete (100 S.F.)	2—4
	— cured concrete (100 S.F.)	4—8
hand	— green concrete (100 S.F.)	6—12
	— cured concrete (100 S.F.)	—
Rubbed	— with burlap and grout (100 S.F.)	1—4
	— with float finish (100 S.F.)	2—6
Exposed aggregrate	— retarder, apply and clean concrete	2—3
Sandblasting	— light penetration	2—4
	— heavy penetration	4—6
Patching	— tie holes and honeycombing (100 S.F.)	1—3
Acid wash	— 100 (S.F.)	1—4
Hardeners and Sealers (spray) — floors (100 S.F.)		$\frac{1}{2}$—1

Figure 9-11 Work Hours for Finishing Concrete Surfaces

EXAMPLE 9–14 — Commercial Building:

Concrete slab
Machine trowel 7,172 s.f.

Rate of work: 1 work hour per 100 s.f.
Labor: $11.75 per hour

$$\text{Time} = 71.72 \times 1 = 71.72 \text{ hours}$$
$$\text{Cost} = 71.72 \text{ hours} \times \$11.75 \text{ per hour} = \$842.71$$

9-7 CURING

Proper *curing* is one of the important factors in obtaining good concrete. The concrete cures with the chemical combinations between the cement and water. This process is referred to as *hydration.* Hydration (chemical combination) requires time, favorable temperatures, and moisture; the period during which the concrete is subjected to favorable temperature and moisture conditions is referred to as the *curing period.* The specified curing period usually ranges from three to 14 days. Curing is the final step in the production of good concrete.

Moisture can be retained by the following methods: leaving the forms in place, sprinkling, ponding, spray mists, moisture retention covers, and seal coats. Sometimes combinations of methods may be used, and the forms left on the sides and bottom of the concrete while moisture retention covers or a seal coat is placed over the top. If the forms are left in place in dry weather, wetting of the forms may be required since they have a tendency to dry out; heat may be required in cold weather.

Sprinkling fine sprays of water on the concrete helps keep it moist, but caution is advised in the drying out of the surface between sprinklings, since the surface will then have a tendency to craze or crack.

Temperature must also be maintained on the concrete. With cold weather construction it may be necessary to begin with heated materials and provide a heated enclosure. While the hydration of cement causes a certain amount of heat (referred to as the *heat of hydration*) and is of some help, additional heat is usually required. Fresh concrete should not be above 70° F. since such temperatures result in lower strengths. The forms may be heated by use of steam, while the enclosed space itself may be heated by steam pipes or unit heaters of the natural gas, LP gas, or kerosene type. The number required varies on the size of the heater used. Heaters are available from about 50,000 to 300,000 BTUH. Since many unions require that an operator work only with the heaters, local union rules should be checked. With operators this heating cost is as much as 20 times the cost per cubic yard as without, depending on the number of operators required.

During mild weather (40-50°F.) simply heating the mixing water may be sufficient for the placement of the concrete; but if temperatures are expected to drop, it may be necessary to apply some external heat. Specifications usually call for a temperature of 60 to 70°F. for the first three days or above 50°F. for the first five days.

Hot weather concreting also has its considerations. High temperatures affect the strength and other qualities of the concrete. To keep the temperature down, it may be necessary to wet down the aggregate stockpiles with cool water and chill the mixing water. Wood forms should be wetted down to avoid the absorption of the moisture, needed for hydration, by the forms.

A continuous spray mist is one of the best methods of keeping an exposed surface moist. There is no tendency for the surface to craze or crack with a continuous spray mist. Special equipment can be set up to maintain the spray mist so it will not require an excessive number of work hours.

Moisture-retaining covers such as burlap are also used. They should thoroughly cover the exposed surfaces and must be kept moist continuously since they dry out (sometimes rapidly). Straw and canvas are also used as moisture-retaining covers. Burlap and canvas may be reused on many jobs, thus spreading their initial cost out considerably.

Watertight covers may be used on horizontal areas such as floors. Materials employed are usually papers and polyethylene films. Seams should be overlapped and taped, and the papers used should be nonstaining. Often these materials may be reused several times with careful handling and storage.

Sealing compounds are usually applied right after the concrete surface has been finished. There are both one- and two-coat applications that are colorless, black, or white. If other materials must be bonded to the concrete surface at a later date, be certain they will bond to the type of compound used.

Estimating the cost of curing means that first a determination must be made of what type of curing will be required and what type of weather the concrete will be poured in. Many large projects have concrete poured throughout the year and thus you must consider the problems involved in cold, mild, and hot weather. Smaller projects have a tendency to fall in one or perhaps two of the seasons, but there is a general tendency to avoid the very coldest of weather.

If a temporary enclosure is required (perhaps of wood and polyethylene film), the size and shape of the enclosure must be determined, a material takeoff made, and the number of work hours determined. For simple enclosures two workers can erect 100 square feet in one-half to one hour, once the materials are assembled in one place. If the wood may be reused, only a portion of the material cost is charged to this portion of the work. Keep in mind that the enclosure must be erected and possibly moved during the construction phases and taken down afterward; each step costs money.

Heating of water and aggregate raises the cost of the concrete by 3 to 10 percent, depending on how much heating is needed and whether ready-mix or job-mix concrete is being used. Most ready-mix plants already have the heating facilities and equipment with the cost spread over their entire production. *Job-mixing* means the purchase and installation of equipment on the job site as well as its operation in terms of fuel, work hours, and upkeep.

The cost of portable heaters used to heat the space is usually based on the number of cubic yards of space to be heated. One worker (not an operator) can service the heaters for 100 cubic yards in about one to two work hours, depending on the type heater and fuel being used. Fuel and equipment costs are also based on the type used.

If approximately 100,000 BTUH are required per 300 to 400 square feet (averaging 8 feet high), estimate the number of units needed (depending on net output of the unit), determine the amount of fuel that will be consumed per hour, the number of hours the job will require, and determine the equipment and fuel costs.

The continuous spray requires purchase of equipment (which is reusable) and the employment of labor to set it up. It will require between one-half and one work hour, to run the hoses and set up the equipment for an area of 100 square feet. The equipment should be taken down and stored when it is not in use.

Moisture-retaining and watertight covers are estimated by the square footage of the surface to be covered and separated into slabs or walls and beams. The materials are estimated as to their initial cost, and this is divided by the number of uses expected of them. Covers over slabs may be placed at the rate of 1,000 square feet per one to three work hours and as many as five uses of the material can be expected (except in the case of canvas, which lasts much longer); wall and beam covers may be placed at the rate of 1,000 square feet per two to six hours. The sealing of watertight covers takes from three-quarters to one work hour per 100 lineal feet.

Sealing compounds are estimated by the number of square feet to be covered divided by the coverage (in square feet) per gallon to determine the number of gallons required. Note whether one or two coats are required. If the two-coat application is to be used, be certain to allow for material to do both coats. For two-coat applications the first coat coverage varies from 200 to 500 square feet, while the second coat coverage varies from 350 to 650 square feet. However, always check the manufacturer's recommendations. Equipment required will vary according to the job size. Small areas may demand the use of only a paint roller, while medium-sized areas may require a pressure-type hand tank or backpack sprayer. Large, expansive areas can best be covered with special mobile equipment. Except for the roller, a great deal of reuse can be expected from the equipment, usually a life expectancy of five to eight years with reasonable care. The cost of the equipment would be spread over the time of its estimated usage. Using the pressure-type sprayer, you can estimate that from one and one-half to three hours will be required per 100 square feet, while in estimating labor for mobile equipment, you should depend more on the methods in which the equipment is mobilized than on anything else, but labor costs can usually be cut by 50 to 80 percent if the mobile equipment is used.

Specifications. The requirements of the architect-engineer regarding curing of the concrete are spelled out in the specifications. Sometimes they detail almost exactly what must be done for each situation, but often simply state that the concrete must be kept at a certain temperature for a given number of days. The responsibility of protecting the concrete during the curing period is that of the general contractor (who may in turn delegate it to a subcontractor).

9-8 TRANSPORTING CONCRETE

The methods used to transport the concrete from its point of delivery by the ready-mix trucks or the field-batching plant include truck chutes, buggies (with and without power), crane and bucket, crane and hoppers, tower and buggies, fork lifts, conveyors, and pumps. Several combinations of methods are available, and no one system is the answer to all the situations. It is likely that several methods will be used on any one project. Selection of the transporting method depends on the type of pour (floor, wall, curb, etc.), the total volume to be poured in each phase (in cubic yards), the distance above or below grade, equipment already owned, equipment available, distance from point of delivery to point of use, and possible methods of getting the concrete to that point.

Once a decision on the method of transporting the concrete has been made, the cost of equipment must be determined, as must the expected quantity of transporting that can be done using that equipment and a given crew of workers. From these the cost of transporting the concrete may be estimated. Generally speaking, the higher the building and the further the point of use from the point of delivery, the more expensive the cost of transporting will be.

9-9 FORMS

This portion of the chapter is not a course in form design, but it is intended to make you aware of the factors involved in forms and the consideration of forms relative to costs. No one design or system will work for all types of formwork. In general, the formwork must be true to grade and alignment, braced against displacement, resistant to all vertical and horizontal loads, resistant to leaking through tight joints, and of a surface finish that produces the desired texture. The pressure on the form is the biggest consideration in the actual design of the forms.

In the design of wall and column forms the two most important factors are the rate of placement of the concrete (feet per hour) and the temperature of the concrete in the forms. From these two variables the lateral pressure (psf) may be determined. Floor slab forms are governed primarily by the actual live and dead loads that will be incurred.

Actual field experience is a big factor in imagining exactly what is required in forming and should help in the selection of the type form to be used. The types of forms, form liners, supports, and methods are many; the preliminary selections must be made during the bidding period. This is one of the phases in which the proposed job superintendent (if you are low bidder) should be included in the discussions of the methods and types of forms being considered as well as in the consideration of what extra equipment and work power may be required.

Engineering data relative to forms and the design of forms is available from the American Concrete Institute (ACI) and Portland Cement Association (PCA) as well as from most manufacturers. This data should definitely be included in your reference library.

The forms for concrete footings, foundations, retaining walls, and floors are estimated by the area (in square feet) of the concrete that comes in contact with the form. The plans should be carefully studied to determine if it is possible to reuse the form lumber on the building and the number of times it may be reused. It may be possible to use the entire form on a repetitive pour item, or the form may have to be taken apart and reworked into a new form.

On higher buildings (eight stories and up), the cost of forms may be reduced and the speed of construction increased if high early strength cements are used instead of ordinary portland cement. This would enable the forms to be stripped in two or three days instead of the usual seven to ten day wait; however, not all architect-engineers will permit this.

Many types of forms and form liners may be rented, and the renting firms often provide engineering services as well as the forms themselves. Often, the cost of the forms required for the concrete work can be reduced substantially.

Wood forms. One of the most commonly used materials of which forms are built is wood. The advantages of wood are that it is readily accessible and easy to work with, and that once used, it may be taken apart and reworked into other shapes. Once it has been decided to use wood, the estimator must work out the quantity of lumber required and the number of uses that can be made of it. This means the construction of the forms must be decided upon in terms of the plywood sheathing, wales, studs or joists, bracing, and ties. The estimator can easily determine all of this if the height of the fresh concrete pour (for columns and walls), the temperature of the placed concrete, and the thickness of the slab (for floors) are known. The manufacturer's brochures or the ACI formwork engineering data may be used.

Complete explanations of the planning, design, materials, accessories, loads, pressure tables, design tables, and more are available in the ACI publication SP-4 "Form Work for Concrete." The address is American Concrete Institute, P.O. Box 19150, Detroit, Michigan 48219. This publication is a must for the estimator and should be in your library (or in the company library).

Metal forms.   Prebuilt systems of metal are used extensively on poured concrete work—not only on large work, but even for foundation walls for homes. Advantages are that these systems are reusable several times, are easily adaptable to the various required shapes, are interchangeable, and require a minimum of hardware and a minimum of wales and ties, which are easily placed. They may be purchased or rented, and several time-saving methods are employed. Curved and battered walls are easily obtained, and while the plastic-coated plywood face liner is most commonly used, other liners are available. Heavy-duty forms are available for heavy construction jobs in which a high rate of placement is desired.

Engineering data and other information pertaining to the uses of steel forms should be obtained from the metal form supplier. They can give information regarding costs (rental and purchase), tie spacing, number of forms required for the project, and labor requirements. Information concerning metal (steel) forms can be obtained from Symons Mfg. Company, 200 East Touhy Avenue, Des Plaines, Illinois 60018, as well as from many other manufacturers.

Miscellaneous forms.   Column forms are available in steel and laminated plies of fiber for round, square, and rectangular columns. Many manufacturers will design custom forms of steel, fiber, and fiberglass to meet project requirements. For columns these would include tapered, fluted, triangular, and half-rounded shapes. Fiber tubes are available to form voids in cast-in-place (or precast) concrete; various sizes are available. Most of these forms are sold by the lineal footage required of a given size. The fiber forms are not reusable, but the steel forms may be reused over and over.

Estimate.   The unit of measurement used for forms is the actual contact area (in square feet) of the concrete against the forms (with the exception of moldings, cornices, sills, and copings, which are taken off by the lineal foot). The forms required throughout the project must be listed and described separately. There should be no deductions in the area for openings of less than 30 square feet.

Materials in the estimate should include everything required for the construction of the forms except stagings and bridging. Thus materials that should appear are struts, posts, bracing, bolts, wire, ties, form liners (unless they are special), and equipment for repairing, cleaning, oiling, and removing.

Items affecting the cost of concrete wall forms are the height of the wall (since the higher the wall, the more lumber that will be required per square foot of contact surface) and the shape of the building, including pilasters.

Items affecting the cost of concrete floor forms include the floor-to-floor height, reusability of the forms, length of time the forms must stay in place, type of shoring and supports used, and the number of drop beams required.

The various possibilities of renting or purchasing forms, using gang forms built on ground and lifted into place, slip forming, and so on should be considered during this phase.

While approximate quantities of materials are given (Fig. 9–12) for work forms,

Type of form	Lumber fbm	Work hours			
		Assemble	Erect	Strip and clean	Repair
Footings	200—400	2—6	2—4	2—5	1—4
Walls	200—300	6—12	3—6	1—3	2—4
Floors	170—300	2—12	2—5	1—3	2—5
Columns	170—350	3—7	2—6	2—4	2—4
Beams	250—700	3—8	3—5	2—4	2—4
Stairs	300—800	8—14	3—8	2—4	3—6
*Moldings	170—700	4—14	2—8	2—6	3—6
*Sills, coping	150—600	3—12	2—6	2—4	2—6

*Values are for 100 lineal feet

Figure 9-12   Wood Forms, Approximate Quantity of Materials, and Work Hours

a complete takeoff of materials should be made. The information contained in Fig. 9–12 is approximate and should be used only as a check.

Estimating wood footing forms.   First determine if the entire footing will be poured at one time or if it will be poured in segments, which would permit the reuse of forms.

EXAMPLE 9–15 — Commercial Building:

Footing forms—single pour, use 1 × 12
l.f. of footings = 397 (from Fig. 9–1)
l.f. of forms (one per side) = 397 × 2 = 794 l.f.
Cutting and waste: 5%

Total l.f. of forms required = 794 × 1.05 = 834 l.f.

$$\frac{1 \times 12 \times 834}{12} = 834 \text{ fbm (foot board measure)}$$

Stakes, 2 × 2 wood, 5'-0" apart

$$\frac{834}{5} = 167 \text{ stakes required}$$

EXAMPLE 9-16 — Residential Building (Appendix G):

Footing forms—single pour, use 2 × 8
l.f. of footings = 146 l.f. (actually 145.33 l.f., from Fig. 9-2)
l.f. of forms = 146 x 2 = 292 l.f.
Cutting and waste: 4%

Total l.f. of forms required = 292 x 1.04 = 303 l.f. (use 300 l.f.)

$$\frac{2 \times 8 \times 300}{12} = 400 \text{ fbm}$$

Stakes, 2 × 2 wood, 5'–0" apart

$$\frac{300}{5} = 60 \text{ stakes required}$$

Labor. The time required for forms has been estimated from the table in Fig. 9–12, The hourly wages are based on local labor and union conditions.

EXAMPLE 9–17 — Commercial Building:

Footing formwork: 834 l.f., 834 fbm (Example 9-15)
Rate of work—erect, strip, and clean: 8 work hours per 200 fbm
Labor: $12.46 per hour

$$\text{Time} = \frac{834 \times 8}{200} = 33.36 \text{ work hours}$$

Cost = 33.36 hours × $12.46 per hour = $415.67

EXAMPLE 9-18 — Residential Building:

Footing formwork: 292 l.f., 400 fbm (Example 9-16)
Rate of work—complete assembly and disassembly: 8 work hours per 300 fbm
Labor: $6.75 per hour

$$\text{Time} = \frac{400 \times 8}{300} = 10.72 \text{ work hours}$$

Cost = 10.72 hours × $6.75 per hour = $72.36

9–10 FORM LINERS

The type of liner used with the form will determine the texture or pattern obtained on the surface of the concrete. Depending on the specified finish, formed concrete surfaces requiring little or no additional treatment can be easily obtained. A variety of patterns and textures may be produced by using various materials as liners. Fiberglass liners, plastic-coated plywood, and steel are among the most commonly used. Textures such as wood grain, rough sawn wood, corrugations (of various sizes and shapes), and all types of specialized designs are available. They have liners that will leave a finish resembling sandblast, acid etch, and bushhammered, as well as others. These textures may be used on floors or walls.

Liners are also used to form waffle slabs and tee beam floor systems; they may be fiberglass, steel, or fiber core. When liners of this type are used, the amount of void must be known so that quantity of concrete may be determined. This information is given in Fig. 9-13. Complete information should be obtained from whichever company supplies the forms for a particular project.

Form liners are often available on a rental or purchase arrangement; specially made form liners of a particular design may have to be purchased. When special designs are required, be certain to get a firm proposal from a manufacturer.

Waffle slab form liners	
Void size and depth	Conc. voided C.F.
19″ x 19″ x 4″	0.77
6″	1.09
8″	1.41
10″	1.90
12″	2.14
30″ x 30″ x 8″	3.85
10	4.78
12	5.53
14	6.54
16	7.44
20	9.16

Figure 9-13 Void Area in Concrete

Specifications. Check the specifications for requirements concerning textures and patterns, liner materials, thicknesses, configurations, and any other liners. It is not uncommon for the specifications to state the type of material the liner must be made of. The drawings should be checked for types of texture, patterns, or other requirements of the form liner. The form liners required may be already in stock at the manufacturer or may need a special order, often requiring months of work including shop drawing approvals.

Estimating. Take off the square footage of surface requiring a particular type of liner. Decide how many liners can be used effectively on the job—this will be the total number of square feet of liner or number of pieces required. Being able to use them several times is what brings the cost down. Divide the total cost of the liners by the square feet of surface to arrive at a cost per square foot for liners. The same approach is used for rental of liners. Different textures and patterns may be required throughout—check your details carefully.

9-11 CHECKLIST

Forms for:
> footings, walls, and columns
> floors
> piers
> beams
> columns
> girders
> stairs
> platforms

ramps
miscellaneous

Form:
erection
removal
repairs
ties
clamps
braces
cleaning
oiling
repairs
liners

Concrete:
footings, walls, and columns
floors
topping
piers
beams
columns
girders
stairs
platforms
ramps
curbing
coping
sills
walks
driveways
slabs
architectural

Materials; Mixes:
cement
aggregates (fine and coarse)
water
color
air entraining
other admixtures
strength required
ready-mix
heated concrete
cooled concrete

Finishes:
hand trowel
machine trowel
bushhammer
wood float
cork float
broom

sand
rubbed
grouted
removing fins

Curing:
admixtures
ponding
spraying
straw
canvas
vapor barrier
heat

Reinforcing:
bars
wire mesh
steel grade
galvanizing
bends
hooks
ties
stirrups
chairs
cutting

REVIEW QUESTIONS

1. Determine the amount of concrete, reinforcing, forms, and accessories required for the building in Appendix E. Use workup sheets and sketches.

2. How are re-bars taken off and how does the required lap affect the quantities?

3. How is vapor barrier taken off and how should the estimator determine the size rolls required to obtain the minimum amount of waste?

4. When taking off expansion joint filler what unit of measure is used?

5. What type of finishes may be required on the concrete?

6. What is the unit of measure for concrete finishes?

7. Why should each finish be listed separately on the workup sheet?

8. Why should the concrete for the various portions of a project (footings, walls, floors, beams, etc.) be listed separately on the workup sheet?

9. What unit of measure is used for concrete forms? How can reuse of forms affect the estimate?

10. What unit of measure is used for form liners? When should form liners be rented instead of purchased?

10
MASONRY

10-1 GENERAL

The term *masonry* encompasses all of the materials made use of by masons in a project, i.e., any block, brick, clay, tile, and stone. The mason is also responsible for the installation of lintels, flashing, metal wall reinforcing, weep holes, precast concrete, stone sills and coping, and manhole and catch basin block.

The tremendous amount of varied material available requires that estimators be certain they are bidding exactly what is required. Read the specifications, check the drawings, and call local suppliers to determine exact availability, costs, and special requirements of the units needed.

10-2 SPECIFICATIONS

Specifications should be carefully checked for size of the unit, type of bond, color, and shape. Simply because a particular unit is specified does not mean it is available in a particular area. The estimator must check with the local manufacturer or supplier on the entire specification, even including aggregates used in the manufacture, the type of curing required, and compliance with the American Society for Materials Testing (ASTM) requirements. Even though concrete blocks are generally manufactured on the local level, it is not uncommon for the local supplier to make arrangements with a manufacturer 300 or more miles away to provide the units required for a particular project—especially if there is a small number of special units involved. If the specifications require a certain fire rating, be sure that the supplier is aware of the requirements before submitting a proposal for the material.

Many specifications require that lintels and flashing, which are built into the wall, be installed by the masons. Check the specifications to see who provides them, who sets them, and what types are required. Large precast concrete lintels may entail the use of special equipment, while steel angle lintels often require cutting of the masonry units.

The type of "tooling" required for the mortar joints should also be noted since the type of tooling will affect the cost of labor. Types of joints are shown in Fig. 10-2.

10-3 LABOR

The amount of time required for a mason (with the assistance of helpers) to lay a masonry unit varies with the size, weight, and shape of the unit, bond (pattern), number of openings, whether the walls are straight or have jogs in them, distance the units must be moved (both horizontally and vertically), and the shape and color of the mortar joint.

The height of the walls becomes very important in estimating labor for masonry units. The masonry work that can be laid up without the use of scaffolding is generally the least expensive. Extra labor costs arise from the erection, moving, and dismantling of the scaffolding as the building goes up. The units and mortar have to be placed on the scaffold, which further adds to the labor costs.

Check union regulations in the locality in which the project will be constructed since unions may require that two masons work together where the units weigh more than 35 pounds each; the unions often also control the amount of work the union members actually lay in the wall each day.

The weather conditions always affect labor costs since a mason will lay more brick on a clear, warm, dry day than on a damp, cold day. Winter construction requires the building and maintenance of temporary enclosures and heating materials.

10-4 BONDS (PATTERNS)

Some types of bonds (Fig. 10-1) required for masonry units can add tremendously to the labor cost of the project. The least expensive bond (pattern) is the *running bond*. Another popular bond is the *stacked bond;* this type of bond will increase labor costs by as much as 50 percent if used instead of the running bond. Various *ashlar patterns* may also be required; these may demand several sizes laid up to create a certain effect. Study the drawings, check the specifications, and keep track of the different bonds that might be required on the project. When doing the quantity takeoff, you must keep the amounts required for the various bonds separately. The types of bonds that may be required are limited only by the limits of the designer's imagination.

The most common shapes of mortar joints are illustrated in Fig. 10-2. The joints are generally first struck off flush with the edge of the trowel. Once the mortar has partially set, the tooled joints are molded and compressed with a rounded or V-shaped joint tool. Some joints are formed with the edge of trowel; the *raked joint* is formed by raking or scratching out the mortar to a given depth; this is generally accomplished with a tool comprised of an adjustable nail attached to two rollers.

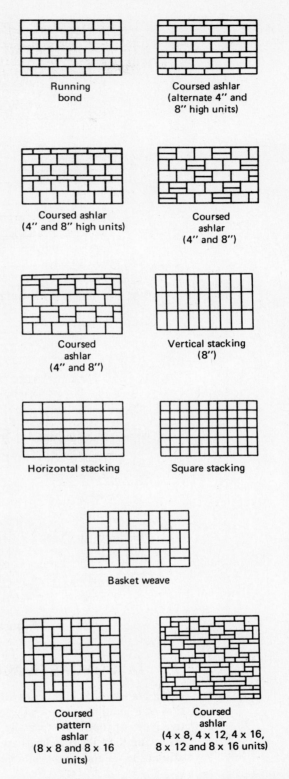

Figure 10-1A Typical Concrete Block Patterns

Running bond

Coursed ashlar
(alternate 4″ and
8″ high units)

Coursed ashlar
(4″ and 8″ high units)

Coursed
ashlar
(4″ and 8″)

Coursed
ashlar
(4″ and 8″)

Vertical stacking
(8″)

Horizontal stacking

Square stacking

Basket weave

Coursed
pattern
ashlar
(8 x 8 and 8 x 16
units)

Coursed
ashlar
(4 x 8, 4 x 12, 4 x 16,
8 x 12 and 8 x 16 units)

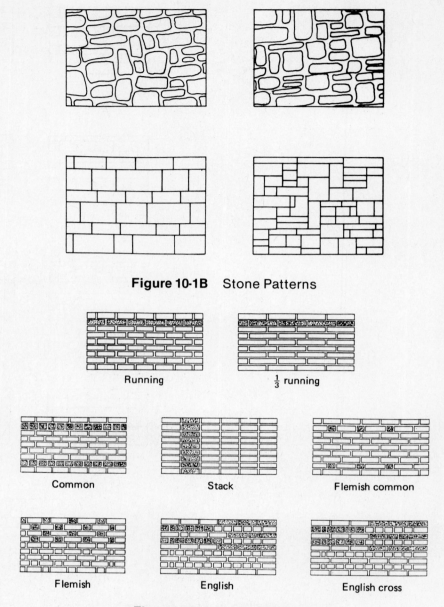

Figure 10-1B Stone Patterns

Running

$\frac{1}{3}$ running

Common

Stack

Flemish common

Flemish

English

English cross

Figure 10-1C Brick Patterns

10-5 CONCRETE MASONRY

Concrete masonry comprises all of the molded concrete units used in the construction of a building and includes concrete brick, hollow and solid block, and decorative types of block. Historically, many of these units are manufactured on the local level, and industry standards are not always adhered to. There is considerable variation in shapes and sizes available.

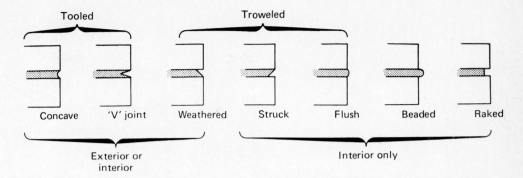

Figure 10-2 Shapes of Mortar Joints

Concrete block. There is no complete standardization of sizes for concrete block. The standard modular face dimensions of the units are 7⅝ inches high and 15⅝ inches long. Thicknesses available are 3, 4, 6, 8, 10 and 12 inches. (These are nominal dimensions; actual dimensions are ⅜ inch less.) A ⅜-inch mortar joint provides a dimension, in the wall, of 8- × 16-inch face dimensions and it requires 112.5 units per 100 square feet of wall. Since there are no industry standards, it is important to check with local suppliers to determine what sizes and shapes are available (Fig. 10–3). If the order requires a large number of special units, it is very possible that the local manufacturer will manufacture the units needed for the project.

Concrete blocks are available either as heavyweight or lightweight units. The heavyweight unit is manufactured out of dense or normal weight aggregates, such as sand, gravel and crushed limestone, cement, and water. Lightweight units use aggregates such as vermiculite, expanded slag, or pumice with the cement and water. The lightweight unit may weigh 30 percent less than the heavyweight unit, while it usually costs a few cents more per unit and usually has a slightly lower compressive strength.

The specifications will have to be checked to determine the size, shape, and color of units, as well as the conformance with some set of standards such as that of the ASTM, which sets up strength and absorption requirements. If a specific fire rating is designated for the units, it may require the manufacture of units with thicker face shells than are generally provided by a particular supplier. Check also for type of bond, mortar, and any other special demands. If you are unfamiliar with what is specified, call local suppliers to discuss the requirements of the specifications with them. Be certain that what is specified can be supplied. Some items, such as units cured with high-pressure steam curing, are unavailable in many areas, and in such a case the estimator (and often the suppliers) may want to call the architect-engineer to discuss the problem of supplying the item with them.

10-6 SPECIFICATIONS—CONCRETE MASONRY

The specifications will state exactly which type of units are required in each location. They specify the size, shape, color, and any requisite features, such as glazed units, strength, and fire ratings. The type, color, thickness, and shape of the mortar joint

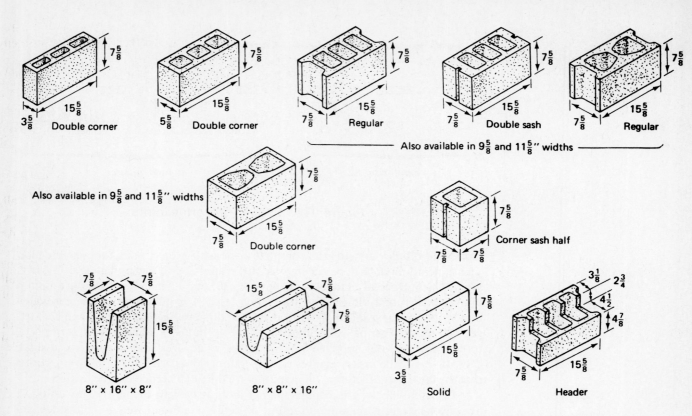

Figure 10-3 Typical Block Sizes (All Sizes Are Actual)

Materials	Materials (per 100 s.f. face area)			
	Concrete block			Brick
	8 x 16	8 x 18	5 x 12	$2\frac{1}{4}'' \times 3\frac{3}{4}'' \times 8''$
No. of units	112.5	100	240	617
Mortar cu. ft.				
Face shell bedding	2.3	2.2	3.6	7.2
Face shell and web bedding	3.2	3.0	5.5	7.2

Values shown are net. Waste for block and brick ranges from 5 to 10%. Waste for mortar may range from 25% to 75% and actual job experience should be considered on this item. It is suggested that 100% waste be allowed by the inexperienced and actual job figures will allow a downward revision. Figures are for a $\frac{3}{4}''$ thick joint.

Figure 10-4 Materials Required for 100 S.F. of Face Area

must be determined as well as the style of bond required. Also to be checked are the reinforcing, control and expansion joints, wall ties, anchors, flashing, and weep holes needed.

If the specifications are not completely clear as to what is required, be certain to call the architect-engineer's office and check. Never guess what they mean—check it out.

10-7 ESTIMATING—CONCRETE MASONRY

Concrete masonry should be taken off from the drawings by the square feet of wall required and divided into the different thickness of each wall. The total square footage of each wall, of a given thickness, is then multiplied by the number of units required per square foot.

While estimating the quantities of concrete masonry, use the exact dimensions shown. Corners should only be taken once, and deductions should be made for all openings in excess of 10 s.f. This area is converted to units, and to this number of units must be added an allowance for waste and breakage.

While making the takeoff, the estimator should note how much cutting of masonry units will be required. Cutting of the units is an expensive item and should be anticipated.

In working up the quantity takeoff, the estimator must separate masonry according to:

1. size of the units
2. shape of the units
3. colors of the units
4. type of bond (pattern)
5. shape of the mortar joints
6. colors of the mortar joints
7. any other special requirements (such as fire rating).

In this manner it is possible to make the estimate as accurate as possible. Since masonry is often a large percentage of the work, accuracy on the order of 99 percent is required. For mortar requirements, refer to Fig. 10-5 and Sec. 10-16.

EXAMPLE 10-1 — Commercial Building: Lengths were calculated in Sec. 9-2.

12-in. block (below grade only):

$$145'-0'' \text{ to a height of } 2'-8'' = 387 \text{ s.f.}$$
$$+ 6\% \text{ waste} = \underline{23} \text{ s.f.}$$
$$410 \text{ s.f.}$$

$$12'' \times 8'' \times 16'' \text{ block} = 112.5 \text{ units per square (from Fig. 10-4)}$$
$$(1 \text{ square} = 100 \text{ s.f.})$$
$$4.1 (112.5) = 462 \text{ units}$$

Mix by volume	Quantities			
	Masonry cement (bags)	Portland cement (bags)	Lime or putty (C.F.)	Sand (C.F.)
1:3 (masonry)	0.33	—	—	0.99
1 : 1 : 6 (portland)	—	0.16	0.16	0.97

Figure 10-5   Mortar Mixes to Mix 1 Cu. Ft. of Mortar

8-in. block:

$$252 \text{ l.f. @ } 10'\text{-}8'' \text{ high} = 2,688 \text{ s.f.}$$
$$145 \text{ l.f. @ } 8'\text{-}0'' \text{ high} = \underline{1,160 \text{ s.f.}}$$
$$3,848 \text{ s.f.}$$

Deduct all openings in excess of 10 square feet and note that all corners have been taken once to get figures as accurate as possible.

Doors:

$$2 \text{ @ } 3'\text{-}0'' \times 7'\text{-}0'' = 42 \text{ s.f.}$$
$$1 \text{ @ } 7'\text{-}0'' \times 10'\text{-}0'' = 70 \text{ s.f.}$$

Windows:

$$2 \text{ @ } 6'\text{-}0'' \times 8'\text{-}0'' = 96 \text{ s.f.}$$
$$4 \text{ @ } 4'\text{-}0'' \times 6'\text{-}0'' = \underline{96 \text{ s.f.}}$$
$$304 \text{ s.f. of openings}$$

Deduct the square footage of openings from the square footage of wall area to determine the net area of 8-in. block required.

$$\text{Gross wall area} = 3,848 \text{ s.f.}$$
$$\text{Opening area} = \underline{- 304 \text{ s.f.}}$$
$$3,544 \text{ s.f. net wall area}$$
$$+ 5\% \text{ waste} = \underline{177 \text{ s.f.}}$$
$$3,721 \text{ s.f.}$$

$$8'' \times 8'' \times 16'' \text{ block} = 112.5 \text{ units per square (from Fig. 10-4)}$$
$$37.21 (112.5) = 4,186 \text{ units}$$

EXAMPLE 10-2 — Residential Building:

$$\text{Height of block} = 2'\text{-}8'' \text{ (from wall section, Appendix G)}$$

NOTE: $2'\text{-}8''$ is 4 full block courses of $8''$ high block

$$\text{Length of foundation wall} = 50' + 22'\text{-}8'' + 50' + 22'\text{-}8'' = 145'\text{-}4''$$
$$\text{Block area} = 145.3 \times 2.67 = 388 \text{ s.f.}$$

$$\text{Units} = 3.88 \times 112.5 = 437 \text{ units}$$
$$+ 5\% \text{ waste} = \underline{22} \text{ units}$$
$$459 \ 8'' \times 8'' \times 16'' \text{ units}$$

Labor. The time required for the concrete masonry has been estimated from the charts in Fig. 10–6. The hourly wages are based on local labor and union conditions.

Work hours per 100 s.f.										
Wall thickness	4''		6''		8''		10''		12''	
Workers	Mason	Laborer	Mason	Laborer	Mason	Laborer	Mason	Laborer	Mason	Laborer
Type of work										
Simple foundation	—	—	—	—	4.5-6.0	5.0-7.5	6.0-9.0	7.0-10.5	7.0-10.0	8.0-11.5
Foundation with several corners, openings	—	—	—	—	5.0-7.5	5.0-7.5	6.5-10.0	7.5-12.0	7.5-10.0	8.5-12.0
Exterior walls, to 4'-0'' high	3.5-5.5	3.5-6.0	4.0-6.0	4.0-6.5	4.5-6.0	5.0-7.5	6.0-9.0	7.0-10.5	7.0-10.0	8.0-11.5
Exterior walls, 4'-0'' to 8'-0'' above ground or floor	3.5-6.0	4.5-7.5	4.0-6.5	4.5-7.0	4.5-6.5	6.0-9.0	6.5-10.0	7.5-12.0	7.5-10.0	8.5-12.0
Exterior walls, more than 8'-0'' above ground or floor	4.5-8.0	6.0-9.5	5.0-9.0	7.0-10.0	5.0-7.0	7.0-10.0	7.0-10.5	7.5-12.0	8.0-12.0	9.0-13.0
Interior partitions	3.0-6.0	3.5-7.0	3.5-6.5	4.5-7.5	4.5-6.0	5.0-7.5	—	—	—	—

Note:
1. The more corners and openings in the masonry wall, the more work hours it requires.
2. When lightweight units are used the work hours should be decreased by 10%.
3. Work hours include simple pointing and cleaning required.
4. Special bonds and patterns may increase work hours by 20% to 50%.

Figure 10-6 Work Hours Required for Concrete Masonry

EXAMPLE 10–3 — Commercial Building:

Concrete block: 12'' wide, 462 units (3.87 squares actual, Example 10-1)
Rate of work per square:

Mason, 7.0 work hours
Laborer, 7.5 work hours

Labor:

Mason, $19.23 per hour
Laborer, $15.57 per hour

Time:

Mason—3.87 squares × 7.0 work hours per square = 27.1 work hours
Labor—3.87 squares × 7.5 work hours per square = 29.0 work hours

Cost:

$$\text{Mason} = 27.1 \times \$19.23 = \$521.13$$
$$\text{Laborer} = 29.0 \times \$15.57 = \underline{451.53}$$
$$\$972.66$$

estimating—concrete masonry

Concrete block: 8″ wide, 4,186 units (35.44 squares net, Example 10-1)
Rate of work per square (Average, 10′-8″ walls):

    Mason, 6.0 work hours
    Laborer, 7.0 work hours

Labor:

    Mason, $19.23 per hour
    Laborer, $15.57 per hour

Time:

$$\text{Mason}—35.44 \times 6.0 = 212.6 \text{ work hours}$$
$$\text{Laborer}—35.44 \times 7.0 = 248.1 \text{ work hours}$$

Cost:

$$\text{Mason}—212.6 \times \$19.23 = \$4,088.30$$
$$\text{Laborer}—248.1 \times \$15.57 = \underline{3,862.92}$$
$$\$7,951.22$$

NOTE: Many firms would use the information from past projects to figure the cost per unit of masonry and then use these costs for jobs being bid. This offers a good check for the estimator, but care must be taken to be certain that the working conditions will be about the same and that there has been no wage or benefits increase since the project was completed.

EXAMPLE 10-4 — Residential Building:

Concrete block: 8″ wide, 460 units (3.9 squares, Example 10-2)
Rate of work per square:

    Mason, 4.5 work hours
    Laborer, 4.5 work hours

Labor:

    Mason, $12.75 per hour
    Laborer, $6.75 per hour

Time:

$$\text{Mason}—3.9 \times 4.5 = 17.6 \text{ work hours}$$
$$\text{Laborer}—3.9 \times 4.5 = 17.6 \text{ work hours}$$

Cost:

$$\text{Mason}—17.6 \times \$12.75 = \$224.40$$
$$\text{Laborer}—17.6 \times \$6.75 = \underline{118.80}$$
$$\$343.20$$

10-8 CLAY MASONRY

Clay masonry includes brick and hollow tile. Brick (Fig. 10-7) is considered a *solid masonry unit,* while tile (Fig. 10-8) is a *hollow masonry unit.* The faces of each type of unit may be given a ceramic glazed finish or a variety of textured faces.

    Brick is a solid masonry unit even if it has cores in it, as long as the core unit

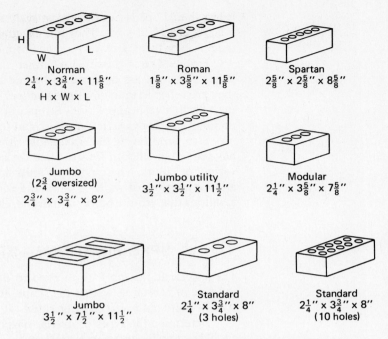

Norman
$2\frac{1}{4}''$ x $3\frac{3}{4}''$ x $11\frac{5}{8}''$
H x W x L

Roman
$1\frac{5}{8}''$ x $3\frac{5}{8}''$ x $11\frac{5}{8}''$

Spartan
$2\frac{5}{8}''$ x $2\frac{5}{8}''$ x $8\frac{5}{8}''$

Jumbo
($2\frac{3}{4}$ oversized)
$2\frac{3}{4}''$ x $3\frac{3}{4}''$ x $8''$

Jumbo utility
$3\frac{1}{2}''$ x $3\frac{1}{2}''$ x $11\frac{1}{2}''$

Modular
$2\frac{1}{4}''$ x $3\frac{5}{8}''$ x $7\frac{5}{8}''$

Jumbo
$3\frac{1}{2}''$ x $7\frac{1}{2}''$ x $11\frac{1}{2}''$

Standard
$2\frac{1}{4}''$ x $3\frac{3}{4}''$ x $8''$
(3 holes)

Standard
$2\frac{1}{4}''$ x $3\frac{3}{4}''$ x $8''$
(10 holes)

Figure 10-7 Brick Sizes (All Dimensions Are Actual)

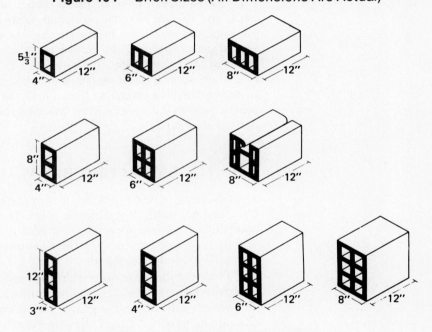

Note: All units shown are horizontal cell tile, also available in vertical cell tile.
*Actual dimension, all other dimensions are nominal ($\frac{1}{2}$ in. less than shown).

Figure 10-8 Structural Clay Tile (Courtesy of Brick Institute of America)

is less than 25 percent of the cross-sectional area of the unit. Bricks are available in a large variety of sizes. They are usually classified by material, manufacture, kind (common or face), size, texture, or finish on the face.

Hollow tile is classified as either *structural clay tile* or *structural facing tile.* When the units are designed to be laid up with the cells in a horizontal position, they are referred to as *side construction;* in *end construction* the units are laid with the cells in a vertical position. Structural clay tiles are available in load-bearing and nonload-bearing varieties, the sizes and shapes of which vary considerably.

Structural facing tile is available either glazed or unglazed. Unglazed units may have rough or smooth finishes, and the ceramic glazed units are available in a wide variety of colors. These types of units are often used on the exteriors of buildings and on interiors of walls of dairy and food producers, laboratories and other such locations that require hard, nonporous, easy-to-clean surfaces.

10-9 SPECIFICATIONS—BRICK

The specifications must be checked to determine exactly the type of material and the type of mortar as well as the shape, thickness, and color of the joint itself. The style of bond must also be determined.

Determine the types of lintels, flashing, reinforcing, and weep holes required, and who supplies and who installs each item.

Glazed-face brick with special shapes for stretchers, jambs, corners, sills, wall ends, and for use in other particular areas is available. Since these special pieces are relatively expensive, be certain that the estimate precisely allows for all of the special pieces specified. Be certain also that the material meets all ASTM requirements of the specification.

Many specifications designate an amount, the *cash allowance,* for the purchase of face brick. The estimator then allows the special amounts in the estimate. This practice allows the owner and architect-engineer to determine the exact type of brick desired at a later date.

10-10 ESTIMATING BRICK

The first thing to be determined in estimating the quantity of brick is the size of the brick and the width of the mortar joint. They are both necessary to determine the number of bricks per square foot of wall area and the quantity of mortar. Brick is sold by the thousand units, so the final estimate of materials required must be in the number of units required.

To determine the number of bricks required for a given project, the first step is to obtain the length and height of all walls to be faced with brick and then calculate the area of wall. Make deductions for all openings so that the estimate will be as accurate as possible. Check the jamb detail of the opening to determine if extra brick will be required for the revel; generally if the revel is over 4 inches deep, extra brick will be required.

Once the number of square feet has been determined, the number of bricks must be calculated. This varies depending on the size of the brick, width of the mortar joint, and style of bond required. The figures must be extremely accurate since actual quantities and costs must be determined. It is only in this manner that you will increase your chances of getting work at a profit—each figure used must be as accurate as possible.

Figure 10–9 shows the number of bricks required per square foot of wall surface for various patterns and bonds. Special bond patterns require that the estimator analyze the style of bond required and determine the number of bricks. One method of analyzing the amount of brick required is to make a drawing of several square feet of wall surface, determine the brick to be used, and divide that into the total area drawn. Sketches are often made right on the workup sheets by the estimator.

Labor costs will be affected by lengths of straight walls, number of jogs in the

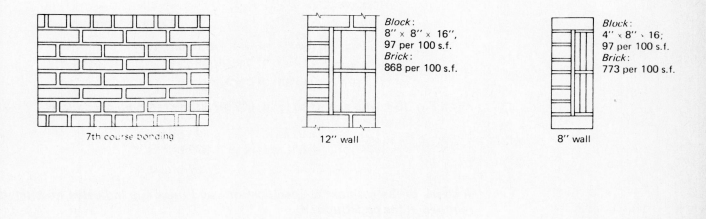

7th course bonding

Block:
8″ × 8″ × 16″,
97 per 100 s.f.
Brick:
868 per 100 s.f.

12″ wall

Block:
4″ × 8″ × 16;
97 per 100 s.f.
Brick:
773 per 100 s.f.

8″ wall

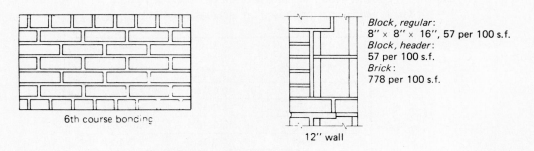

6th course bonding

Block, regular:
8″ × 8″ × 16″, 57 per 100 s.f.
Block, header:
57 per 100 s.f.
Brick:
778 per 100 s.f.

12″ wall

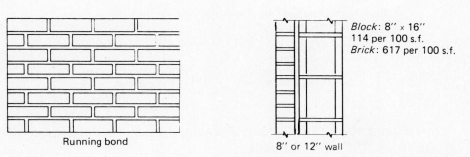

Running bond

Block: 8″ × 16″
114 per 100 s.f.
Brick: 617 per 100 s.f.

8″ or 12″ wall

Figure 10-9 Bricks Required per S.F. of Wall Surface

wall, windows, piers, pilasters, and anything else that might slow the mason down, such as the weather conditions.

Also to be calculated are the amount of mortar required, and any lintels, flashing, reinforcing, and weep holes that may be specified. Note any special requirements such as colored mortar, shape of joint, and type of flashing on the workup sheet.

EXAMPLE 10-5 — Commercial Building: Standard size brick, with a 3/8-inch joint, is specified for the 12-inch wall of the building being estimated.
The following openings are deducted (refer to Appendix A):

Doors: 1 @ 3'-0" × 7'-0" = 21 s.f.
Windows: 2 @ 6'-0" × 8'-0" = 96 s.f.
117 s.f. openings

1,160 s.f. of wall area (Sec. 10-7)
− 117 s.f. of openings
1,043 s.f. of brick area

From Fig. 10-4, with a 3/8-inch joint, 617 bricks are required per square.

10.43 (617) = 6,435 bricks
+ 5% waste = 322 bricks
6,757 bricks

A check of the residential specifications and drawings indicates no brick is required in the construction.

Labor.   The time required for the brick has been estimated from the table in Fig. 10-10. The hourly wages are based on local labor and union conditions.

Type bond	Mason	Laborer
Common	10.0–15.0	11.0–15.5
Running	8.0–12.0	9.0–13.0
Stack	12.0–18.0	10.0–15.0
Flemish	11.0–16.0	12.0–16.0
English	11.0–16.0	12.0–16.0

Note:

1. The more corners and openings in the wall, the more work hours it will require.

Norman brick:	+30%
Roman brick:	+25%
Jumbo brick:	+25%
Modular brick:	Same as standard
Jumbo utility brick:	+30%
Spartan brick:	+25%

Figure 10-10 Work Hours Required to Lay 1,000 Standard Size (8" × 2¾" × 3¾") Bricks

EXAMPLE 10-6 — Commercial Building:

Brick (standard size): 6,757 bricks (6.76 m, or thousands)
Rate of work:
  Mason, 12 work hours per 1,000 bricks
  Laborer, 13 work hours per 1,000 bricks
Labor:
  Mason, $19.23 per hour
  Laborer, $15.57 per hour
Time:

$$\text{Mason—}6.76 \times 12 = 81.12 \text{ work hours}$$
$$\text{Laborer—}6.76 \times 13 = 87.88 \text{ work hours}$$

Cost:

$$\text{Mason—}81.12 \times \$19.23 = \$1,559.94$$
$$\text{Laborer—}87.88 \times \$15.57 = \underline{1,368.29}$$
$$\$2,928.23$$

10-11 SPECIFICATIONS—TILE

Check the specifications for the type, size, and special shapes of the tile. The type, thickness, and color of mortar must be noted as well as the type of lintels, flashing, and reinforcing.

A wide variety of glazed finishes and special shapes is available. The specifications and drawings must be studied to determine the shapes required for the project. Since the special shapes tend to be expensive, estimate them carefully. Be certain that the materials specified meet all ASTM specifications.

10-12 ESTIMATING TILE

Hollow masonry units of clay tile are estimated first by the square feet of wall area required, each thickness of wall being kept separate. Then the number of units required is determined.

To determine the amount of tile required for a given project, the first step is to take off the length, height, and thickness of units required throughout. All openings should be deducted. Check the details for any special shapes or cuts that might be specified.

When the total square footage of a given thickness has been determined, calculate the number of units. Figure the size of the unit plus the width of mortar joint; this total face dimension is divided into the square feet to determine the number of units.

The labor costs will be affected by jogs in the wall, openings, piers, pilasters, weather conditions, and the height of the building. Items that need to be included are mortar, lintels, flashing, reinforcing, weep holes, and any special shapes.

10-13 STONE MASONRY

Stone masonry is primarily used as a veneer for interior and exterior walls; it is also used for walkways, riprap, and trim on buildings. Stone masonry is usually broken down into that which is laid up dry with no mortar being used such as on some low walls, sloping walls, walkways, and riprap; and that which is wet masonry in which mortar is used.

Stone is used in so many sizes, bonds, and shapes that a detailed estimate is required. The types of stone most commonly used are granite, sandstone, marble, slate, limestone, and trap. The finishes available include various split finishes, and tooled, rubbed, machine, cross-broached, and brushed finishes. Stone is generally available in random, irregular sizes, sawed-bed stone, and cut stone, which consists of larger pieces of cut and finished stone pieces. Random irregular sized stone is used for rubble masonry, and rustic and cobblestone work; it is often used for chimneys, rustic walls, and fences. Sawed-bed stones are used for veneer work on the interior and exterior of buildings. Patterns used include random and coursed ashlar. Coursed ashlar has regular courses while random ashlar has irregular-sized pieces and generally will require fitting on the job.

10-14 SPECIFICATIONS—STONE

Check exactly the type of stone required, the coursing, thickness, type, and color of mortar as well as wall ties, flashing, weep holes, and other special requirements.

Not all types and shapes of stone are readily available. Be certain that the supplier of stone is involved early in the bidding period. For large projects the stone required for the job will actually be quarried and finished specifically in accordance with the specifications. Otherwise, the required stone may be shipped in from other parts of the country or from other countries. In these special cases, be prepared to place a purchase order as soon as the bid is awarded.

10-15 ESTIMATING STONE

Stone is usually estimated by the area in square feet with the thickness being given. In this manner, the total may be converted to cubic feet and cubic yards easily, and it still gives the estimator the basic square-foot measurement to work with. Stone trim is usually estimated by the lineal foot.

Stone is sold in various ways, sometimes by the cubic yard, often by the ton. Cut stone is often sold by the square foot; of course, the square foot price goes up as the thickness increases. Large blocks of stone are generally sold by the cubic foot. It is not unusual for suppliers to submit lump-sum proposals whereby they will supply all of a certain type of stone required for a given amount of money. This is especially true for cut stone panels.

In calculating the quantities required, note that the length times height equals the square footage required; if the number must be in cubic feet, multiply the square footage by the thickness. Deduct all openings but usually not the corners. This will give you the volume of material required. However, the stone does not take up all the space: the volume of mortar must also be deducted. The pattern in which the stone is laid and the type of stone used will greatly affect the amount of mortar required. Cut stone may have 2 to 4 percent of the total volume as mortar, ashlar masonry 6 to 20 percent, and random rubble 15 to 25 percent.

Waste is equally hard to anticipate; cut dressed stone has virtually no waste, while ashlar patterns may have 10 to 15 percent waste.

Dressing a stone involves the labor required to provide a certain surface finish to the stone. Dressing and cutting stone require skill on the part of the mason, and this varies considerably from person to person. There is an increased tendency to have all stone dressed at the quarry or supplier's plant rather than on the job site.

The mortars used should be nonstaining mortar cement mixed in accordance with the specifications. When cut stone is used, some specifications require that the mortar joints be raked out and that a specified thickness of caulking be used. The type and quantity of caulking must then be taken off. The type, thickness, and color of the mortar joint must also be taken off.

Wall ties are often used for securing random, rubble, and ashlar masonry to the back-up material. The type of wall tie specified must be noted as must the number of wall ties per square foot. Divide this into the total square footage to determine the total number of ties required.

Cleaning must be allowed for as outlined in Sec. 10-19. Flashing should be taken off by the square foot and lintels by the lineal foot. In each case note the type required, the supplier, and the installer.

Stone trim is used for door and window sills, steps, copings, and moldings. This item is usually priced by the supplier by the lineal foot or as a lump sum. Some type of anchor or dowel arrangement is often required for setting the pieces. Check with the supplier as to who is supplying the anchors and dowels, and who will provide the anchor and dowel holes. Be certain the holes are larger than the dowels being used.

Large pieces of cut stone may require cranes or other lifting devices to move the stone and set it in place. Where large pieces of stone are being used as a facing on a building, special inserts and attachments to hold the stone securely in place will be required.

All of the above mentioned items will also affect the work hours required to put the stone in place.

10-16 MORTAR

The *mortar* used for masonry units may consist of portland cement, mortar, sand, and hydrated lime; or of mortar, cement, and sand. The amounts of each material required vary depending on the proportions of the mix selected, the thickness of the mortar joint (⅜ inch is a common joint thickness, but ¼- and ½-inch joints are also used), and the color of the mortar. There has been an increase in the use of colored mortar on the work; colors commonly used are red, brown, white, and black, but almost any color may be specified. Pure white mortar may require the use of white cement and white mortar sand; the use of regular mortar sand will generally result in a creamy color. The other colors are obtained by adding color pigments to the standard mix. Considerable trial and error may be required before a color acceptable to the owner and architect-engineer is found.

When both colored and gray mortars are required on the same project, keep in mind that the mixer used to mix the mortar must be thoroughly cleaned between mixings. If both types will be required, it may be most economical to use two mixers: one for colored mortar, the other for gray. White mortar should never be mixed in the mixer that is used for colored or gray mortar unless the mixer has been very thoroughly cleaned.

EXAMPLE 10-7 — Commercial Building: Using the masonry calculations in Secs. 10-7 and 10-10, calculate the mortar required. Assume 3/8-in. joints, gray, with face shell and web bedding.

Total s.f. of block area (less waste):

$$\begin{array}{r} 387 \text{ s.f.} \\ + \underline{3{,}544} \text{ s.f.} \\ 3{,}931 \text{ s.f.} \end{array}$$

Mortar required (from past job files): 4.8 c.f. per square

$$39.3\,(4.8) = 189 \text{ c.f. of mortar for blockwork}$$

Total square footage of brick area (less waste): 1,043 s.f.
Mortar required (from past job files): 9.4 c.f. per square

$$10.43\,(9.4) = 98 \text{ c.f. of mortar for brickwork}$$

Mortar required:

$$\begin{array}{r} 189 \text{ c.f.} \\ + \underline{98} \text{ c.f.} \\ 287 \text{ c.f. for block and brick} \end{array}$$

EXAMPLE 10-8 — Residential Building:
Total s.f. of block area (less waste): 388 s.f.
Mortar required: 4.8 c.f. per square

$$3.9 \times 4.8 = 18.7 \text{ c.f. of mortar}$$

Labor. The amount of labor required for mortar mixing is usually considered a part of the labor to lay the masonry unit.

10-17 ACCESSORIES

Masonry wall reinforcing. Steel reinforcing, in a wide variety of styles and wire gauges, is placed continuously in the mortar joints (Fig. 10–11). It is used primarily to minimize shrinkage, temperature, and settlement cracks in masonry, as well as to provide shear transfer to the steel. The reinforcing is generally available in lengths up to 20′. The estimator must determine the lineal footage required. The drawings and specifications must be checked to determine the spacing required (sometimes every course, often every second or third course). The reinforcement is also used to tie the outer and inner wythes together in cavity wall construction. The reinforcing is available in plain or corrosion-resistant wire. Check the specifications to determine what is required. Then check your local supplier to determine prices and availability of the specified material.

In calculating this item, you will find that the easiest method for larger quantities (over 1,500 sq. ft.) is to multiply the square footage by the factor given:

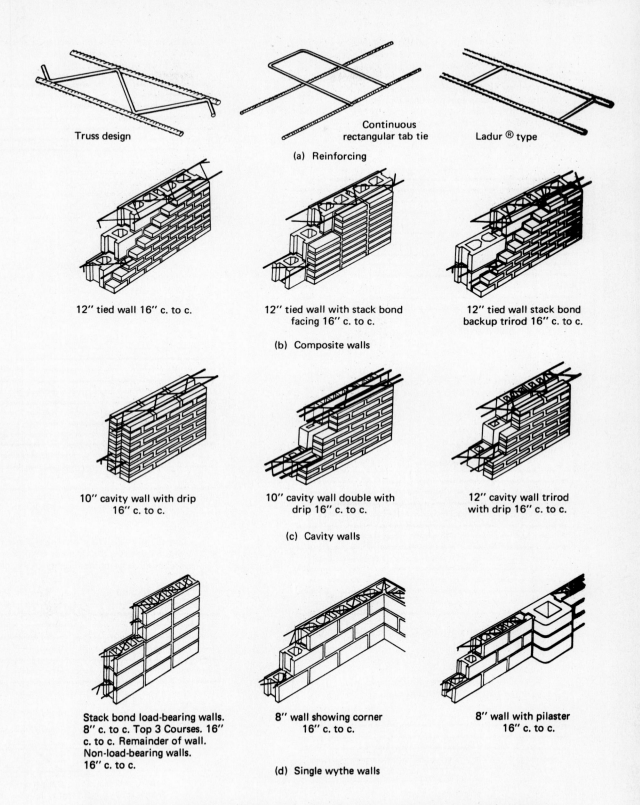

Truss design

Continuous
rectangular tab tie

Ladur ® type

(a) Reinforcing

12″ tied wall 16″ c. to c.

12″ tied wall with stack bond
facing 16″ c. to c.

12″ tied wall stack bond
backup trirod 16″ c. to c.

(b) Composite walls

10″ cavity wall with drip
16″ c. to c.

10″ cavity wall double with
drip 16″ c. to c.

12″ cavity wall trirod
with drip 16″ c. to c.

(c) Cavity walls

Stack bond load-bearing walls.
8″ c. to c. Top 3 Courses. 16″
c. to c. Remainder of wall.
Non-load-bearing walls.
16″ c. to c.

8″ wall showing corner
16″ c. to c.

8″ wall with pilaster
16″ c. to c.

(d) Single wythe walls

Figure 10-11A Horizontal Masonry Reinforcement (Courtesy of Dur-O-WaL)

accessories

Wall openings. Reinforcing should be installed in the first and second bed joints, 8 inches apart immediately above lintels and below sills at openings and in bed joints at 16-inch vertical intervals elsewhere. Reinforcement in the second bed joint above or below openings shall extend two feet beyond the jambs. All other reinforcement shall be continuous except it shall not pass through vertical masonry control joints.

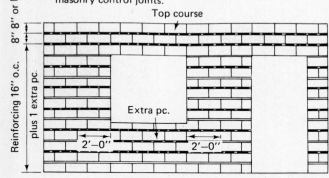

Wall with openings—running bond

Single wythe walls. Exterior and interior. Place reinforcing 16" o.c. and in bed joint of the top course.

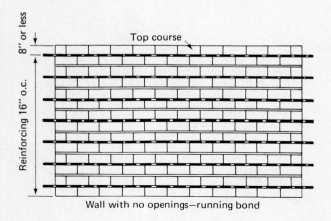

Wall with no openings—running bond

Foundation walls. Place reinforcing 8" o.c. in upper half to two-thirds of wall.

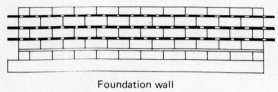

Foundation wall

Basement walls. Place reinforcing in first joint below top of wall and 8" o.c. in the top 5 bed joints below openings.

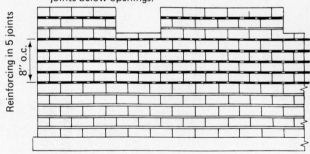

Basement wall

Stack bond. Reinforcing should be placed 16" o.c. vertically in walls laid in stack bond except it shall be placed 8" o.c. for the top 3 courses in load bearing walls

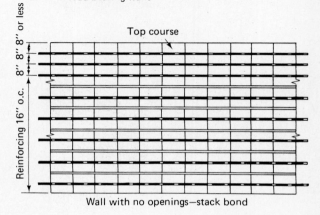

Wall with no openings—stack bond

Control joints. All reinforcement should be continuous except it shall not pass through vertical masonry control joints.

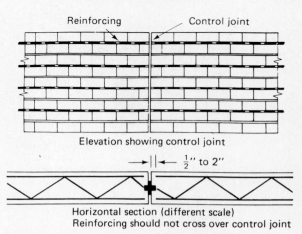

Elevation showing control joint

Horizontal section (different scale)
Reinforcing should not cross over control joint

Figure 10-11B Reinforcement Layout (Courtesy of Dur-O-WaL)

Reinforcement every course 1.50
second course 0.75
third course 0.50

To this amount, add 5% waste and 5% lap.

For small quantities the courses involved and the length of each course must be figured. Deduct only openings in excess of 50 square feet.

Control joints.   A *control joint* is a straight vertical joint that completely cuts the masonry wall from top to bottom. The horizontal distance varies from ½ to about 2 inches. The joint must also be filled with some type of material; materials usually specified are caulking, neoprene and molded rubber, and copper and aluminum. These materials are sold by the lineal foot and in a variety of shapes. Check the specifications to find the types required and check both the drawings and the specifications for the locations of control joints. Extra labor is involved in laying the masonry, since alternate courses utilizing half-size units will be required to make a straight vertical joint.

Wall ties.   Wall ties (Fig. 10–12) are used to tie the outer wythe with the inner wythe. They allow the mason to construct one wythe of wall to a given height before working on the other wythe, resulting in increased productivity. Wall ties are available in a variety of sizes and shapes including corrugated strips of metal 1¼ inches wide and 6 inches long (about 22 gauge), and wire bent to a variety of shapes. Adjustable wall ties are among the most popular, since they may be used where the coursing of the inner and outer wythes is not lined up. Noncorrosive metals or galvanized steel may be used. Check the specifications for the type of ties required and their spacing. To determine the amount required, take the total square footage of masonry and divide it by the spacing. A spacing of 16 inches vertically and 24 inches horizontally requires one tie for every 2.66 square feet; a spacing of 16 inches vertically and 36 inches horizontally requires one tie for every 4.0 square feet. Often, closer spacings are required. Also allow for extra ties at control joints, wall intersections, and vertical supports as specified.

Flashing.   The flashing (Fig. 10–13) built into the walls is generally installed by the mason. It is installed to keep moisture out and to divert any moisture that does get in back to the outside of the building. Flashing may be required under sills and copings, over openings for doors and windows, at intersections of roof and masonry wall, at floor lines, and at the bases of the buildings (a little above grade) to divert moisture out. Materials used include copper, aluminum, copper-backed paper, copper and polyethylene, plastic sheeting (elastomeric compounds), wire and paper, and copper and fabric. Check the specifications to determine the type required. The drawings and specifications must also be checked to determine all of the locations in which the flashing must be used. Flashing is generally sold by the square foot or by the roll. A great deal of labor may be required to bend metal flashings into shape. Check carefully whether the flashing is to be purchased and installed under this section of the estimate, or if it is to be purchased under the roofing section and installed under the masonry section.

Weep holes.   In conjunction with the flashing at the base of the building (above grade level), weep holes are often provided to drain out any moisture that might have got through the outer wythe. Weep holes may also be required at other loca-

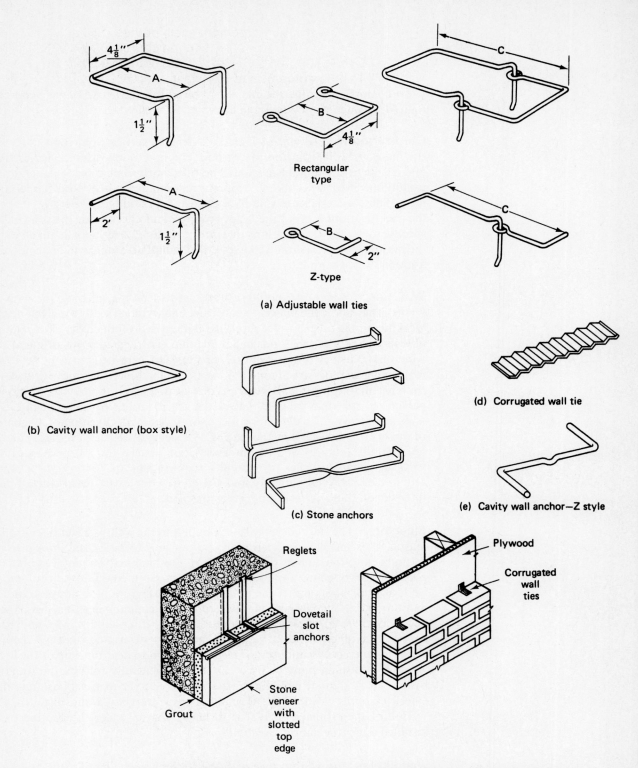

4 1/8"

A

1 1/2"

B

4 1/8"

C

A

2'

1 1/2"

B

2"

Rectangular type

Z-type

(a) Adjustable wall ties

(b) Cavity wall anchor (box style)

(c) Stone anchors

(d) Corrugated wall tie

(e) Cavity wall anchor—Z style

Reglets

Dovetail slot anchors

Grout

Stone veneer with slotted top edge

Plywood

Corrugated wall ties

Figure 10-12 Typical Wall Ties (Courtesy of Dur-O-WaL) and Wall Tie Insulation

masonry

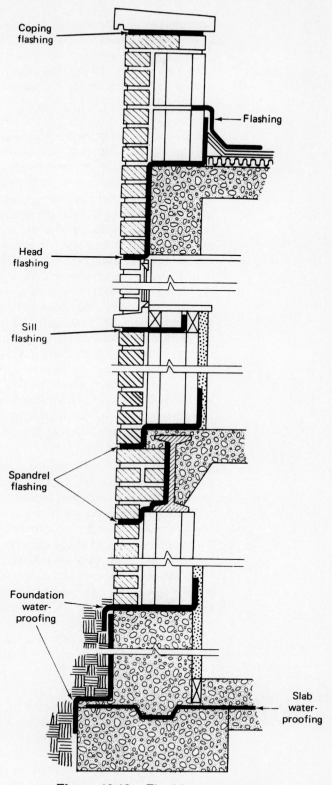

Figure 10-13 Flashing Locations

tions in the construction. The maximum horizontal spacing for weep holes is about 3'-0", but specifications often require closer spacing. The holes may be formed by using short lengths of cord inserted by the mason or they may be formed by well-oiled rubber tubing. The material used should extend up into the cavity for several inches to provide a drainage channel through the mortar droppings that accumulate in the cavity.

Lintels. A *lintel* is the horizontal member that supports the masonry above a wall opening. It spans from one side of the opening to the other. Materials used for lintels include steel angle iron, composite steel sections, lintel block (shaped like a "U") with reinforcing bars and filled with concrete, and precast concrete lintels. The lintels are usually set in place by masons as they lay up the wall. Some specifications require that the lintel materials be supplied under this section while other specifications require the steel angles and composite steel section to be supplied under "structural steel" or "miscellaneous accessories." Precast lintels may be supplied under "concrete"; the lintel block will probably be included under "masonry," as will the reinforcing bars and concrete used in conjunction with it.

It is not unusual for several types of lintels, in a variety of sizes, to be required on any one project. They must be broken down into the types, sizes, and lengths for each material used. Steel lintels may require extra cutting on the job so that the masonry will be able to fit around it. If the lintel is heavy, it may be necessary to use equipment (such as a lift truck or a crane) to put it in place. In determining the length, be certain to take the full masonry opening and add the required lintel bearing on each end. Lintel bearing for steel is generally a minimum of 4 inches, while lintel block and precast lintels are often required to bear 8 inches on each end. Steel is purchased by the pound, precast concrete by the lineal foot, and lintel block by the unit (note the width, height, and length).

Sills. *Sills* (Fig. 10-14) are the members at the bottoms of window or door openings. Materials used are brick, stone, tile, and precast concrete. These types of sills are installed by the mason, although the precast concrete may be supplied under a different portion of the specifications. The brick and tile sills are priced by the number of units required, while the stone and precast concrete sills are sold by the lineal foot. The estimator should check the maximum length of stone and precast concrete sill required and note it on the takeoff.

Also to be checked is the type of sill required (Fig. 10-13): a *slip sill,* which is slightly smaller than the width of the opening and can be installed after the masonry is complete, or a *lug sill,* which extends into the masonry at each end of the wall and must be built into the masonry as the job progresses.

Some specifications require special finishes on the sill and will have to be checked. Also, if dowels or other inserts are required, it should be noted.

Coping. The coping (Fig. 10-15) covers the top course of a wall to protect it from the weather. It is most often used on parapet walls. Masonry materials used include coping block, stone, tile, and precast concrete. Check the specification for the exact type required and who supplies it. The drawings will show the locations in which it is used, its shape, and how it is to be attached. The coping block and tile are sold by the unit, while the stone and precast coping are sold by the lineal foot.

Special colors, finishes, dowels, dowel holes, and inserts may be required. Check the drawing and specifications for these items and note all requirements on the workup sheet.

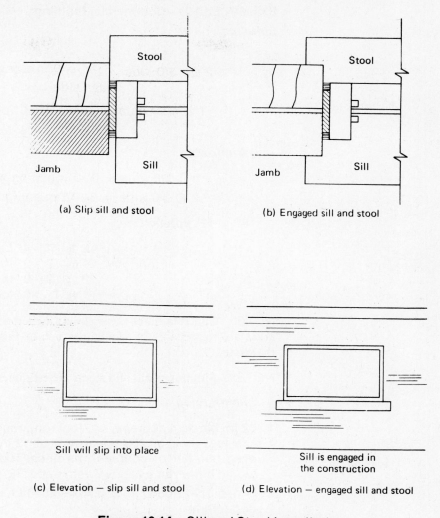

(a) Slip sill and stool

(b) Engaged sill and stool

Sill will slip into place

Sill is engaged in
the construction

(c) Elevation — slip sill and stool

(d) Elevation — engaged sill and stool

Figure 10-14   Sill and Stool Installation

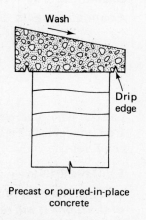

Figure 10-15   Masonry Coping

EXAMPLE 10-9 — Commercial Building:

Lintels:

   Doors:

$$2 @ \quad 3/0 \text{ wide } + \text{ frame } = 3'\text{-}4'' \text{ opening } + 6'' \text{ bearing}$$
$$each \text{ side } = 4'\text{-}4'' \text{ long}$$

$$1 @ 10/0 \text{ wide} \qquad 10'\text{-}0'' \text{ opening } + 6'' \text{ bearing}$$
$$each \text{ side } = 11'\text{-}0'' \text{ long}$$

Windows:

$$4 @ 6/0 \text{ wide } = 6'\text{-}0'' \text{ opening } + \text{ bearing } = 7'\text{-}0''$$
$$2 @ 8/0 \text{ wide } = 8'\text{-}0'' \text{ opening } + \text{ bearing } = 9'\text{-}0''$$

8" × 12" lintels:

$$2 @ \quad 4'\text{-}4'' \text{ long}$$
$$4 @ \quad 7'\text{-}0'' \text{ long}$$
$$2 @ \quad 9'\text{-}0'' \text{ long}$$
$$1 @ 11'\text{-}0'' \text{ long}$$

NOTE: Concrete lintel areas should be subtracted from the block area if these areas are significant. In a building such as this, the block are considered part of the waste.

NOTE: Aluminum sills and stools are estimated as part of the windows.

   Reinforcing:

1. 12-in. block, 4 courses, 145 feet long requires two courses of reinforcing 145 feet long or 290 lineal feet of 12-in. reinforcing. Since there is an even number of courses, the lineal footage works out the same as it would if the values above were used (387 s.f. × 0.75 = 290 l.f.).
2. 8-in. block: 2,688 s.f. × 0.75 = 2016 l.f. of 8-in. reinforcing (no openings deducted).
3. 8-in. block and 4-in. brick: 1,160 s.f. × 0.75 = 870 l.f. of 12-in. cavity wall reinforcing (no openings deducted).

EXAMPLE 10-10 — Residential Building:

Foundation vents, adjustable, 8" × 16" aluminum: Maximum 12'-0" o.c.

$$\text{Perimeter of residence } = 145.33 \text{ l.f. (from Example 10-2)}$$

$$\frac{145.33}{12} = 12.1 = 13 \text{ vents required}$$

Labor. The time required for the installation of accessories has been estimated from the table in Fig. 10-16. The hourly wages are based on local labor and union conditions.

EXAMPLE 10-11 — Commercial Building:

Lintels: The cost of the installation balances out against the installation cost of the block it will replace.

Accessory	Unit	Work hours
Wall reinforcing	1000 l.f.	1.0 to 1.5
Lintels (up to 300 lbs) precast conc.	100 l.f.	4.5 to 7.0
Coping, sills (up to 150 lbs) precast conc.	100 l.f.	5.0 to 8.0
Wall ties	100 pieces	1.0 to 1.5

Figure 10-16 Work Hours Required for Installing Accessories

Wall reinforcing: 3,176 l.f. = 3.2 mlf (thousand lineal feet)
Rate of work per 1,000 l.f. (mlf):
  Mason, 0.75 work hours
  Laborer, 0.50 work hours
Labor:
  Mason, $19.23 per hour
  Laborer, $15.57 per hour
Time:

$$\text{Mason}—3.2 \times 0.75 = 2.4 \text{ work hours}$$
$$\text{Laborer}—3.2 \times 0.50 = 1.6 \text{ work hours}$$

Cost:

$$\text{Mason}—2.4 \times \$19.23 = \$46.15$$
$$\text{Laborer}—1.6 \times \$15.57 = \underline{\quad 24.91}$$
$$\$71.06$$

EXAMPLE 10-12 — Residential Building: *Vents:* The cost of installation balances out against the installation cost of the block it will replace.

10-18 REINFORCING BARS

Reinforcing bars are often used in masonry walls to create multistory bearing wall construction; they are also used in conjunction with bond beam block and grout in bond beams used to tie the building together.

Reinforcing bars are sold by weight; the various lengths required of each size are taken off, and the total weight of each size required is multiplied by the price.

The specifications should be checked to determine the type of steel required. If galvanized re-bars are required, the cost for materials will easily double. Also galvanized re-bars must be specially ordered, so be certain they are ordered quickly when you are low bidder. (Sec. 9–3 also discusses reinforcing.)

10-19 CLEANING

The specifications must be checked to determine the amount of cleaning required and the materials that must be used to clean with. The materials exposed inside and outside the building will probably require cleaning, while the concealed masonry, such as block used as a backup, generally receives no cleaning.

Clay masonry. For brickwork, no attempt to clean should be made for a minimum of 48 hours after completion of the wall. After the minimum time, soap powder (or other mild solutions) with water and stiff brushes may be tried. When cleaning unglazed brick and tile, you should make your first attempt with plain water and stiff brushes. If these solutions do not work, the surface should be thoroughly wetted down with clear water, scrubbed with a solution of acid and water, and thoroughly rinsed down. Always try the acid solution on an inconspicuous area prior to using it on the entire wall. Acids should not be used on glazed facing tile.

Concrete masonry. Acid is not used on concrete masonry. If mortar droppings fall on the units, the droppings should be allowed to dry before removal to avoid smearing the face of the unit. When the droppings have dried, they can be removed with a trowel, and a final brushing will remove most of the mortar.

The estimator must determine the type of materials required to clean with, the area of the surfaces to be cleaned, the equipment required, the amount of cleaning that will actually have to be done, and the many work hours that will be required. The better the workmanship on the job, the less money that has to be allowed for cleaning. When the color of the mortar is different from that of the masonry unit, the cost for cleaning will be higher because all of the mortar droppings must be cleaned off to get an unblemished facing.

Stone masonry. Clean stone masonry with a stiff fiber brush and clean water (soapy water may be used if necessary). Then rinse with clean water. This will remove construction and mortar stains. Machine cleaning processes should be approved by the stone supplier before they are used. Wire brushes, acids, and sandblasting are not permitted for cleaning stonework.

10-20 EQUIPMENT

The equipment required for laying masonry units includes the mason's hand tools, mortar boxes, mortar mixer, hoes, hoses, shovels, wheelbarrows, mortar boards (tubs), pails, scaffolding, power hoist, hand hoist, elevator tower, hoisting equipment, and lift trucks.

The estimator must decide what equipment is required on the project, how much of each type is required, and the cost that must be allowed. While always remembering to include the costs of ownership (or rental), operating the equipment as well as mobilization to and from the project, erection, and dismantling.

Items to be considered in determining the amount of equipment required are the height of building, the number of times the scaffolding will be moved, the number of masons and helpers needed, and the type of units being handled.

10-21 COLD WEATHER

Cold weather construction is more expensive than warm weather construction. The increased costs stem from the construction of temporary enclosures so the masons can work, higher frequency of equipment repair, thawing materials, and the need for temporary heat.

Masonry should not be laid if the temperature is 40°F. and falling or less than 32°F. and rising at the place where the work is in progress, unless adequate precautions against freezing are taken. The masonry must be protected from freezing for at least 48 hours after it is laid. Any ice on the masonry materials must be thawed before use.

Mortar also has special requirements: its temperature should be between 70° and 120°F. During cold weather construction, it is common practice to heat the water used in order to raise the temperature of the mortar. Moisture present in the sand will freeze unless heated; upon freezing, it must be thawed before it can be used.

10-22 SUBCONTRACTORS

In most localities masonry subcontractors are available, and the estimator will have to decide whether it is advantageous to use a subcontractor on each individual project.

The decision to use a subcontractor does not mean the estimator does not have to prepare an estimate for that particular item; the subcontractor's bid must be checked to be certain it is neither too high nor too low. Even though a particular contractor does not ordinarily subcontract masonry work, it is possible the subcontractor can do the work for less money. There may be a shortage of masons, or the contractor's masonry crews may be tied up on other projects.

If the decision is made to consider the use of subcontractors, the first thing the estimator should decide is which subcontractors he wants to submit a proposal for the project. The subs should be notified as early in the bidding period as possible to allow them time to make a thorough and complete estimate. Often, the estimator will meet with the sub to discuss the project in general and go over exactly which items are to be included in the sub's proposal. Sometimes, the proposal is for materials and labor, other times for labor only. Be certain that both parties clearly understand the items that are to be included.

10-23 CHECKLIST

Masonry:
     type (concrete, brick, stone, gypsum)
     kind
     size (face size and thickness)
     load-bearing
     nonload-bearing
     bonds (patterns)
     colors
     special facings
     fire ratings
     amount of cutting
     copings
     sills
     steps
     walks

Reinforcing:
    bars
    wall reinforcings
    galvanizing (if required)

Mortar:
    cement
    lime (if required)
    fine aggregate
    water
    admixtures
    coloring
    shape of joint

Miscellaneous:
    inserts
    anchors
    bolts
    dowels
    reglets
    wall ties
    flashing
    lintels
    expansion joints
    control joints
    weep holes

REVIEW QUESTIONS

1. Determine the amount of masonry, mortar, accessories, and any other related items required for the building in Appendix E. Use workup sheets and sketches.

2. How may the type of bond (pattern) specified affect the amount of materials required?

3. Why is very high accuracy required with an item such as masonry?

4. When special colors or shapes are specified, why should local suppliers be contacted early in the bidding process?

5. What unit of measure is used to take off masonry units and how are they converted to units required?

6. Why is it important to note the finish required on the masonry units?

7. What is a cash allowance? How does it work?

8. List the equipment required for the work to be performed in problem 1.

9. How may cold weather construction affect the cost of a building?

10. What variables will affect the work hours required for the labor required for masonry items?

11

WATERPROOFING AND DAMPPROOFING

11-1 WATERPROOFING

Waterproofing is designed to resist the passage of water and usually to resist the hydrostatic pressures to which a wall or floor might be subjected. Dampproofing resists dampness, but is not designed to resist water pressure.

Waterproofing can be effected by the use of an admixture mixed with the concrete, the *integral method;* by placing layers of waterproofing materials on the surface, *membrane waterproofing;* and by the *metallic method.*

11-2 MEMBRANE WATERPROOFING

Membrane waterproofing (Fig. 11-1) consists of a buildup of tar or asphalt and membranes (plies) into a strong impermeable blanket. This is the only system that is a dependable method of waterproofing against hydrostatic head. In floors, the floor waterproofing must be protected against any expected upward thrust from hydrostatic pressure.

The actual waterproofing is provided by an amount of tar or asphalt applied between the plies of reinforcement. The purpose of the plies of reinforcement is to build up the amount of tar or asphalt that meets the waterproofing requirements and to provide strength and flexibility to the membrane.

Reinforcement plies are of several types, including a woven glass fabric with an open mesh, saturated cotton fiber, and tarred felts.

Newly applied waterproofing protection membranes should always be protected against rupture and puncture during construction and backfilling. Materials applied

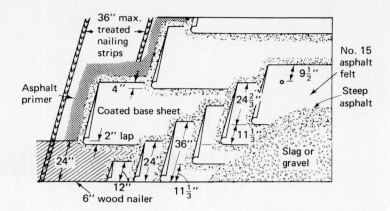

Figure 11-1 Nailers for Back Nailing (Courtesy of Celotex, Division of Jim Walter Corp.)

over the membrane include building board and rigid insulation for protection on exterior walls against damage during construction and backfill.

Another type of waterproofing is effected by the use of a sprayed-on asphalt with chopped fiberglass sprayed on simultaneously. The fiberglass reinforcement helps the resultant film bridge hairline cracks that may occur in the wall. The sprayed-on waterproofing requires that one to two days of good weather pass before backfilling can begin. Take care during backfill so that the film is not damaged.

11–3 ESTIMATING MEMBRANE WATERPROOFING

Membrane waterproofing.   The unit of measurement is the square, 1 square equals 100 square feet. To estimate the quantity of reinforcement felt required, first determine the square footage of the walls and floors that require membrane waterproofing; keep walls and floors separate.

The specifications must be checked to determine the number of membrane (plies) required, the kind and weight of reinforcement ply, the number of coatings, type of coating (tar or asphalt), the pounds of coating material required to complete each 100 square feet of waterproofing. Check also to determine what industry requirements must be met and whether the specifications limit the manufacturer's materials that may be used.

If one layer of reinforcement is specified, the amount must allow for laps over the footings and at the top of the wall, as well as the lap required over each strip of reinforcement. A 4-inch side lap of layers plus the top and bottom laps will require about 20 percent additional, a 6-inch lap will require about 25 percent additional to the actual area being waterproofed. Lap is generally listed in the specifications.

The number of plies used usually ranges from two to five, and the amount to be added for laps and extra material drops to about 10 percent of the actual wall or floor area.

Most felt and cotton fabric reinforcement is available in rolls of 432 square feet. Glass yarn reinforcement is generally available in rolls of 450 square feet. Different manufacturers may have various size rolls and sheets available of certain types of reinforcements.

waterproofing and dampproofing   *182*

The amount of tar required varies depending on the number of plies, and the amount used per square (one square = 100 square feet) should be in the specifications.

Figure 11–2 shows the amount of reinforcing and coating required. Membrane waterproofing is similar to applying built-up roofing, but the cost is greater for vertical surfaces than for horizontal surfaces and there is generally less working space on the vertical.

Material	Plies					
	1 ply 4" lap	1 ply 6" lap	2 ply	3 ply	4 ply	5 ply
Reinforcement S.F.	120	125	230	330	440	550
Moppings of bituminous coatings	2	2	3	4	5	6
Lbs of bituminous coatings	60–70	60–70	90–100	120–135	150–170	180–205

Figure 11-2   Materials Required for Membrane Waterproofing

The square footage of protective coating must be also determined; this is the same as the actual area to be waterproofed. The type of protective coating, thickness, and method of installation are included in the specifications.

The sprayed-on asphalt and chopped fiberglass require special equipment to apply, and most manufacturers specify that only if these are applied by approved applicators, will they be responsible for the results on the job. Materials required per square are 8 to 10 gallons of asphalt and 7 to 10 pounds of fiberglass. The surface that is being sprayed must be clean and dry.

Labor.   This item is almost always installed by a subcontractor (usually a roofing subcontractor). A typical subcontractor's bid (to the contractor) is shown in Fig. 11-3. Typical installation times are shown in Fig. 11-4.

Agree to furnish and install 3-ply built-up membrane waterproofing in accordance with the contract documents for the sum of $8,225.00

Figure 11-3   Subcontractor's Bid for Membrane Waterproofing

Type of work	Work hours per 100 s.f.
Cleaning	0.25–3.5
Painting brush, per coat	0.5–2.0
Parging, per coat	1.0–2.5
Metallic	1.5–3.0
Waterproofing, asphalt or tar, per coat felt or reinf., per layer	0.7–2.0 1.2–2.5

Figure 11-4   Work Hours Required for Applying Waterproofing and Dampproofing

11-4 INTEGRAL METHOD

The admixtures added to the cement, sand, and gravel generally are based on oil-water repellent preparations. Calcium chloride solutions and other chemical mixtures are used for liquid admixtures and stearic acid for powdered admixtures. These admixtures are added in conformance with manufacturer's recommendations. Most of these admixtures enhance the water resistance of the concrete, but offer no additional resistance to water pressure.

Estimating. The amount to be added to the mix varies considerably, depending on the manufacturer and composition of the admixture. The powdered admixtures should be mixed with water before they are introduced into the mix, while the liquid admixtures may be measured and added directly, which is a little less trouble. The costs involved derive from delivering the admixture to the job site, moving it into a convenient place for use, and adding it to the mix. The containers may have to be protected from the weather, particularly the dry admixture containers.

11-5 METALLIC METHOD

Metallic waterproofing employs a compound that consists of graded fine-iron aggregate combined with oxidizing agents. The principle involved is that, when the compound is properly applied, it provides the surface with a coating of iron that fills the capillary pores of the concrete or masonry and, as the iron particles oxidize, the compound creates an expanding action that fills the voids in the surface and becomes an integral part of the mass. Always apply in accordance with the manufacturer's specifications and always apply the compound to the inside of the wall.

Estimating. Check the specification of the type of compound to be used, and the thickness and method of application specified. From the drawings determine the square footage to be covered and, using the manufacturer's information regarding square feet coverage per gallon, determine the number of gallons required.

Labor required will depend on the consistency of the compound, the thickness, and method of application required as well as the convenience afforded by the working space. Equipment required includes scaffolds, planks, ladders, mixing pails, and trowels.

11-6 DAMPPROOFING

Dampproofing is designed to resist dampness and is used on foundation walls below grade and exposed exterior walls above grade. It is not intended to resist water pressure.

The methods used to dampproof include painting the wall with bituminous materials below grade and transparent coatings above grade. *Parging* with a rich cement base mixture is also used below grade, often with a bituminous coating over it.

11-7 PAINTING METHOD

Among the most popular paints or compounds applied by brush or spray are tar, asphalt, cement washes, and silicone-based products. This type of dampproofing may be applied with a brush or mop, or may be sprayed on. The type of application, and the number and thickness of coats required will be determined by the specifications and manufacturer's recommendations.

Below grade on masonry or concrete, the dampproofing material is often a mastic black coating that is applied to the exterior of the foundation. It is applied with a brush, mop, or in a spray. The walls must be thoroughly coated with the mastic filling all voids or holes. The various types available require from one to four coats to do the job, and square footage per gallon varies from 30 to 100. The more porous the surface, the more material it will require.

Transparent dampproofing of exterior masonry walls is used to make them water-repellent. The colorless liquids must be applied in a quantity sufficient to completely seal the surface. Generally, one or two coats are required by the manufacturers with the square footage per gallon ranging from 50 to 200.

11-8 ESTIMATING PAINTING

To estimate these types of dampproofing, it is necessary to first determine the types of dampproofing required, the number of coats specified (or recommended by the manufacturer), and the approximate number of square feet that one gallon of material will cover so that the amount of material required may be determined. If two coats are required, twice the material is indicated.

The amount of labor required to do the work varies with the type of compound being used, the number of coats, the height of work, and the method of application. When the surface to be dampproofed is foundation walls below grade and the work is in close quarters, it will probably take a little longer. Sprayed-on applications require much less labor time but more expensive equipment. High buildings require scaffolds that will represent an added cost factor.

Equipment that may be required includes ladders, planks, mixing cans, brushes, mops, spraying equipment, and scaffolding.

EXAMPLE 11-1 — Commercial Building: The block below grade, on the building, must be brushed with a bituminous product within 2 inches of the finished grade. If a one-coat product is used and if the coverage is estimated at 60 square feet per gallon, how many gallons are required?

First, the square footage to be covered must be figured.

$$\text{Area below grade} = 400 \text{ ft } (2.5 \text{ ft}) = 1{,}000 \text{ s.f.}$$

$$1000/60 = 16.7 \text{ gallons} = 17 \text{ gallons required}$$

This product is available in 5-gallon containers, so 20 gallons or 4 containers are required.

Labor. The time required for the application of the dampproofing has been estimated from the table in Fig. 11-4. The hourly wages are based on local labor and union conditions.

EXAMPLE 11-2 — Commercial Building:

Dampproofing: 1,000 s.f. (10 squares)
Rate of work: 1.25 work hours per square (100 s.f.)
Laborer: $15.57 per hour
Time:

$$10 \times 1.25 = 12.5 \text{ work hours}$$

Cost:

$$12.5 \times \$15.57 = \$194.63$$

11-9 PARGING (PLASTERING)

The material used to parge on exterior portions of foundation walls below grades is a mixture of portland cement, fine aggregates (sand), and water. A water-repellent admixture is often added to the mix and is troweled or sprayed in place, the most common thickness being about ½ inch. Compounds with a cement base are also available from various manufacturers.

On some projects the specifications require parging and the application of a coating of bituminous materials in a liquid form. The bituminous type of coating is discussed in Secs. 11-7.

11-10 ESTIMATING PARGING

To determine the amount of material required for parging, first check the specifications to determine exactly the type of materials required. They may be blended and mixed on the job or they may be preblended compounds. Information concerning the specially formulated compounds must be obtained from the manufacturer. The job-mixed parging requires that the mix proportions be determined. The next step is to determine the number of square feet to be covered and the thickness. With this information the cubic feet required can be calculated, or the number of gallons determined if it is preblended.

The amount of labor will depend on the amount of working space available and the application technique. Labor costs will be higher when the materials are mixed on the job, but the material cost will be considerably lower.

Equipment required includes a mixer, trowel or spray accessories, scaffolding, planks, shovels, and pails. Water must always be available for mixing purposes.

11-11 CHECKLIST

Waterproofing:
    integral
    membrane
    paints
    plaster
    foundation walls

    slabs
    sump pits
    protective materials
    exterior
    interior
    admixtures
    drains
    pumps
Dampproofing:
    integral
    parge
    vapor barriers
    bituminous materials
    drains
    foundation walls
    slabs

REVIEW QUESTIONS

1. Determine the amount of dampproofing required for the building in Appendix E. Use workup sheets and sketches.

2. What is the difference between waterproofing and dampproofing?

3. Assume that membrane waterproofing is required for the building in Appendix E. Determine the amount of materials required if a 3-ply system is specified under the floor slab and up the walls to within 2″ of grade. Assume the floor has been dropped down to just above the footing.

4. What is parging and what unit of measure is used?

5. Assume ¾″ of parging is required from the top of the footing (Appendix E) to within 2″ of grade. Determine how many c.f. of material are required.

12

STRUCTURAL SYSTEMS

12-1 PRECAST CONCRETE

The term *precast concrete* is applied to individual concrete members that are cast in separate forms and then placed in the structure. They may be cast in a manufacturing plant or on the job site by the general contractor or supplier. The most common types of structural precast concrete are double and single tees, floor planks, columns, beams, and wall panels (Fig. 12-1).

Depending on the requirements of the project, precast concrete is available reinforced or prestressed. Reinforced concrete utilizes reinforcing bars encased in concrete; it is limited in its spans, with 40'-0" being about the maximum for a roof. When longer spans are required, prestressed concrete is used, it may be either pretensioned or posttensioned. Prestressing generally utilizes high-strength steel or wire or wire strands as the reinforcing. In pretensioning the longitudinal reinforcing is put in tension before the concrete is cast. The reinforcing is stretched between anchors and held in this state as the concrete is poured around the steel and cured. With posttensioning the longitudinal reinforcement is not bonded with the concrete. The reinforcement may be greased or wrapped to avoid bonding with the concrete, or conduits of some type (tubes, hoses) may be cast in the concrete and the reinforcement added later. The reinforcement is then stretched and anchored at the ends and, when released, the stretched wires tend to contract and in this manner compress the concrete.

Most precast concrete is priced by the square foot, lineal foot, or in a lump sum. It is important to determine exactly what is included in the price. Most suppliers of precast concrete items price them delivered to the job site and installed, especially if they are structural items. When the specifications permit it, some contractors will

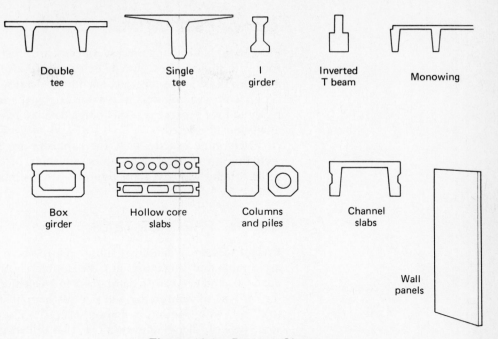

Figure 12-1 Precast Shapes

precast and install the pieces required themselves. Unless experienced personnel are available, doing your own precasting may cost more than subcontracting the work out to others.

The aggregates used for precast concrete may be heavyweight or lightweight; however, it should be noted that some types of lightweight concrete are not recommended by some consultant engineers for use on posttensioning work, and care should be taken regarding all materials used in the concrete.

Hole cutting in tees is subject to the distance between the structural tee portions of the members. The lengths of holes are not as rigidly controlled, but should be approved by the structural engineer.

12-2 SPECIFICATIONS

The specifications (Fig. 12-20) must be checked to determine if a particular manufacturer is specified. Be sure all manufacturers who can supply materials that will meet the specifications bid the project. Often there is only limited competition in bidding on precast concrete items. Check also to determine the strength, reinforcing, and inserts as well as any special accessories required.

If the project entails any special finish, such as sandblasting, filling air holes, colored concrete, special aggregates, or sand finish, this should be noted on the workup sheets. Be certain that there is a clear understanding of exactly what the manufacturer is proposing to furnish. For example, it should be understood who will supply any required anchor bolts, welding, cutting of required holes, and filling of joints. Always check to see if the manufacturer (or subcontractor) will install the precast. If not, the estimator will have to calculate the cost of the installation.

12-3 ESTIMATING

Floor, wall, and ceiling precast concrete are most commonly taken off by the square foot with the thickness noted. Be certain to also note special requirements such as insulation cast in the concrete, anchorage details, and installation problems.

Beams and columns are taken off by the lineal foot with each different size required kept separately. Anchorage devices, inserts, and any other special requirements must be noted.

Determine exactly what the suppliers are proposing to do. If they are not including the cutting of *all* holes, finishing of concrete, welding, or caulking, all of these items will have to be figured by the estimator.

12-4 PRECAST TEES

Precast tees are available as simply reinforced or prestressed. The shapes available are double and single tee. For the double tee the most common widths available are 4'-0" and 6'-0" with depths of 8 to 26 inches. Spans range up to 75'-0" depending on type of reinforcing and size of the unit. Single tee widths vary from 6'-0" to 12'-0" with lengths up to about 112'-0". The ends of the tees must be filled with some type of filler as shown in Fig. 12-4. Fillers may be concrete, glass, plastic, etc.

Most manufacturers who bid this item will bid it delivered and installed either on a square-foot or lump-sum basis. One accessory item often overlooked is the special filler block required to seal the end of the tees.

Concrete fill is often specified as a topping for tees; it may be used for floors or sloped on a roof to direct rainwater. The fill is usually a minimum of 2 inches thick and should have at least light reinforcing mesh placed in it. Due to the camber in tees, it is sometimes difficult to pour a uniform 2-inch thick topping and end up with a level floor.

Specifications. Determine what manufacturers are specified, the strength of concrete, type and size of reinforcing, type and size of aggregate, the finish required, and whether topping is specified. For reinforcing bars you must determine the type of chairs to be used to hold them in place. The bars should be corrosion-resistant. Who cuts the holes and caulks the joints should also be checked as well as what type of caulking to be used.

Estimating. Take off the square footage required. If the supplier made a lump-sum bid, check out the square-foot price. The square footage required should have all openings deducted. If the project is bid by the square foot, call the supplier to check your square footage against their takeoff. This provides a check for your figures. (However, it is only a check. If they don't agree, recheck the drawings. Never use anyone else's quantities when working up an estimate.)

When an interior finish is required, keep in mind the shape of the tee; a tee 4'-0" wide and 40'-0" long does not have a bottom surface of 160 square feet to be finished. The exact amount of square footage involved varies, depending on the depth and design of the unit; it should be carefully checked.

The installation of mechanicals sometimes takes some special planning with precast tees in regard to where the conduit, heating, and plumbing pipes will be located and how the fixtures will be attached to the concrete. All of these items must be checked.

The concrete fill *(topping)* is placed by the general contractor in most cases. The type of aggregate must be determined and the cubic yards required must be taken off. The square footage of reinforcing mesh and the square footage of concrete that must be finished must be determined. From the specifications you may determine the aggregate, strength, type of reinforcing, surface finish, and any special requirements. This operation will probably be done after the rest of the concrete work is completed. The decision must be made as to whether ready-mixed or field-mixed concrete will be used and how it will be moved to the floor and the particular spot on which it is being placed.

EXAMPLE 12-1 — Using the commercial building in Appendix A, assume specifications Alternate 1, with accompanying details (Fig. 12-2), requires a bid using double tees instead of the hollow core plank shown in the working drawings. This portion of the estimate must be clearly noted Alternate 1 and not be included in with the estimate for the base bid.

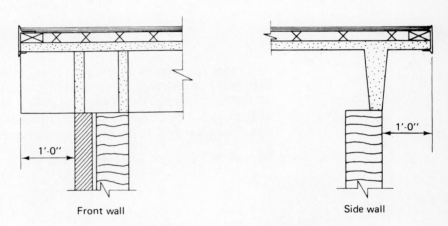

Front wall   Side wall

Figure 12-2   Filler in Precast Tees

Using a roof system of flat, hollow core, precast slabs, determine the square footage required for the building, if there is to be a 1′-0″ overhang all around. The area of the roof must be calculated and a sketch (Fig. 12-3) is the easiest way to ensure accuracy. Refer also to Appendix A.

Basic area:

$$67'-0'' \times 127'-0'' = 8{,}509 \text{ s.f.}$$

Recesses and offsets:

Southwest offset = 12 × 20 =	240 s.f.
North recess = 10 × 38 =	380 s.f.
Total Deductions =	620 s.f.
Net area of roof = 8,509 − 620 =	7,889 s.f.

The precast concrete fillers must also be estimated.

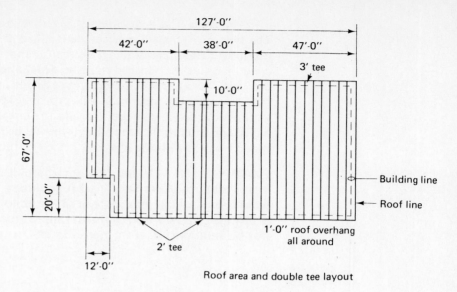

Roof area and double tee layout

Figure 12-3   Roof Area

From the sketch of the roof area, there are 30 full-width tees, 2'-width tees and 3'-width tees. For estimating the number of fillers, assume there are 32 full-width fillers. From the wall detail, in Fig. 12-2, two fillers are required at each wall. Also to be considered is that each end of a full double tee requires two fillers (see Fig. 12-4). So, in this estimate the number of fillers equals:

32 tees × 2 ends × 2 per wall thickness × 2 filler areas per end = 256 fillers

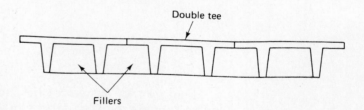

Figure 12-4   Fillers in Double Tees

Labor.   Two typical subcontractors' bids are included here. In Fig. 12-5 the subcontractor is bidding on a square-foot basis, while in Fig. 12-6 a lump-sum bid is used. Typically, the contractor will have to have the concrete fillers installed by masons. For these fillers the estimate is based on Fig. 10-6 and past experience. Labor rates area based on local labor and union conditions.

Double tees, 16' deep × 4'0" wide . . . . 6.75 s.f.
   (including installation)
Fillers (delivered to job site) . . . . 4.45 each

Figure 12-5   Subcontractor's Bid for Tees and Fillers (S.F.)

Furnish and install all 16" deep x 4'-0"
double tees for a lump sum price of · · · · · · · · $53,250.75

Furnish double tee fillers · · · · · · · · · · · · · · · · · · · $1,139.20

Figure 12-6 Subcontractor's Bid for Tees and Fillers (Lump Sum)

EXAMPLE 12-2 — Concrete fillers:
Rate of work: 8 units per 1 mason work hour and 1 laborer work hour
Labor:

Mason, $19.85 per hour
Laborer, $15.57 per hour

Time:

$$\text{Mason}—256 \div 8 = 32 \text{ work hours}$$
$$\text{Laborer}—256 \div 8 = 32 \text{ work hours}$$

Cost:

$$\text{Mason}—32 \times \$19.85 = \$635.20$$
$$\underline{\text{Laborer}—32 \times \$15.57 = \$498.24}$$
$$\$1,133.44$$

NOTE: Whenever you are considering an alternate, it must be carefully analyzed to determine what other work may be affected; e.g., Are the same number of courses of block required? Is a concrete block the beam required or perhaps some extra accessories? In the case of double tees an important factor may be that the ceiling area to be painted is large. Also, a different fascia detail is often used for this type of roof assembly. Check carefully and be certain to discuss alternates with all subcontractors.

12-5 PRECAST SLABS

Precast slabs are available in hollow, cored, and solid varieties for use on floors, walls, and roofs. For short spans, various types of panel and channel slabs with reinforcing bars are available in both concrete and gypsum. Longer spans and heavy loads most commonly involve cored units with prestressed wire.

The solid panel and channel slabs are available in heavyweight and lightweight aggregates. The thicknesses and widths available vary considerably, but the maximum span is generally limited to about 10'-0". Some slabs are available tongue-and-grooved, some with metal-edged tongue-and-groove. These types of slabs use reinforcing bars or reinforcing mesh for added tension strengths. These lightweight, easy-to-handle, nail, drill, and saw pieces are easily installed on the job over the supporting members. A clip or other special fastener should be used in placing the slabs.

Cored units with prestressed wire are used on roof spans up to about 44'-0". Thicknesses available range from 4 to 16 inches with various widths available, 40 and 48 inches being the most common. Each manufacturer must be contacted to determine the structural limitations of each individual product. The units generally have high fire resistance ratings and are available with an acoustical finish. Some types are available with exposed aggregate finishes for walls.

Specifications. The type of material used and the manufacturer specified are the first items to be checked. The materials used to manufacture the plank, type and size of reinforcing, and required fire rating and finish must be checked.

Note also who cuts required holes in the planks and who caulks the joints and the type of caulking to be used. If topping is required, the thickness, reinforcing, aggregates, and strength specified must also be noted. Any inserts, anchors, or special requirements must be noted as well.

Estimating. The precast slabs are generally quoted by the square foot or in a lump sum. Solid panels and channel slabs can often be purchased from the manufacturer and installed by the general contractor. The manufacturers of cored units generally furnish and install the planks themselves. When calculating the square footage, deduct all openings. Also determine who will cut the holes, do any welding and special finishing, and provide anchors and inserts. Concrete topping is commonly used over the cored units, the most commonly specified thickness being 2 inches. Determine the cubic yards of concrete, square feet of reinforcing mesh, and square feet of surface to be finished.

Check the drawings and specifications to determine how the planks will be held in place, special anchorage details, inserts, and any other items that may present a cost or problem on the job. Mechanical and electrical requirements should also be checked.

EXAMPLE 12-3 — Commercial Building: A review of the wall sections for Alternate 1 (Fig. 12–2) and the base bid wall sections (Appendix A) shows that the same roof overhang was used in both sections. This means that the roof area determined for the double tees in Example 12-1 can be used here.

Roof area = 7,889 s.f.

Labor. Two typical subcontractors' bids are included here. Figure 12–7 shows a square-foot (unit price) bid, and Fig. 12–8 shows a lump-sum bid.

8″ thick prestressed hollow core slabs, furnished
and installed .... $6.85 s.f.

Figure 12-7   Subcontractor's Bid for Slabs (S.F.)

8″ thick prestressed hollow core slabs, furnished and
installed $54,039.65

Figure 12-8   Subcontractor's Bid for Slabs (Lump Sum)

12–6 PRECAST BEAMS AND COLUMNS

Precast beams and columns are available in square, rectangular, "T," and "I" shaped sections. They are available simply reinforced, with reinforcing bars, or prestressed with high-strength wire. The sizes and spans depend upon the engineering requirements of each particular project, and the beams and columns are not poured in any one particular size or shape. Special forms can easily be made out of wood to form the size and shape required for a particular project.

Specifications. The manufacturers' specified strengths and materials, reinforcing, connection devices, anchors, inserts, and finishes required should all be noted. The different shapes required throughout and any other special requirements should also be specified.

Estimating. If the contractor intends to precast the concrete, the costs involved are indicated in Sec. 12–8. Manufacturers who bid this item will bid it per lineal foot or in a lump sum. In doing a takeoff, keep the various sizes separate. Take special note of the connection devices required.

EXAMPLE 12-4 — Commercial Building: Lintels are required over all openings in the building. The precast lintels are 8″ × 12″ and bear 6 inches on the wall. The details for door openings should be checked to determine their actual sizes.
If the door is 3′-0″ wide and if a hollow metal frame is used, it is possible for the actual masonry opening (with a 2-inch frame) to be 3′-4″. If a 6-inch bearing is required at each end, the minimum lintel length would be 4′-4″.

	NO.	M.O.*	LENGTH REQUIRED	L.F. REQUIRED
Doors	2	3′-4″	4′-4″	8′-8″
	1	10′-0″	11′-0″	11′-0″
Windows	2	8′-0″	9′-0″	18′-0″
	4	6′-0″	7′-0″	28′-0″

(*M.O. indicates masonry opening.)

Reinforcing required: Four #8 bars
Color: Gray
Finish: Rubbed

65′-8″ of 8″ × 12″ concrete lintels

Wherever the concrete lintels are placed, they take up area that was originally figured as concrete block. If the area the lintels will occupy is significant, the concrete block figures should be revised.

Labor. Precast beams (lintels) for door and window openings are generally installed by the mason. If the weight of an individual lintel exceeds about 300 lbs., it may be necessary to have a small lift or crane to put them in place. A typical subcontractor's bid is shown in Fig. 12–9. The estimate of time is based on Fig. 12–10. Labor rates are based on local labor and union conditions.

Furnish 8″ × 12″ concrete lintels in required lengths · · $8.40 l.f.

Figure 12-9 Subcontractor's Bid for Lintels

Item	Unit	Work hours
Lintels (up to 300 lbs.)	100 l.f.	4.5 to 7.0
Coping (up to 150 lbs.)	100 l.f.	5.0 to 8.0

Figure 12-10 Work Hours Required for Lintels and Coping

EXAMPLE 12-5 — Commercial Building: Lintels: When the lintel areas have not been deducted from the block wall area, the labor cost balances out against the installation cost of the block it will replace.

Lintels: 65.67 l.f.

Rate of work:

Mason, 3.5 work hours per 100 l.f.
Laborer, 3.0 work hours per 100 l.f.

Labor:

Mason, $19.85 per hour
Laborer, $15.57 per hour

Time:

$$\text{Mason} - .66 \times 3.5 = 2.3 \text{ work hours}$$
$$\text{Laborer} - .66 \times 3.0 = 2.0 \text{ work hours}$$

Cost:

$$
\begin{aligned}
\text{Mason} &- 2.3 \times \$19.85 = \$45.66 \\
\text{Laborer} &- 2.0 \times \ \ 15.57 = \underline{\$31.14} \\
& \$76.80
\end{aligned}
$$

12-7 MISCELLANEOUS PRECAST

Precast panels for the exterior walls of homes, warehouses, apartment, and office buildings are available. Their sizes, thicknesses, shapes, designs, and finishes vary considerably. Each individual system must be analyzed carefully to determine the cost in place. Particular attention should be paid to the attachment details at the base, top, and midpoints of each panel, how attachments will be handled at the job site, how much space is required for erection, how much bracing is required for all panels to be securely attached, and how many men are required.

Some of the various methods involve the use of panels 4'-0" wide, panels the entire length and width of the house, and precast boxes which are completely furnished before they are installed on the job. There is a tremendous amount of research going into precast modules. The higher the cost of labor in the field, the more research there will be to arrive at more economical building methods.

The estimator must analyze carefully each individual system, consider the fabrication costs and time, space requirements, how mechanicals will relate and be installed, and try to determine any hidden costs. New systems require considerable thought and study.

structural systems

12-8 PRECAST COSTS

If the specifications allow the contractor to precast the concrete shapes required for a project, or if the contractor decides to at least estimate the cost for precasting and compare it with the proposals received, the following considerations will figure in the cost:

1. Precasting takes quite a lot of space: Is it available on the job site or will the material be precast off the site and transported to the job site? If precast off the site, whatever facilities are used must be charged off against the items being made.

2. Determine whether the types of forms will be steel, wood, fiberglass, or a combination of materials, who will make the forms, how long will manufacture take, and how much it will cost. The cost of forms must be charged to the precast items being made.

3. Will a specialist be required to supervise the manufacture of the items? Someone will have to coordinate the work and the preparation of shop drawings. This cost must also be included.

4. Materials required for the manufacture must be purchased:
   a. Reinforcing
   b. Coarse and fine aggregates
   c. Cement
   d. Water
   e. Anchors and inserts

5. Allowance must be made for the actual cost of labor required to:
   a. Clean the forms.
   b. Apply oil or retarders to the forms.
   c. Place the reinforcing (including pretensioning if required).
   d. Mix and pour the concrete (including troweling the top off).
   e. Cover the concrete and apply curing method.
   f. Uncover the concrete after the initial curing.
   g. Strip from form and stockpiling to finish curing.
   h. Erect the concrete on the job.

   If poured off the site, it will also be necessary to load the precast concrete on trucks, transport it to the job site, unload it, and then erect it.

6. Equipment required may include mixers, lift trucks, and cranes. Special equipment is required to prestress concrete. Equipment to cure the concrete may be required and miscellaneous equipment such as hoes, shovels, wheelbarrows, hammers, and vibrators will also be required.

7. Shop drawings should be prepared either by a company draftsman or a consultant.

12-9 PRECAST CHECKLIST

shapes
bearing requirements

accessories

walls

floors

ceilings

beams

joists

girders

lintels

strength requirements

inserts

attachment requirements

finish

color

special requirements

12-10 STRUCTURAL STEEL

General contractors generally handle structural steel in one of two ways: either they purchase the steel fabricated and erect it with their own construction crew; or they have the steel company fabricate and erect the steel or arrange erection by another company. Many contractors do not have the equipment and skilled personnel required to erect the steel. The structural steel includes all columns, beams, lintels, trusses, joists, bearing plates, girts, purlins, decking, bracing, tension rods, and any other items required.

When estimating structural steel, you should estimate each item such as column bases, columns, trusses, lintels, etc., separately. Structural steel is purchased by the ton, and the cost per ton varies depending on the type and shape of steel required. Also, labor operations are different for each type.

The estimate of the field cost of erecting structural steel will vary depending on weather conditions, prompt delivery of all materials, equipment available, size of building, and amount of riveting and welding required.

12-11 STRUCTURAL STEEL FRAMING

The structural steel used for the framing of a structure includes primarily wide flange beams, light beams, I beams, plates, channels, and angles. Special shapes and composite members are often specified; they require a great deal of fabrication and should be listed separately from the standard mill pieces. Keep a careful list of each size and length. Check not only the structural drawings, but also the sections and details. Note in the specifications the fastening method required and check for fastening details on the drawings.

Estimating.   Steel is sold by weight, so the takeoff is made in pounds and converted into tons. The takeoff should be first a listing of all the steel required for the structure. A definite sequence for the takeoff should be maintained; a commonly used sequence is columns and details, beams and details, and bracing and flooring (if required). Floor by floor, a complete takeoff is required.

Structural drawings, details, and specifications do not always show all of the required items. Among the items that may not specifically be shown and yet are required for a complete job are various field connections, field bolts, ties, beam separators, anchors, turnbuckles, rivets, bearing plates, welds, setting plates, and templates. The specification may require conformance with AISC standards, with the exact methods left up to the fabricator and erector. When this is the situation, a complete understanding is required of the AISC and code requirements to do a complete estimate. Without this thorough understanding the estimator can make only an approximate estimate for a check on subcontractor prices, but should not attempt to use those figures for compiling a bid price.

Once a complete takeoff of the structural steel has been made, the steel should be grouped according to the grade (A36, A242, A374A, 375, etc.) required and the shape of the structural piece. The special built-up shapes must be listed separately and in each area the shapes should be broken down into small, medium, and large weights per foot. For standard mill pieces the higher weight per foot, the lower the cost per pound will be. The weights of the various standard shapes may be obtained from any structural steel handbook or from the manufacturing company. To determine the weight of built-up members, total up each of the component shapes used to make the special shape.

12-12 STEEL JOISTS

Steel joists, also referred to as *open web steel joists,* are prefabricated lightweight trusses. Most trusses follow the Warren or modified Warren type of design (Fig. 12–11). There are six basic series of steel joists.

1–2. J and H series—short-span series, spans to 48 ′-0 ″

3–4. LA and LH series—long-spans joists, spans to 96 ′-0 ″

5–6. DLJ and DLH series—deep long-span joists, spans to 144 ′-0 ″

The members will be called out on the drawings as "12 H 4. "

12⌐   H   4◄———designation within series
depth of joist   └series specified

In order to find the weight per foot, refer to the manufacturer's catalog, Sweets Catalog, or AISC manual for the appropriate series and check the listing; for example, a 12 H 4 weighs 6.2 pounds per foot. Calculate the lineal footage of joists required and multiply it by its weight per foot; the product represents the total weight. The cost per pound is considerably greater than that for some other types of steel (such as reinforcing bars or wide flange shapes) due to the sophisticated shaping and fabrication required.

Steel joists by themselves do not make an enclosed structure—they are one part of an assembly. The other materials included in this assembly are varied, and selection of them may be made based on economy, appearance, sound control, fire-rating requirements, or any other criterion. When doing the takeoff of joists, be aware of the other parts in the assembly; they must all be included somewhere in the estimate.

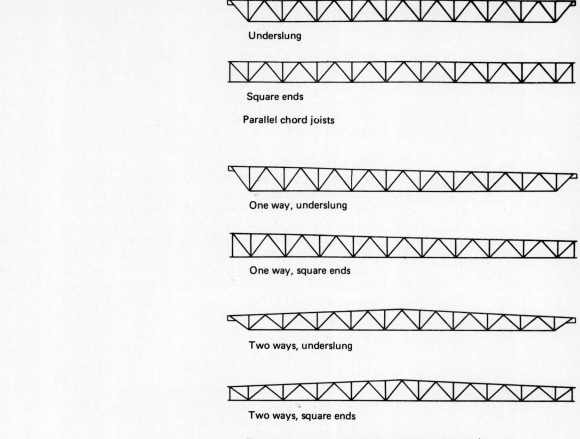

Underslung

Square ends

Parallel chord joists

One way, underslung

One way, square ends

Two ways, underslung

Two ways, square ends

Pitched top chord joists (Note: Standard pitch is $\frac{1}{8}''$ per ft)

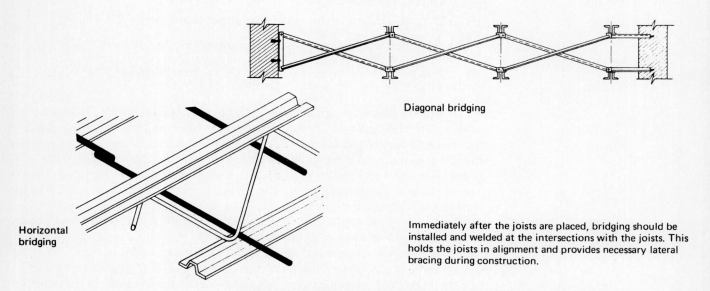

Diagonal bridging

Horizontal bridging

Immediately after the joists are placed, bridging should be installed and welded at the intersections with the joists. This holds the joists in alignment and provides necessary lateral bracing during construction.

Figure 12-11A Common Steel Joist Shapes and Joist Bridging

structural systems

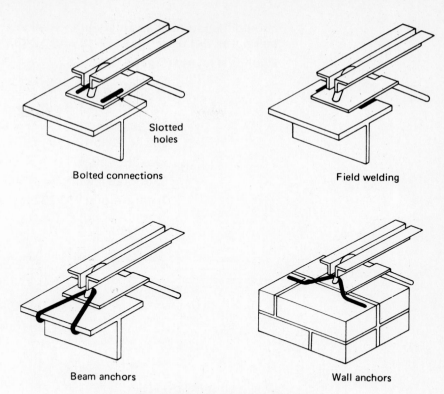

Slotted
holes

Bolted connections

Field welding

Beam anchors

Wall anchors

Figure 12-11B   End Anchorage

The mechanicals and electrical requirements are generally quite compatible with steel joists; only large ducts have to be carefully planned for job installation.

Specifications.   Specifications will list the type of joist required, the type of attachment they will have to the rest of the structure, and also the accessories and finish. Many specifications will enumerate industry standards that must be met for strength and for the type of steel used in the joists, erection and attachment techniques, and finishes. Read the standards carefully. Accessories that may be specified are bridging, ceiling extensions, masonry wall anchors, bridging anchors, and header angles.

Estimating.   Since structural steel is sold by the pound, the total number of pounds required must be determined. First, take off the lineal feet of each different type of joist required; then multiply the lineal footage of each type times the weight per foot to determine the total weight of each separate type. Also, estimate the accessories required, both their type and the number needed.

If the contractor is to erect the joists, an equipment list, costs, and the labor work hours are required. On small jobs, it is not unusual for the contractor to use his own forces to set the joists. Remember that all accessories must also be installed.

If the joist erection is to be subbed out, check to see what the general contractor's responsibilities are and who will install the accessories.

EXAMPLE 12-6 — Commercial Building: Alternate 2: Contractors shall state in their Proposal Form the amount to be added or deducted from the base bid,

should they furnish and install the steel joist roof with suspended acoustical ceiling as detailed (Figs. 12-12 and 12-13).

Roof joist layout (Fig. 12-13):

$$27 - 36\ LA15,\ 67'-0''\ long = 1,890$$
$$1,890\ l.f. \times 43\ lbs./l.f. = 77,787\ lbs.$$
$$5 - 36LA08,\ 47'-0''\ long = \quad 235\ l.f.$$
$$235 \times 23\ lbs./l.f. = 5,405\ lbs.$$
$$12 - 36LA09,\ 57'-0'' = 684\ l.f.$$
$$684 \times 25\ lbs./l.f. = 17,100\ lbs.$$

Total weight: 100,292 lbs. or 50.15 tons

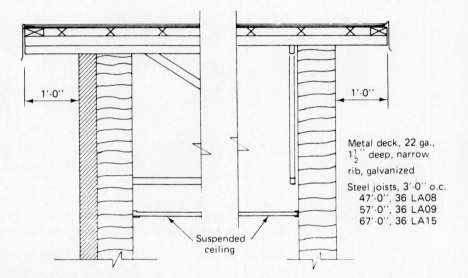

Metal deck, 22 ga.,
$1\frac{1}{2}''$ deep, narrow
rib, galvanized

Steel joists, 3'-0" o.c.
47'-0", 36 LA08
57'-0", 36 LA09
67'-0", 36 LA15

Suspended ceiling

Figure 12-12   Steel Joist Roof with Suspended Acoustical Ceiling

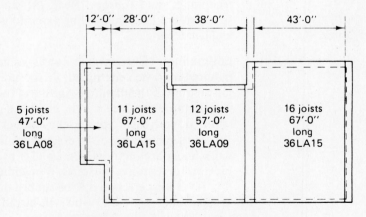

12'-0"	28'-0"	38'-0"	43'-0"

5 joists
47'-0"
long
36LA08

11 joists
67'-0"
long
36LA15

12 joists
57'-0"
long
36LA09

16 joists
67'-0"
long
36LA15

Figure 12-13   Roof Joist Layout

Labor. A typical subcontractor's bid for this work is shown in Fig. 12-14. The time required for the installation of steel joists may be taken from Fig. 12-15. The hourly wages are based on local labor and union conditions. A crane would be usually required for joists over 300 lbs. The crane would reduce installation time on large projects, and justify the cost of the crane, even if the joist weight was less than 300 lbs.

Furnish and install all steel joists required for
Alternate 2 . . . . $29,747.00

Figure 12-14 Subcontractor's Bid for Steel Joists

Joist type	Work hours per ton
J & H, up to 30 ft.	3.0 to 5.0
J & H, over 30 ft.	4.0 to 6.0
LJ & LH	4.0 to 6.0
DLJ & DLH	4.0 to 6.0

Bolted cross bridging (for welded cross bridging, add 25%)

Figure 12-15 Work Hours Required for Installing Joists

12-13 STEEL DECKING

Steel decking is used for floor and roof applications. Depending on the particular requirements of the job, a wide selection of shapes, sizes, thicknesses, and accessories are available.

For roof applications the simplest types are simply decks over which insulation board and built-up roofing are applied, or forms and reinforcing over which concrete may be poured, ranging to decking that can receive recessed lighting and has acoustical properties. Depending on the type used and design of the deck, allowable spans range from 3'-0" to about 33'-0".

Decking for floor applications is equally varied from the simplest type used as a form and reinforcing for concrete to elaborate systems that combine electrical and telephone outlets, electrical raceways, air ducts, acoustical finishes, and recessed lighting.

Decking is generally available either unpainted, primed, painted, or galvanized. Accessories available include flexible rubber closures to seal the flutes, clips that fasten the decking to the purlins, lighting, and acoustical finish.

Specifications. Determine the type of decking required, thickness, gauge of metal, finish required on the decking as received from the supplier, method of attachment, accessories required, and manufacturers specified.

Items that are necessary for the completion of the decking and that you should check to be certain that they are included in the specifications include painting of the underside of the deck, acoustical treatment, openings, and insulation.

Estimating. Since steel decking is priced by the square, the first thing for you to do is to determine how many squares are required. Again, a systematic plan should be used: start on the floor on which the decking is first used and work up through the building, keeping the estimates for all floors separate.

Decking is usually installed by welding directly through the bottom of the rib, usually a maximum of 12 inches on center, with side joints mechanically fastened not more than 3′-0″ on center. The estimator will have to determine approximately how many weld washers will be required and how long it will take to install them. Otherwise fastening is sometimes effected by clips, screws, and bolts.

The estimator should consult local dealers and suppliers for material prices to be used in preparing the estimate. Materials priced f.o.b. dealer will require that the estimator add the cost of transporting the materials to the job site. Once at the job site, they must be unloaded, perhaps stored, and then placed on the appropriate floor for use. In most cases one or two workers can quickly and easily position the decking and prepare them for the welder to make the connections.

Be especially careful on multifloor buildings to count the number of floors requiring the steel deck, and keep the roof deck, and any possible poured concrete first floor separate. Most estimators make a small sketch of the number of floors to help avoid errors.

In checking the number of floors, also be aware that many times there may also be a lobby level, lower lobby level, and basement.

EXAMPLE 12-7 — Commercial Building: A review of Alternate 2 (Fig. 12-12) indicates that a 22-gauge, 1½″-deep, narrow-rib galvanized steel deck is specified.

The roof area required is the same as that used for precast double tees in Example 12-1: 7,889 s.f.

Roof area = 7,889 s.f. + 5% waste = 8,180 s.f.

Labor. The time required for the installation of steel decking may be taken from Fig. 12-16. The hourly wages used are based on local labor and union conditions.

Steel decking, gauge	Work hours per square
22	1.0 to 2.5
18	2.0 to 4.0
14	2.5 to 4.5

For 5,000 s.f. or more (for less than 5,000 s.f., add 30%)

Figure 12-16 Work Hours Required for Installing Steel Decking

EXAMPLE 12-8:

Steel decking: 7,889 s.f. (area covered)
Rate of work: Steelworker, 1.0 work hour per square
Labor: Steelworker, $20.65 per hour

Time:

$$78.8 \text{ squares} \times 1 = 78.8 \text{ work hours}$$

Cost:

$$78.8 \times \$20.65 = \$1,627.22$$

12-14 MISCELLANEOUS STRUCTURAL STEEL

Other types of structural steel framing are sometimes used. They include structural steel studs, joists, purlins, and various shapes of structural pipe and tubes available. The procedure for estimating each of these items is the same as that outlined previously for the rest of the structural steel.

1. Take off the various types and shapes.
2. Determine the pounds of each type required.
3. The cost per ton times the tonnage required equals the material cost.
4. Determine the work hours and equipment required and their respective costs.

12-15 STEEL ERECTION SUBCONTRACTORS

Most of the time these subcontractors can erect structural steel at a considerable saving, compared to the cost to the average general contractor. They have specialized, well-organized workers to complement the required equipment that includes cranes, air tools, rivet busters, welders, and impact wrenches. These factors, when combined with the experienced organization that specializes in one phase of construction, are hard to beat.

Using subs never lets the estimator off the hook; an estimate of steel is still required because cost control is difficult to maintain and improve upon.

12-16 STEEL CHECKLIST

Shapes:
    sections
    lengths
    quantities
    weights
    locations
    fasteners

Engineering:
    fabrication
    shop drawings
    shop painting
    testing
    inspection
    unloading, loading

erection
plumbing up

Installation:
    riveting
    welding
    bracing (cross and wind)
    erection
    bolts

Miscellaneous:
    clips
    ties
    rods
    painting
    hangers
    plates
    anchor bolts

12-17 WOOD SYSTEMS

Wood is used as a component in quite a number of structural systems, among them wood trusses, laminated beams, wood decking, and box beams. Wood is used for a variety of reasons: the wood trusses are economical, while the laminated beams and box beams are both economical and used for appearance.

12-18 WOOD TRUSSES

Wood joists with spans in excess of 150 feet are readily available. The cost savings of these types of trusses compared to steel, range as high as 25 percent: less weight is involved and the trusses are not as susceptible to transportation and erection damage. Decking is quickly and easily nailed directly to the chords. Trusses of almost any shape (design) are possible. Ducts, piping, and conduits may be easily incorporated into the trusses. The trusses are only part of the system and must be used in conjunction with a deck of some type. Typical truss shapes are shown in Fig. 12-17.

Specifications.   Check the type of truss required. If any particular manufacturer is specified, note how the members are to be attached (to each other and to the building) and what stress-grade-marked lumber is required. Note also any requirements regarding the erection of the trusses and any finish requirements.

Estimating.   Determine the number, type, and size required. If different sizes are required, keep them separate. Note any special requirements for special shapes; then get a written proposal from the manufacturer or supplier. If they are not familiar with the particular project, make arrangements for them to see the contract documents so that they will provide a complete price. Check to see how the truss is attached to the wall or column supporting it.

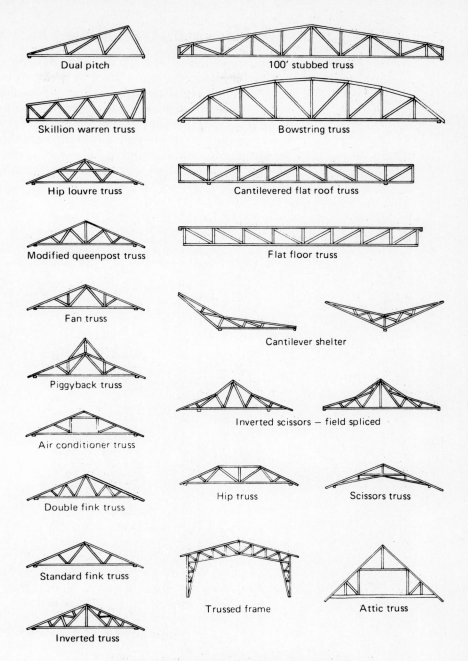

Figure 12-17 Typical Wood Truss Shapes

Dual pitch

Skillion warren truss

Hip louvre truss

Modified queenpost truss

Fan truss

Piggyback truss

Air conditioner truss

Double fink truss

Standard fink truss

Inverted truss

100' stubbed truss

Bowstring truss

Cantilevered flat roof truss

Flat floor truss

Cantilever shelter

Inverted scissors — field spliced

Hip truss

Scissors truss

Trussed frame

Attic truss

If the trusses are to be installed by the general contractor, an allowance for the required equipment (booms, cranes) and workers must be made. The type of equipment and number of workers will depend on the size and shape of the truss, and the height of erection. Trusses 100 feet long may be placed at the rate of 1,000 to 1,500 square feet of coverage per hour, using two mobile cranes mounted on trucks and seven workers.

wood trusses

207

12-19 LAMINATED BEAMS AND ARCHES

Laminated structural members are pieces of lumber glued under controlled temperature and pressure conditions. The glue used may be either interior or exterior. They may be rectangular beams or curved arches, such as parabolic, bowstring, "V," or "A." The wood used includes Douglas fir, southern pine, birch, maple, and redwood (Fig. 12-18). Spans just over 300 feet have been built by using laminated arches in a cross vault. They are generally available in three grades: industrial appearance, architectural appearance, and premium appearance grade. The specifications should be carefully checked so that the proper grade is used. They are also available prefinished.

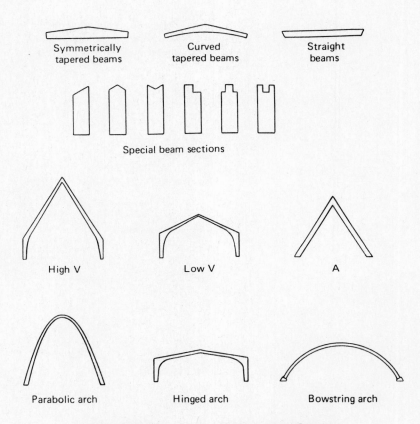

Figure 12-18   Typical Laminated Shapes

Specifications.   Check for quality control requirements, types of adhesives, hardware, appearance grade, finish, protection, preservative, and erection requirements.

Estimating.   The cost of materials will have to come from the manufacturer or supplier, but the takeoff is made by the estimator. The takeoff should list the lineal footage required of each type, and the size or style of beam or arch. Note also the type of wood, appearance grade, and finishing requirements. If the laminated shapes are not prefinished, be certain that finishing is covered somewhere in the specifications.

If the laminated beams and arches are to be installed by the general contractor, consider who will deliver the material to the site, how it will be unloaded, where it will be stored, and how much equipment and how many workers will be required to erect the shapes. Erection time varies with the complexity of the project. In a simple erection job beams may be set at the rate of six to eight lengths per hour, while large, more complex jobs may require four hours or more for a single arch. Check carefully the fastening details required, total length of each piece, how they will be braced during construction, and whether it is one piece or is segmental.

12-20 WOOD DECKING

Wood decking is available as a solid-timber decking, plank decking, and laminated decking. Solid-timber decking is available either in natural finish or prefinished; the most common sizes are 2 × 6, 3 × 6, and 4 × 6. The most commonly used woods for decking are southern pine, western red cedar, inland white fir, western white spruce, and redwood.

Plank decking is tongue-and-groove decking fabricated into panels. The most common panel size is 21 inches wide and lengths are up to 24 feet. Installation costs are reduced by using this type of deck. Note wood species, finish, size required, and appearance grade.

Laminated decks are available in a variety of thicknesses and widths. Installation costs are reduced substantially. Note the wood species, finish, and size required. Appearance grades are also available.

Estimating.   No matter what type of wood deck is required, the takeoff is the square foot area to be covered. Since the decking is sold by the board measure, the square footage is multiplied by the appropriate factor shown in Fig. 12-19. Take particular notice of fastening details and size of decking required. (See board measure, Sec. 13-2).

| Block decking | | Laminated decking | | |
nominal dimensions		actual dimensions		
3 x 6	4 x 6	$2\frac{1}{4} \times 5\frac{1}{2}$	$3\frac{1}{16} \times 5\frac{1}{2}$	$3\frac{13}{16} \times 5\frac{1}{2}$
3.43	4.58	3.33	4.4	5.5

Multiply the square foot area to be covered by the factor listed to arrive at the required board measure. These factors include no waste.

Figure 12-19   Wood Deck Estimating Factors

Consider also if the decking is priced delivered to the job site or must be picked up, where the decking will be stored, how it will be moved onto the building, and the number of workers required.

Solid decking requires between 12 and 16 work hours for a M.B.F.M. (thousand board feet of measure), while plank decking may save 5 to 15 percent of installation costs on a typical project. Laminated decking saves 15 to 25 percent of the installation costs since it is attached by nailing instead of with heavy spikes.

Specifications. Check for wood species, adhesives, finishes, appearance grade, fastening, size requirements, and construction of the decking. Size of decking should also be noted.

12-21 PLYWOOD SYSTEMS

The strength and versatility of plywood has recommended it for use in structural systems. The systems presently in use include box beams, rigid frames, folded plates, and space planes for long-span systems. For short-span systems, stressed skin panels and curved panels are used.

The availability, size, and ease of shaping of plywood makes it an economical material to work with. Fire ratings of one hour can be obtained also.

Technical information regarding design, construction, estimating and cost comparison may be obtained through the American Plywood Association.

Plywood structural shapes may be supplied by a local manufacturer, shipped in, or built by the contractor's work forces. The decision will depend on local conditions such as suppliers available, workers, and space requirements.

Specifications. Note the thickness and grade of plywood, whether interior or exterior glue is required, fastening requirements, and finish that will be applied. All glue used in assembling the construction should be noted, and any special treatments such as with fire or pressure preservatives, must also be taken off.

Estimating. Take off the size and shape of each member required. Total up the lineal footage of each different size and shape. If the material is being supplied by a manufacturer, it will probably be priced by the lineal foot of each size or as a lump sum. If the members are to be built by the contractor's work force, either on or off the job site, a complete takeoff of lumber and sheets of plywood will be required.

12-22 WOOD CHECKLIST

Wood:
  species
  finish required
  solid or hollow
  laminated
  glues
  appearance grade
  special shapes
  primers

Fastening:
  bolts
  nails
  spikes
  glue
  dowels

Erection:
  cranes

GENERAL

All provisions of the "Sections 1A and 1B" form a part of this section.

SCOPE

The Contractor shall furnish and install all precast reinforced concrete double tee roof slabs and filler blocks as indicated on the Drawings.

WORKMANSHIP

Skilled mechanics shall install slabs in strict conformity with approved shop drawings and details.

The Contractor shall place no warped, cracked, or broken tees.

MATERIALS

Concrete for the slabs shall be vibrated concrete with a minimum strength of 5,000 psi at 28 days, using a maximum aggregate size of 3/4". The manufacturer shall furnish the Architect with laboratory test results verifying the strength of concrete.

Reinforcing for the slabs shall be two (2) No. Nine (9) bars with 6" x 6" #10/#10 wire mesh as shown on the Drawings. Reinforcing shall conform to the requirements of ASTM A15 for billet-steel bars for concrete reinforcement.

All roof slabs shall be cast using plastic chairs to hold the reinforcing in place. Galvanized chairs with plastic coated ends are also acceptable.

FILLER BLOCKS

Filler blocks shall be 2" thick and shall be made of the same material as the masonry block units.

Allow 1/2" mortar joint on all sides of the filler block.

INTERIOR FINISH OF SLABS

All "air holes" and "air pockets" in the roof slabs shall be filled with grout and wiped smooth. Grout to be one part Portland Cement to two parts clean, fine sand.

SHOP DRAWINGS

The Contractor shall submit complete shop drawings for approval by the Architect prior to the manufacturer of the roof planks.

CAULKING

Caulking shall be installed upon completion of erection on longitudinal joints between the double tee units exposed in finish construction. Apply on dry, clean, and sufficiently warm surfaces to insure lasting adhesion, with a uniform bead netaly laid into joint.

Caulking shall be Tremco Manufacturing Company's "Dura Bilt" or approved equal.

LINTELS

Precast lintels shall be made to sizes and shapes shown on the drawings with reinforcing noted. Concrete strength shall be 5,000 psi at 28 days. The exposed finish shall be rubbed with all "air holes" filled.

Figure 12-20 Typical Concrete Specification

wood checklist   211

booms
scaffolds
hoists

Engineering:
fabrication
shop drawings
unloading, loading
inspection
bracing

REVIEW QUESTIONS

1. Determine the amounts required for the structural system for the building in Appendix E. Use workup sheet and sketches.

2. What unit of measure may be used by suppliers in pricing precast concrete structural systems?

3. Assume 8″-thick hollow core slabs are required for the roof deck of the building (Appendix E). How many s.f. are required?

4. Many suppliers of precast concrete take the responsibility of installing the units. Why may this be desirable?

5. Under what conditions might it be desirable for the contractor to precast the concrete required?

6. In taking off structural steel, why are the various shapes, such as wide flange shapes and steel joists, listed separately?

7. What are the advantages of using steel erection subcontractors?

8. What unit of measure is most likely to be used for laminated beams?

9. What unit of measure is most likely to be used for wood decking? How is it arrived at?

10. Assume 3″-thick wood decking (roof) is required on the building in Appendix E. What board measure must be ordered if 12% waste is anticipated?

13

WOOD FRAME CONSTRUCTION

13-1 FRAME CONSTRUCTION

The wood frame construction discussed in this chapter relates primarily to light construction of ordinary wood buildings. It covers the rough carpentry work, which includes framing, sheathing, subfloors, and insulation. Flooring, roofing, drywall, and wetwall construction are all included in their respective chapters and discussions of them are not repeated here.

The lumber most commonly used for framing is yard lumber; the size classification of lumber most commonly used is dimensional (2 to 5 inches thick and any width). The allowable spans of wood depend on the loading conditions and the allowable working stresses for the wood. As higher working stresses are required, it may become necessary to go to a stress-grade lumber that has been assigned working stresses. This type of lumber is graded by machine testing and is stamped at the mill. This information is available in pamphlet form from your local supplier and is generally included in the building code manual put out by the governing body having jurisdiction.

Rough lumber is wood that has been sawed, edged, and trimmed, but has not been dressed. When lumber is surfaced by a planing member, it is referred to as *dressed*. This process gives the piece uniform size and a smooth surface. Before the lumber is dressed, it has the full given dimension (2 × 4 inches), but once it has been surfaced on all four sides (S4S), it will actually measure about 1½ by 3½ inches.

When calculating board measure, you must use the full dimension, since lumber is marketed at its original size. The 2- by 4-inch size is considered the nominal size.

The only safe way to estimate the quantity of lumber required for any particular job is to do a take-off of each piece of lumber needed for the work. Since the time to do such an estimate is excessive, tables are included to provide as accurate material quantities as are necessary in as short a time as possible, and these are the methods generally used.

The estimator should possess a good working knowledge of the trade, as well as a familiarity with the possible job conditions, in order to estimate quantities and costs accurately.

13-2 BOARD MEASURE

When any quantity of lumber is purchased, it is priced and sold by the thousand-feet-board measure, abbreviated to *mfbm*. The number of board feet required on the job must be calculated by the estimator.

One board foot is equal to the volume of a piece of wood 1 inch thick and 1 foot square (Fig. 13–1). By the use of the following formula, the number of board feet can be quickly determined. The nominal dimensions of the lumber are used even though the actual dimensions are smaller, once the lumber is dressed.

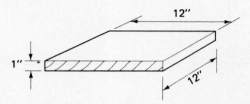

A. One foot of board measure

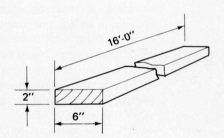

B. Sixteen feet of board measure

FIGURE 13-1   Board Measures, Examples

$$N = P \frac{T(W)}{12} L$$

N = Number of feet (board measure)
P = Number of pieces of lumber
T = Thickness of the lumber (in inches)
W = Width of the lumber (in inches)
L = Length of the pieces (in feet)

EXAMPLE 13-1 — Estimate the fbm of ten pieces of lumber 2 × 6 inches and 16′-0″ long. (It would be written 10/2 × 6-16′-0″.) (See Fig. 13-1B.)

$$N = 10\frac{2\,(6)}{12}16 = 160 \text{ fbm}$$

EXAMPLE 13-2 — Estimate the fbm in 24/2 × 8-16′-0″.

$$N = 24\frac{2\,(8)}{12}16 = 512 \text{ fbm}$$

This formula may be used for any size order and any number of pieces that are required. The table (Fig. 13–2) shows the fbm, per piece, for some typical lumber sizes and lengths. Multiply the fbm by the number of pieces required for the total fbm.

Size of lumber inches*	Length of piece in feet						
	8	10	12	14	16	18	20
2 x 4	$5\frac{1}{3}$	$6\frac{2}{3}$	8	$9\frac{1}{3}$	$10\frac{2}{3}$	12	$13\frac{1}{3}$
2 x 6	8	10	12	14	16	18	20
2 x 8	$10\frac{2}{3}$	$13\frac{1}{3}$	16	$18\frac{2}{3}$	$21\frac{1}{3}$	24	$26\frac{2}{3}$
2 x 10	$13\frac{1}{3}$	$16\frac{2}{3}$	20	$23\frac{1}{3}$	$26\frac{2}{3}$	30	$33\frac{1}{3}$
2 x 12	16	20	24	28	32	36	40
2 x 14	$18\frac{2}{3}$	$23\frac{1}{3}$	28	$32\frac{2}{3}$	$37\frac{1}{3}$	42	$46\frac{2}{3}$
2 x 16	$21\frac{1}{3}$	$26\frac{2}{3}$	32	$37\frac{1}{3}$	$42\frac{2}{3}$	48	$53\frac{1}{3}$

*nominal

Figure 13-2 Typical Board Feet

13-3 FLOOR FRAMING

In beginning a wood framing quantity takeoff the first portion estimated would be the floor framing. As shown in Fig. 13–3, the floor framing generally consists of a girder, sill, floor joist, joist headers, and subflooring.

The first step in the estimate is to determine the grade (quality) of lumber required and to check the specifications for any special requirements. Make note of this information on the workup sheets.

In the examples of wood framing the residence in Appendix G is used.

GIRDERS

When the building is of a width greater than that which the floor joists can span, a beam of some type is required. A built-up wood member, referred to as a *girder,* is often used. The sizes of the pieces used to build up the girder must be listed and the length of the girder noted.

From the foundation plan determine the length of the girder. This length is

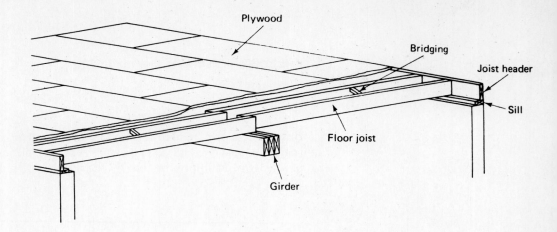

FIGURE 13-3   Floor Framing

equal to the distance between the block walls *plus* the distance that the girder rests or "bears" on each wall (Fig. 13–4).

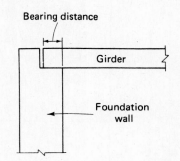

Figure 13-4   Bearing Distance

EXAMPLE 13-3 — Residential Building (Appendix G): Girder (Fig. 13-5):

SIZE	LENGTH	L.F.	B.F.
3-2 × 10	49′-8″	149	248.33

$$\text{b.f.} = 149 \times \frac{2 \times 10}{12} = 248.33$$

SILLS

Sill plates are most commonly 2 × 6, 2 × 8, and 2 × 10 inches. They are placed on the foundation so the length of sill plate required is the distance around the perimeter of the building. Lengths ordered will depend on the particular building. Not all

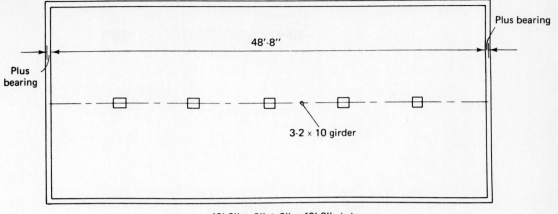

48'-8" + 6" + 6" = 49'-8" girder

Figure 13-5 Girder Layout

frame buildings require a sill plate so the details should be checked; but generally, where there are floor joists, there are sill plates.

The length of sill is often taken off as the distance around the building (perimeter)—in this case, 148 lineal feet (50 + 24 + 50 + 24) as shown in Fig. 13-6.

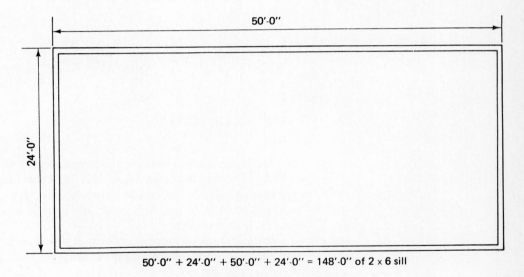

50'-0" + 24'-0" + 50'-0" + 24'-0" = 148'-0" of 2 x 6 sill

Figure 13-6 Sill Layout

An exact takeoff would require that the estimator allow for the overlapping of the sill pieces at the corners. In this case there is a 5½ in. actual overlap at each corner. As shown in Fig. 13-7, the actual length required would be 146'-2", meaning that 148 l.f. would have to be ordered. This type of accuracy is required only when the planning of each piece of wood is involved on a series of projects, such as mass-produced housing.

floor framing

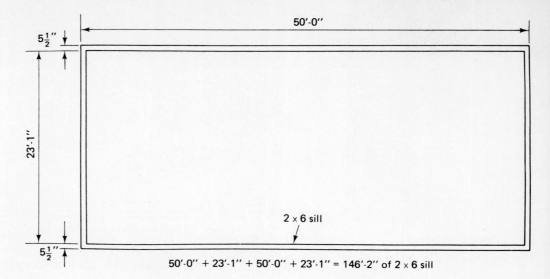

50'-0" + 23'-1" + 50'-0" + 23'-1" = 146'-2" of 2 x 6 sill

Figure 13-7 Actual Sill Length

EXAMPLE 13-4 — Residential Building: Sill plate:

SIZE	L.F.	B.F.
2 × 6	148	148

$$\text{b.f.} = 148 \times \frac{2 \times 6}{12} = 148$$

WOOD FLOOR JOISTS

The wood joists should be taken off and separated into the various sizes and lengths required. The spacing most commonly used for joists is 16 inches on center, but spacings of 12, 20, 24, 30, and 36 inches on center are also found. The most commonly used sizes are 2 × 6, 2 × 8, 2 × 10, and 2 × 12 inches, although material 3 inches wide is sometimes used.

To determine the number of joists required for any given area, the length of the floor (in feet) is divided by the joist spacing (in feet), and then one joist is added for the extra joist that is required at the end of the span. If the joists are to be doubled under partitions, or if headers frame into them, one extra joist should be added for each occurrence. Factors for various spacings are given in Fig. 13-8.

The length of the joist is taken as the inside dimension of its span plus 4 to 6 inches at each end for bearing on the wall or bearing beam.

EXAMPLE 13-5 — The area to be covered is 32 feet long and 32 feet wide (Fig. 13-9), with a bearing beam down the center of the building.

It will require joists 16 feet long to allow for a wall bearing on one end and a beam bearing on the other end. With a joist spacing of 16 inches on center, one joist will be required every 1⅓ feet; or 32 divided by 1⅓; or ¾ as

O.C. spacing inches	Divide length to be framed by	OR	Multiply length to be framed by
12	1		1
16	1.33		0.75
20	1.67		0.60
24	2		0.50
30	2.5		0.40
36	3		0.33

*Answer gives number of spaces involved, add one (1) to obtain the number of framing members required. This table makes no allowance for waste, doubling of members or intersecting walls (for stud take off).

Figure 13-8 Number of Framing Members Required for a Given Spacing

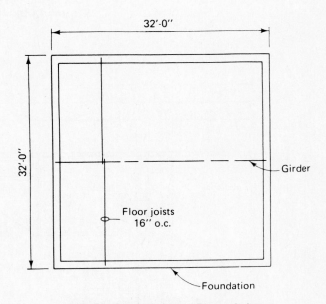

Figure 13-9 Wood Floor Joists

many joists as the length of the building (the ¾ from Fig. 13-8). There are 24 joists required plus 1 extra joist for the end, making a total of 25 joists times two (two sides to be framed).

Estimating steps.

1. From the foundation plan and wall section (Appendix G), determine the size of the floor joists required.
2. Determine the number of floor joists required by first finding the number of spaces; then add one extra joist to enclose the last space. A sketch of the residence joist layout is shown in Fig. 13–10.

floor framing

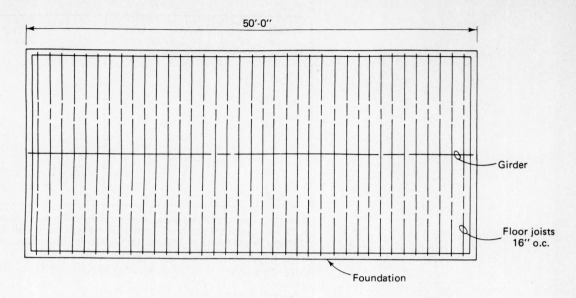

Figure 13-10 Residence Joist Layout

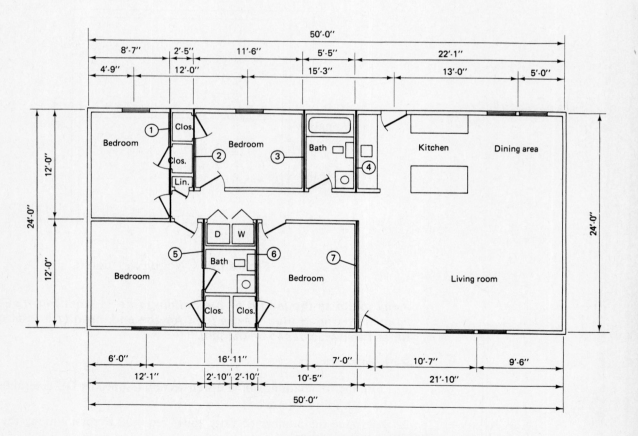

Figure 13-11 Floor Plan

wood frame construction

3. Since there are two rows of joists, double the number of joists.

4. Add one extra for partitions which run parallel to the joists (Fig. 13–11).

5. Determine the required length of the floor joists. This is done by gathering some information from the drawings:

   a. The total width of the building is given on the floor plan as 24'–0" (Fig. 13–12).

NOTE: This dimension is from the exterior face of stud to exterior face of stud.

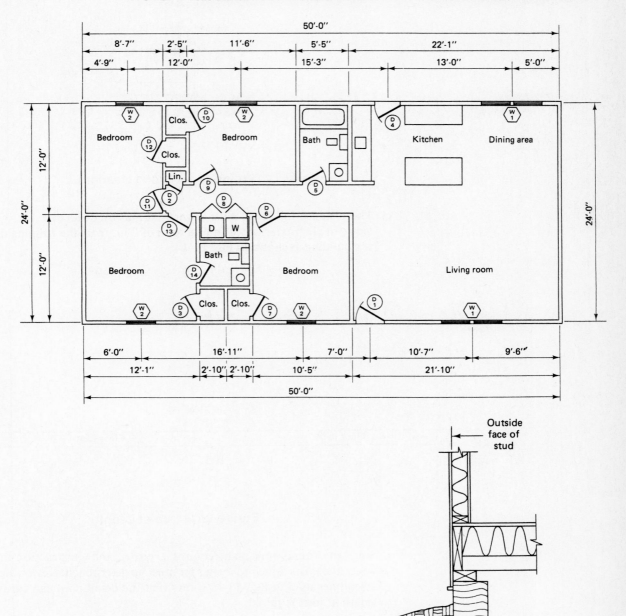

Figure 13-12   Floor Plan

b. Determine the horizontal distance from the outside face of the stud to the edge of the floor joist by carefully checking the wall section. Typically this distance will be the thickness of the joist header, or 1½ in. (Fig. 13–13).

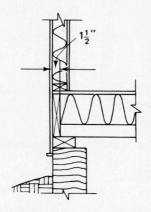

Figure 13-13 Joist Header

c. The distance the joists must cover is found by deducting 1½ in. from each side (for each exterior wall), or a total of 3 in. from the framing dimension of the building as shown in Fig. 13–14.

$$24'-0'' - 0'-3'' = 23'-9''$$

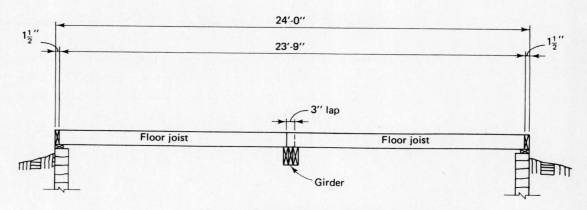

Figure 13-14 Joist Length

d. Since the foundation plan requires a girder, the construction method used where the joists attach to the girder must be determined. Reviewing the typical assemblies used in Fig. 13-15 shows that the detail used can easily affect the length of joist required.

If there is no detail showing the assembly required, recheck the specifications. If not in the specifications or on the drawings, it will be necessary to call the architect-engineer (or contractor, owner, etc.) and determine what method assembly is to be used.

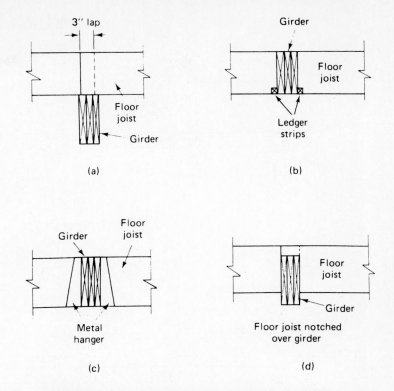

Figure 13-15 Typical Joist–Girder Assemblies

NOTE: For competitive bidding it would be necessary to get this information in the form of an addendum.

One last point should be made—don't assume that the assembly will be done any one particular way. Any of the assemblies shown in Fig. 13-15 may be required, or even some variation of the assemblies shown. Never assume that you know what is required.

In this estimate the detail shows that the joists pass over top of the girder and lap a minimum of 3 in. as shown in Fig. 13-15A. This information is then sketched on the workup sheet and shown in Fig. 13-16.

e. Since the girder is located in the center of the house (from the foundation plan), the distance from the edge of the joist to the center line of the girder is one-half the inside dimension found. This is added to the sketch and shown in Fig. 13-16.

f. Add one-half of the lap to the center line dimension to find the length of floor joist required (Fig. 13–17).

6. List the required floor joists on the workup sheet.

Floor joists: 2 × 10, 16″ o.c., #2 KD

	NO	LENGTH
Joists	78	12′
Extra joists under partition	7	12′

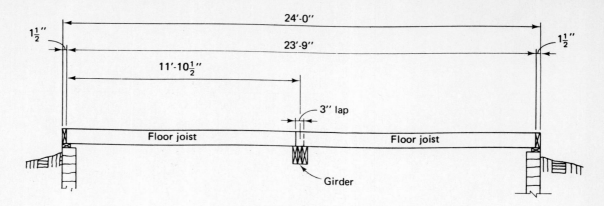

Figure 13-16   Joist Sketch

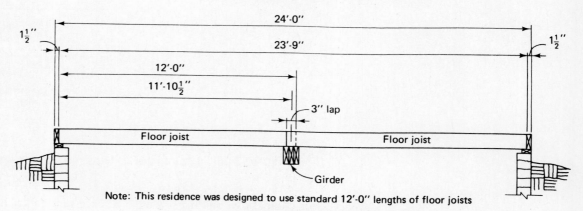

Note: This residence was designed to use standard 12'-0" lengths of floor joists

Figure 13-17   Joist Length

TRIMMERS AND HEADERS

Openings in the floor, such as for stairs or fireplaces, are framed with trimmers running in the direction of the joists and headers which support the cut-off "tail beams" of the joist (Fig. 13–18).

Unless the specifications say otherwise, when the header length (Fig. 13–19) is 4 ft. most codes allow single trimmers and headers to be used.

For header lengths greater than 4 ft., codes usually require double trimmers and headers (Fig. 13–20) and for certain special conditions some codes even require them to be tripled.

To determine the extra material required for openings we will investigate two situations:

1. a 3'–0" × 4'–0" opening as shown in Fig. 13–21.
2. a 3'–0" × 8'–0" opening as shown in Fig. 13–21.

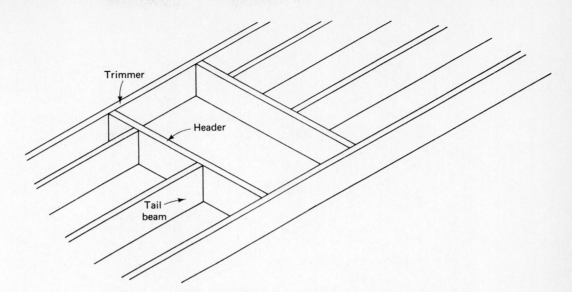

Figure 13-18 Framing for Floor Openings

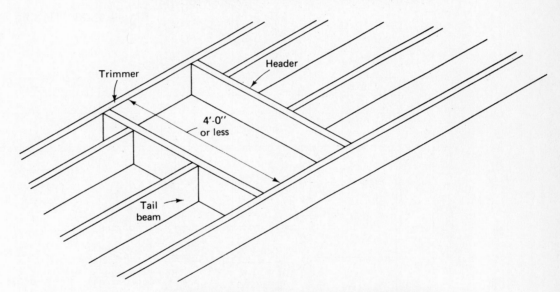

Figure 13-19 Single Trimmers and Headers

The material is determined by:

1. Sketch the floor joists without an opening (Fig. 13–22).

2. Locate the proposed opening on the floor joist sketch (Fig. 13-23).

3. Sketch what trimmer and header pieces are required and how the cut pieces of the joist may be used as headers and trimmers (Fig. 13-24).

floor framing

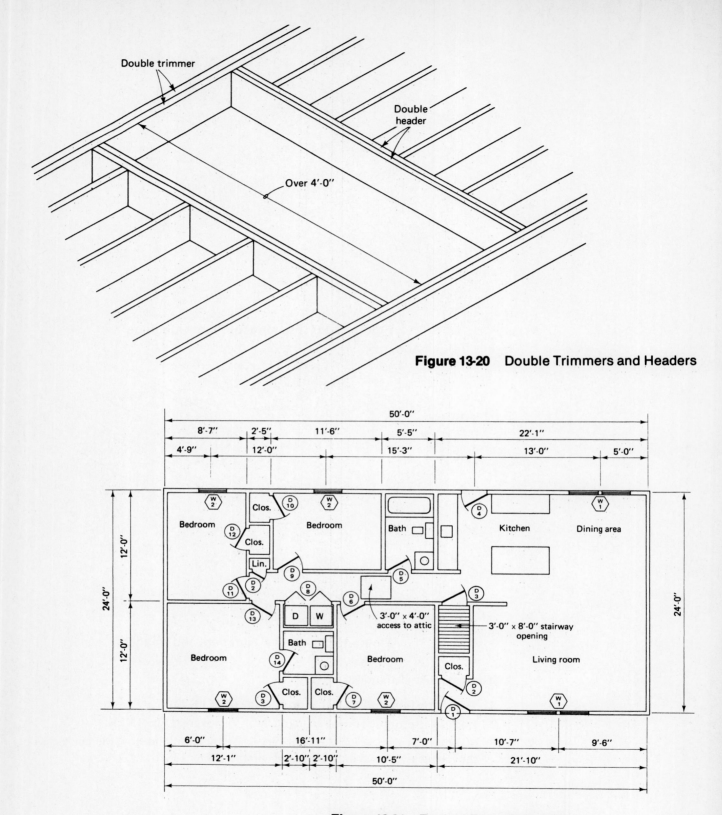

Figure 13-20 Double Trimmers and Headers

Figure 13-21 Typical Floor Openings

wood frame construction

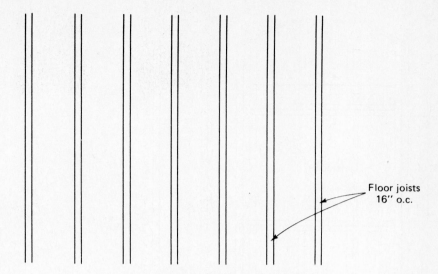

Figure 13-22 Floor Joist Sketch

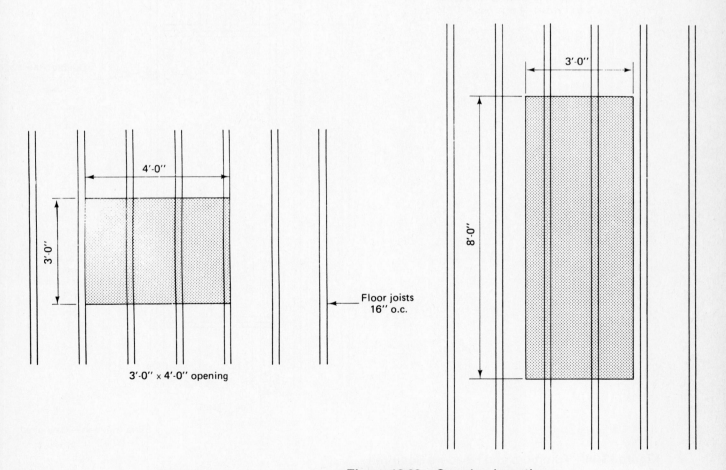

Figure 13-23 Opening Location

floor framing

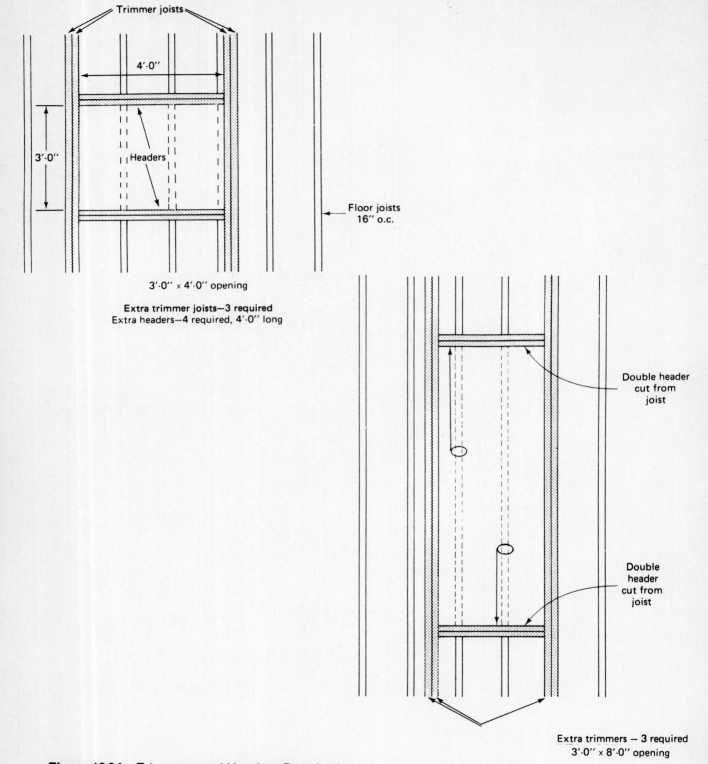

Trimmer joists

4'-0"

3'-0"

Headers

3'-0" × 4'-0" opening

Extra trimmer joists—3 required
Extra headers—4 required, 4'-0" long

Floor joists
16" o.c.

Double header
cut from
joist

Double
header
cut from
joist

Extra trimmers — 3 required
3'-0" × 8'-0" opening

Figure 13-24 Trimmers and Headers Required

Joist headers are taken off next. The header runs along the ends of the joist to seal the exposed edges (Fig. 13–25). The headers are the same size as the joists, and the length required will be two times the length of the building (Fig. 13–26).

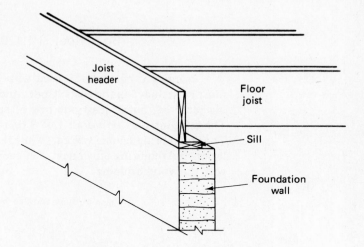

Figure 13-25   Joist Header

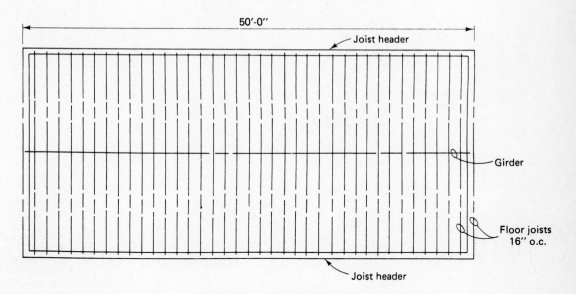

Figure 13-26   Header Length

Board feet of framing material for the floor assembly is now determined and noted on the workup sheets.

NOTE: There are *no* floor openings in the residence.

Floor joists: 2 × 10, 16″ o.c., #2KD

	NO.	LENGTH	LF.
Joists	78	12'-0"	936
Extra under partition	7	12'-0"	84
Headers	2	50'-0"	100
			1,120 l.f.

$$\text{Board feet} = 1,120 \times \frac{2 \times 10}{12} = 1,867 \text{ b.f.}$$

Bridging is customarily used with joists (except for glued nailed systems) and must be included in the costs. Codes and specifications vary on the amount of bridging required, but at least one row of cross-bridging is required between the joists. The bridging may be wood, 1 × 3 or 1 × 4 inches, or metal bridging (Fig. 13–27). The wood bridging must be cut, while the metal bridging is obtained ready for installation and requiring only one nail at each end, half as many nails as would be needed with the wood bridging.

Bridging for floor and flat roof joists and beams

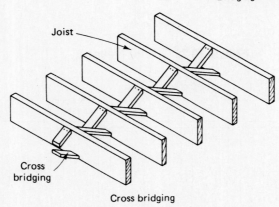

Cross bridging

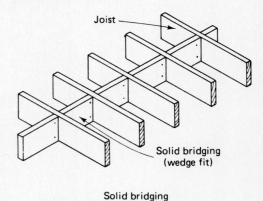

Solid bridging

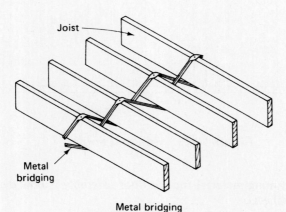

Metal bridging

Minimum sizes—wood cross bridging, 1 inch by 3 inches. Solid wood bridging, 2 inches thick, and same depth as members bridged. Metal bridging, minimum 18 gage.

Figure 13-27 Bridging

In this estimate the specifications require metal bridging (Fig. 13-28) with a maximum spacing of 8 ft. between bridging or bridging and bearing. A check of the joists' length shows they are about 12 ft. long and that one row of bridging will be required for each row of joists.

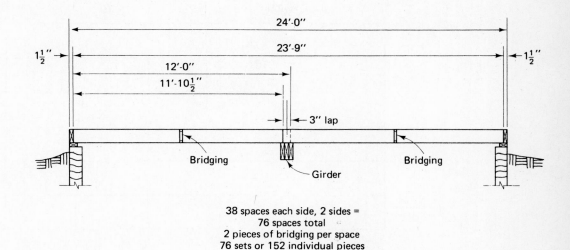

38 spaces each side, 2 sides =
76 spaces total
2 pieces of bridging per space
76 sets or 152 individual pieces

Figure 13-28   Metal Bridging

Since each space requires two pieces of metal bridging, determine the amount required by multiplying the joist *spaces* (not the number of joists) times two (2 pieces each space) times the number of rows of bridging required.

Sheathing is taken off next. This sheathing is most commonly plywood but diagonal sheathing is still occasionally (infrequently) used. First, a careful check of the specifications should provide the plywood information required. The thickness of plywood required may be given in the specifications or in the wall sections (Fig. 13-29). In addition, the specifications will spell out any special installation requirements, such as the glued-nailed system. The plywood is most accurately estimated by

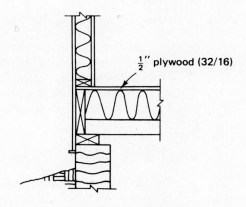

Figure 13-29   Sheathing

doing a sketch of the area to be covered and planning the plywood sheet layout. Using 4 × 8 sheets a layout for the residence is shown in Fig. 13-30.

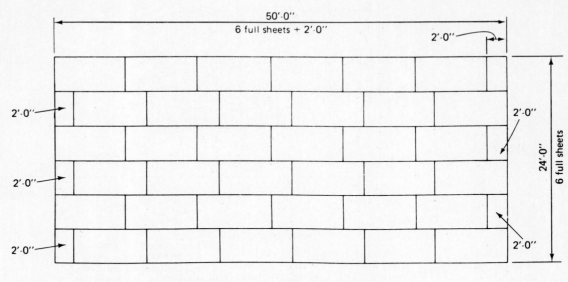

36 full sheets plus 6-2'-0'' pieces = $37\frac{1}{2}$ sheets
order 38 sheets

Figure 13-30   Residence Sheathing Layout

Another commonly used method of estimating plywood is to determine the square footage to be covered and divide the total square footage to be covered by the area of one sheet of plywood (Fig. 13-31).

While both methods give almost the same answer, the use of square feet alone does not allow planning of sheet layout on irregular plans (Fig. 13-32) and allowing properly for waste.

List the quantity of plywood required for the sheathing on the workup sheet, being certain to list all related information.

Plywood subflooring: ½ '', 32/16, 32-4 × 8 sheets, C-D.

Subflooring is sometimes used over the sheathing. Subflooring may be a pressed particle board (where the finished floor is carpet) or plywood (where the finished floor is resilient tile, ceramic tile, slate, etc.). A careful review of the specifications and drawings will show whether subflooring is required, the type and thickness and location it is required. It is taken off in sheets the same as plywood.

13-4 WALL FRAMING

In this section a quantity takeoff of the framing required for exterior and interior walls is done. Since the exterior and interior walls have different finish materials on them, they will be estimated separately. The exterior walls are taken off first, then the interior.

$$\begin{array}{r} 24 \\ \underline{\times\ 50} \\ 1200\ \text{s.f.} \end{array} \qquad \frac{1200}{32} = 37.5 \text{ sheets (order 38 sheets)}$$

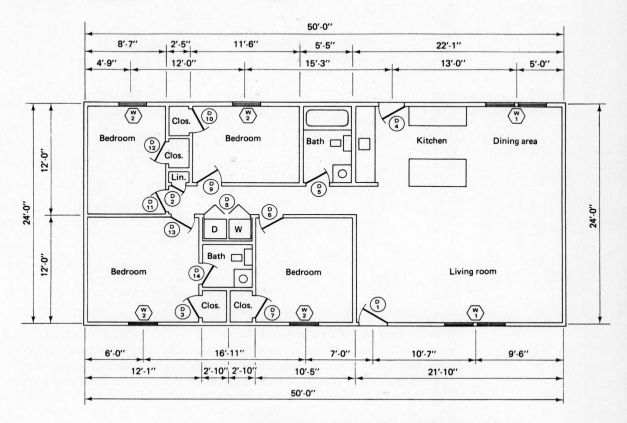

Figure 13-31 Sheathing Requirements

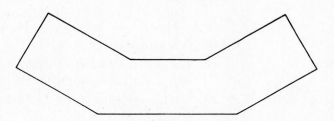

Figure 13-32 Irregular Plan

EXTERIOR WALLS

Basically, most of the wall framing consists of sole plates, studs, double plates, headers, and finish materials (Fig. 13–33).

Plates. The most commonly used assembly incorporates a double-top plate and a single-bottom plate, although other combinations may be used. Begin by first review-

wall framing

233

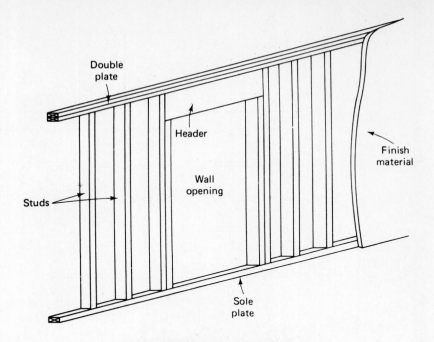

Figure 13-33　Wall Framing

ing the specifications and drawings for the thickness of materials (commonly 2 × 4 or 2 × 6) grade of lumber to be used and information on the number of plates required. The commonly used assembly in Fig. 13-33 provides an 8'-0" ceiling height which is most commonly used and works economically with 4' × 8' sheets of plywood and gypsum board. When an 8'-2" ceiling height is required, a double-top and bottom plate is used (Fig. 13-34).

Estimating steps.

1. Plates are required around the perimeter of the building. Since this is the same perimeter used to determine the sill material in Sec. 13-3 and Example 13-4, the perimeter is already known.
2. The total lineal feet of exterior plates is determined by multiplying the lineal feet of wall times the number of plates.
3. List this information on the workup sheet and calculate the board feet required.

Wall Framing:

　Exterior wall—2 × 4

　Plate—2 × 4

L.F.	PLATES	TOTAL L.F.	B.F.
148	3	444	296

$$\text{Board feet} = 444 \times \frac{2 \times 4}{12} = 296 \text{ b.f.}$$

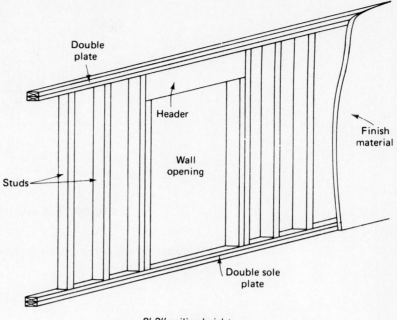

Double
plate

Header

Finish
material

Wall
opening

Studs

Double sole
plate

8'-2'' ceiling height

Figure 13-34   Double Top and Bottom Plates

NOTE: Interior plates will be done with the interior walls later in this portion of the takeoff, but the same basic procedure shown here will be used.

Studs.   The stud takeoff should be separated into the various sizes and lengths required. Studs are most commonly 2 × 4, 16 in., and 24 in. on center and 2 × 6, 24 in. on center (providing a greater width for extra wall insulation).

Estimating steps.

1. Review the specifications and drawings for the thickness and spacing of studs and lumber grade.
2. The exterior studs will be required around the perimeter of the building; the perimeter used for the sills and plates may be used.
3. Divide the perimeter (length of wall) by the spacing of the studs to determine the number of spaces. Then add one to close off the last space.

Studs: 2 × 4, 16″ o.c., 148 l.f.

$$\frac{148}{1.33} = 111 \text{ spaces} = 112 \text{ studs}$$

4. Add extra studs for corners, wall intersections, (where two walls join) and wall openings.
    a. Corners—using 2 × 4 studs—a corner is usually made up of 3 studs (Fig. 13-35). This requires 2 extra studs at each corner. Estimate the extra material

wall framing   235

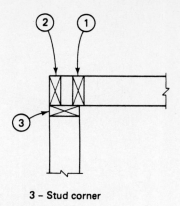

3 – Stud corner

Figure 13-35 Corners—2 × 4 Studs

required by counting up the number of corners and multiplying the number of corners by two.

STUDS	NO.
2 × 4, 16″ o.c., 148 l.f.	112
4 corners—4 × 2	8

b. Corners—using 2 × 6 studs—a corner is usually made up of 2 studs (Fig. 13-36), requiring 1 extra stud. Drywall clips are used to secure the drywall panel to the studs. Some specifications and details require that 3 studs be used for 2 × 6 corners (similar to the 2 × 4 corner).

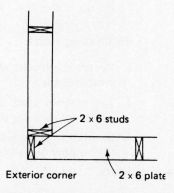

Exterior corner   2 x 6 plate

Figure 13-36 Corners—2 × 6 Studs

c. Wall intersections—using 2 × 4 studs the wall intersection is made up of 3 studs (Fig. 13-37). This requires 2 extra studs at each intersection. The extra material is estimated by counting up the number of wall intersections (on the exterior wall and multiplying the number of corners by two).

wood frame construction

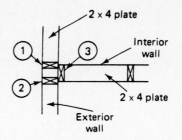

Figure 13-37 Wall Intersection

STUDS (FIG. 13-38)	NO.
2 × 4, 16″ o.c., 148 l.f.	112
4 corners—4 × 2	8
9 intersections—9 × 2	18

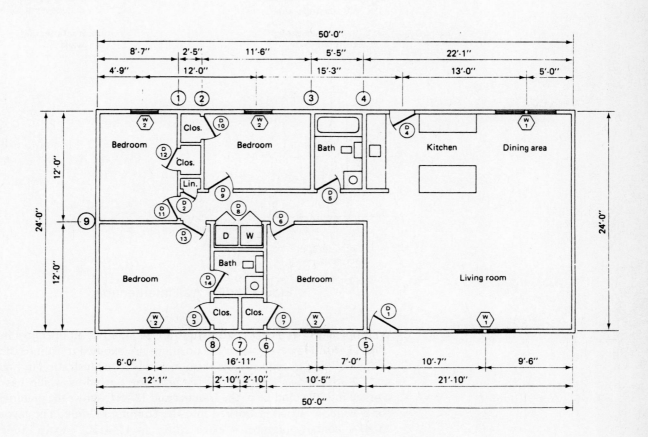

Figure 13-38 Floor Plan

d. Wall intersections—using 2 × 6 exterior studs the wall intersection may be made up of 1 or 2 studs (Fig. 13-39) with drywall clips used to secure the gypsum board (drywall). Some specifications and details require that 3 studs be used for 2 × 6 corners (similar to the 2 × 4 corner).

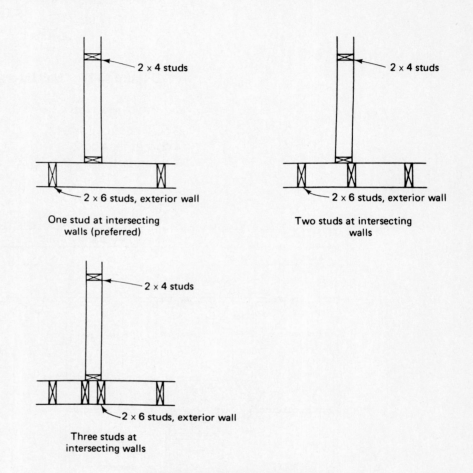

Figure 13-39   Wall Intersections

e. Wall openings—typically, extra studs are required at all openings in the wall (Fig. 13–40). However, when the openings are planned to fit into the framing of the building, it requires less extra material. The illustration in Fig. 13–41A and 13–41B shows the material required for a window which has not been worked into the module of the framing and 13–41C shows the material required for a window which is worked into the framing module. The layout in Fig. 13–41A actually requires 4 extra studs; in 13–41B, 3 extra studs; and in 13–41C, 2 extra studs.

In this example there is no indication that openings have been worked into the framing module. For this reason, an average of 3 extra studs were included for all window and door openings.

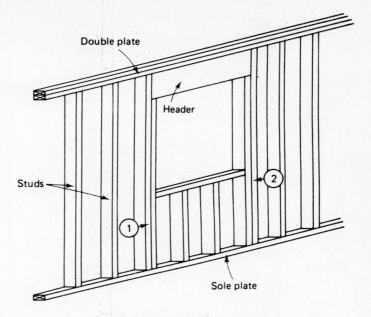

Figure 13-40   Wall Opening

STUDS	NO.
2 × 4, 16″ o.c.	112
4 corners—4 × 2	8
9 intersections—8 × 2	16
8 openings—8 × 3	24

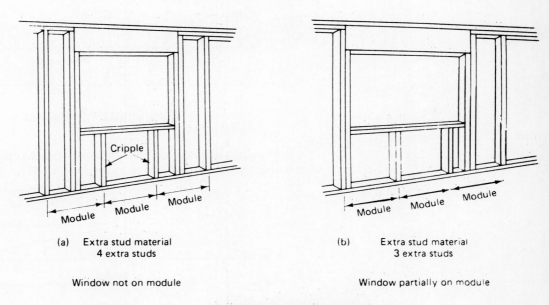

(a)   Extra stud material
4 extra studs

Window not on module

(b)   Extra stud material
3 extra studs

Window partially on module

Figure 13-41   Wall Openings

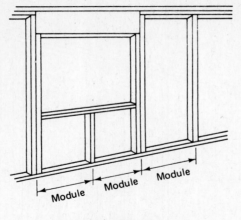

(c) Extra stud material
2 extra studs

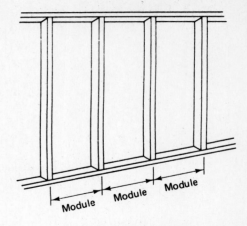

(d) Framing without window

Figure 13-41   Wall Openings (Continued)

5. Additional studs are also required on the gable ends of the building (Fig. 13–42) which will have to be framed with studs between the double plate and the rafter (unless trusses are used, Sec. 13–7). The specifications and elevations should be checked to determine whether the gable end is plain or has a louver.

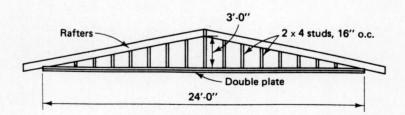

Figure 13-42   Gable End

Estimating the studs for each gable end is accomplished by first drawing a sketch of the gable end and noting its size (Fig. 13-43).

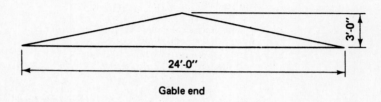

Gable end

Figure 13-43   Gable End Sketch

wood frame construction

Find the number of studs required by dividing the length by the stud spacing. Then, from the sketch find the approximate average height of the studs required and record the information on the workup sheet. Multiply the quantity by two since there are two gable ends on this building (Fig. 13-44).

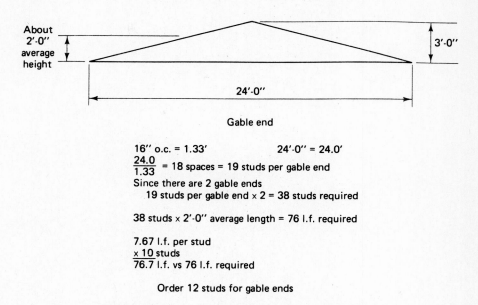

16″ o.c. = 1.33′   24′-0″ = 24.0′

$\frac{24.0}{1.33}$ = 18 spaces = 19 studs per gable end

Since there are 2 gable ends
      19 studs per gable end x 2 = 38 studs required

38 studs x 2′-0″ average length = 76 l.f. required

7.67 l.f. per stud
x 10 studs
76.7 l.f. vs 76 l.f. required

Order 12 studs for gable ends

Figure 13-44 **Studs Required for Gable End**

6. The total stud material has now been estimated and is converted into board feet.

STUDS	NO.
2 × 4, 16″ o.c.	112
4 corners—4 × 2	8
9 intersections—9 × 2	18
8 openings—8 × 3	24
Gable ends	12
Total exterior studs	174

$$174 \times 7'\text{-}8'' \times \frac{2 \times 4}{12} = 890 \text{ b.f.}$$

Headers. Headers are required to support the weight of the building over the openings. A check of the specifications and drawings must be made to determine if the headers required are solid wood, headers and cripples, or plywood sheathing (Fig. 13–45). For ease of construction many carpenters and home builders feel that a solid header provides best results and they use 2 × 12's as headers throughout the project, even in nonload-bearing walls. Shortages and higher costs of materials have increased the usage of plywood and smaller-sized headers.

The header length must also be considered. As shown in Fig. 13–46, the header extends over the top of the studs and it is wider than the opening. Most specifications

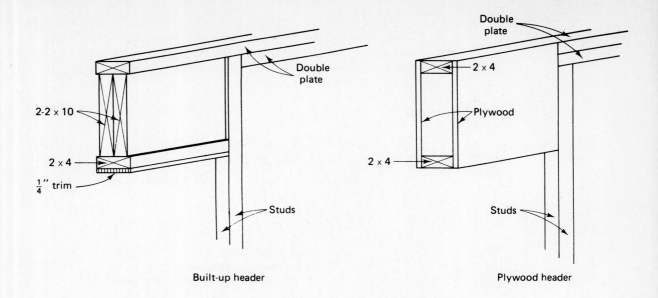

Built-up header

Plywood header

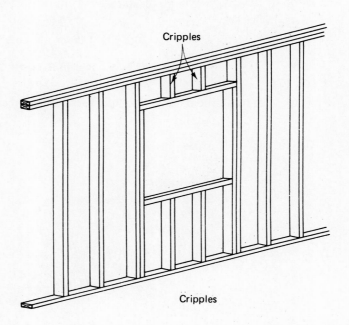

Figure 13-45 Headers

and building codes require that headers for openings up to 6 ft. wide must extend over 1 stud at each end (Fig. 13–46), and headers for openings 6 ft. and wider must extend over 2 studs at each end (Fig. 13-46).

List the number of openings, their width, the number of headers, their length, and the lineal feet and board feet required.

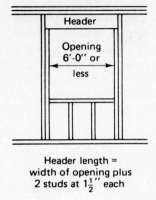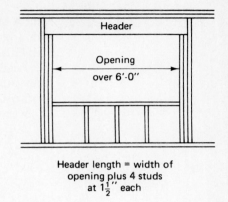

Header length =
width of opening plus
2 studs at $1\frac{1}{2}''$ each

Header length = width of
opening plus 4 studs
at $1\frac{1}{2}''$ each

Figure 13-46 Header Length

Headers: 2 × 12 (Fig. 13-47)

	OPENING	PIECES REQUIRED	LENGTH REQUIRED	L.F.
9 openings	36″ wide	18	3′–3″	58′–6″
1 opening	2′–8″ wide	2	2′–11″	5′–10″
				63′–16″ = 64′–6″

Four 2 × 12's 10′ long will give twelve 3′–3″ pieces. One 2 × 12 14′ long will give two 3′–3″ pieces, two 2′–11″ pieces, and 1′–8″ of waste.

$$66 \times \frac{2 \times 12}{12} = 132 \text{ b.f.}$$

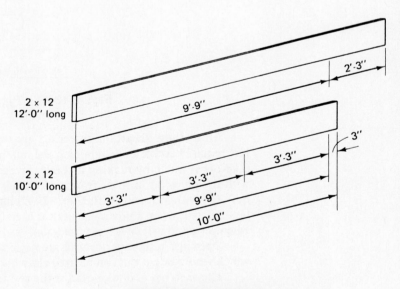

Figure 13-47 Headers

Finish Materials. Exterior wall sheathing may be a fiberboard material soaked with a bituminous material, insulation board (often a urethane insulation covered with an aluminum reflective coating), or plywood (Fig. 13–48). Carefully check the specifications and working drawings to determine what is required (insulation requirements, thickness). Fiberboard and insulation board sheathing must be covered by another material (such as brick, wood, or aluminum siding), while the plywood may be covered or left exposed.

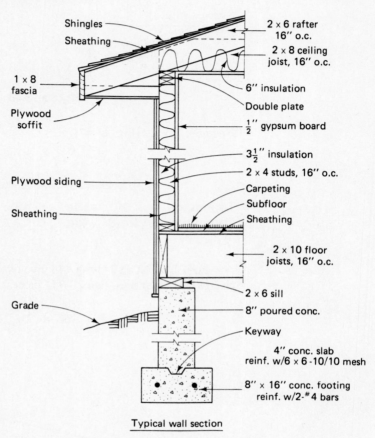

Typical wall section

Figure 13-48   Typical Wall Section

All of these sheathing materials are taken off first by determining the square feet required and then determining the number of sheets required. The most accurate take-off is made by making a sketch layout of the material required (as was done with the plywood in floor framing). Check the height of plywood carefully since a building with a sloped soffit (Fig. 13–49) may require a 9-ft. length, while 8 ft. may be sufficient when a boxed-in soffit is used (Fig. 13–50). This is especially important when plywood will be left exposed and as few joints as possible are desired.

Openings in the exterior wall are neglected unless they are large and the plywood which would be cut out can be used elsewhere, otherwise it is considered waste.

The sketches and take-off for the residence are shown in Fig. 13–51. Sheathing is also required to cover the gable ends.

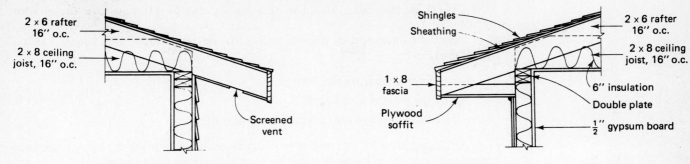

Figure 13-49 Sloped Soffit

Figure 13-50 Boxed-In Soffit

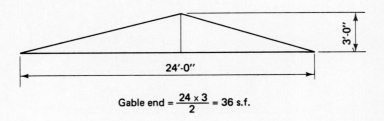

Gable end = $\dfrac{24 \times 3}{2}$ = 36 s.f.

Figure 13-51 Gable end

Exterior plywood: ½", 32/16, A–C, 4 × 8 sheets

148 l.f. perimeter = 148 l.f. × 8 ft. high = 1,184 s.f.

$$\frac{148 \text{ l.f.}}{4 \text{ ft.}} = 37 \text{ sheets required (neglect openings)}$$

Gable ends = 36 s.f. (Fig. 13-51) × 2 ends = 72 s.f.

The six 32" × 36" pieces of plywood left from the openings can be used on part of the gable ends. These pieces amount to 6 × 2.67' × 3.0' = 48 s.f.

Total s.f. = 1,184 s.f.
Gable (72 − 48) = 24 s.f.
1,208 s.f.

Allowing for waste, order 40 sheets, 1,280 s.f.

NOTE: The inside of the exterior wall is covered with gypsum board (drywall). The gypsum materials will be taken off in Chapter 14.

INTERIOR WALLS

Interior walls are framed with studs, top and bottom plates, and a finish on both sides of the wall. When estimating the material for interior walls, the first step is to determine from the specifications and drawings when thickness(es) of walls is (are) required. Most commonly, 2 × 4 studs are used, but 2 × 3 studs and metal studs

(Sec. 14–2) are also used. Also, the stud spacing may be different from the exterior walls.

Next, the lineal feet of interior walls must be determined. This length is taken from the plan by:

1. using dimensions from the floor plan
2. scaling the lengths with a scale
3. using a distance measurer over the interior walls.

On large projects extreme care must be used when using a scale or distance measurer since the drawing may not be done to the exact scale shown.

Any walls which are different thicknesses (such as a 6-in.-thick wall, sometimes used where plumbing must be installed) and of special construction (such as a double or staggered wall, Fig. 13–52) may require larger stud or plate sizes.

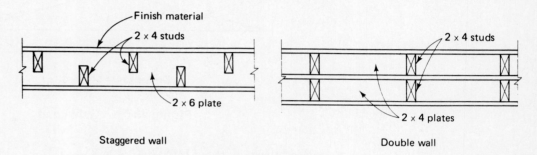

Figure 13-52   Special Wall Construction

Plates.   Refer to the discussion under "Exterior Walls" earlier in this section.

Interior walls: 2 × 4, 16″ o.c., 149 l.f.
Plate: 2 × 4

L.F.	PLATES	TOTAL L.F.	B.F.
149	3	447	298

Studs.   Refer to the discussion under "Exterior Walls" earlier in this section. As in exterior walls, deduct only where there are large openings and take into account all corners, wall openings and wall intersections.

L.F.	STUDS
149 l.f. of plate	113
17 intersections	34
12 openings × 3	36
Interior studs	183 = 936 b.f.

wood frame construction

Headers. Refer to the discussion under "Exterior Walls" earlier in this section.

OPENINGS	HEADER LENGTH	NO. OF PIECES	L.F. REQUIRED
One 1'-6" wide	1'-9"	2	3'-6"
Two 2'-4" wide	2'-7"	4	10'-4"
Eight 2'-6" wide	2'-9"	16	44'-0"
One 5'-0" wide	5'-3"	2	10'-6"
			68'-4"

One 12' length will give four 2'-9" pieces and 1'-0" of waste—you will need four 12' lengths for sixteen 2'-9" pieces.

One 8' length will give one 5'-3" piece, one 2'-7" piece and 2" of waste—you will need two 8' lengths for two 5'-3" pieces and two 2'-7" pieces.

One 10' length will give two 2'-7" pieces, two 1'-9" pieces and 1'-4" of waste.

Order:

$$\begin{array}{lll}
\text{Four } 2 \times 12 & 12' \text{ long} & = 48 \text{ l.f.} \\
\text{Two } 2 \times 12 & 8' \text{ long} & = 16 \text{ l.f.} \\
\text{One } 2 \times 12 & 10' \text{ long} & = \underline{10 \text{ l.f.}} \\
& & 74 \text{ l.f.} = 148 \text{ b.f.}
\end{array}$$

At this point, the entire materials' takeoff for the framing of the interior and exterior walls has been estimated. Finishes for the interior and exterior wall (drywall, interior plywood, etc.) will be covered in Chapter 14.

13-5 CEILING ASSEMBLY

In this section a quantity takeoff of the framing required for the ceiling assembly of a wood frame building is done. The ceiling assembly will require a takeoff of ceiling joists, headers, and trimmers, quite similar to the takeoff done for the floor assembly (Sec. 13-3).

First, a careful check of the specifications and drawings must be made to determine if the ceiling and roof are made up of joists and rafters (Fig. 13-53), often called "stick construction" or prefabricated wood trusses (Fig. 12-17) which are discussed in Sec. 13-7.

Ceiling joists, from the contract documents determine:

1. Size, spacing, and grade of framing required
2. The number of ceiling joists required by dividing the spacing into the length and adding one (the same as done for floor joists)
3. The length of each ceiling joist. (Don't forget to add one-half of any required lap.)

   This information is compiled on the workup sheets.

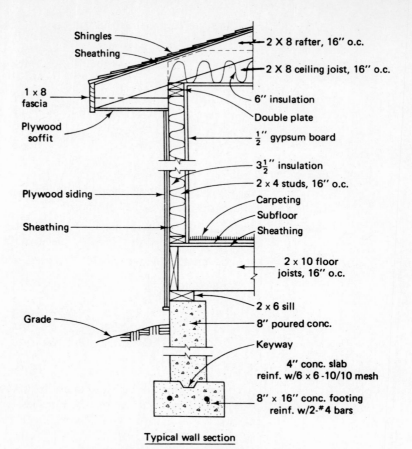

Typical wall section

Figure 13-53 Typical Wall Section

Ceiling joist: 2×8, 16" o.c., #2KD

	NO.	LENGTH	B.F.
Joists	78	12'	1,248

The *headers* and *trimmers* required for any openings (such as stairways, fireplace and attic access openings) are considered next. This is taken off the same as done for floor framing (Sec. 13–3).

The ceiling finish material (drywall or wetwall) is estimated in Chapter 14.

NOTE: A $3'$–$0'' \times 3'$–$9''$ attic access is required in the residence (Fig. 13-54).

	NO.	LENGTH	B.F.
Joists	78	12'	1,248
Trimmers	2	12'	32
Header	1	16'	22
			1,302

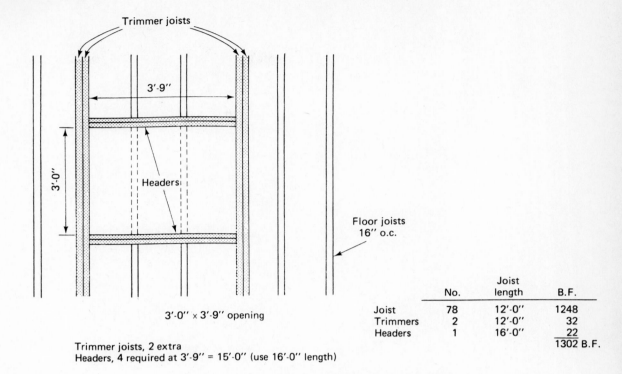

	No.	Joist length	B.F.
Joist	78	12'-0"	1248
Trimmers	2	12'-0"	32
Headers	1	16'-0"	22
			1302 B.F.

3'-0" x 3'-9" opening

Trimmer joists, 2 extra
Headers, 4 required at 3'-9" = 15'-0" (use 16'-0" length)

Figure 13-54 Headers and Trimmers

13-6 ROOF ASSEMBLY

In this section a quantity take-off of the framing required for the roof assembly of a wood frame building is done. The roof assembly will require a take-off of rafters, ridge, lookouts, collar ties (or supports), plywood sheathing, and the felt which covers the sheathing.

If trusses are called for, then a separate take-off for roof and ceiling assembly will not be made and it would be estimated as discussed in Sec. 13-7, "Trusses."

RAFTERS

Roof rafters should be taken off and separated into the sizes and lengths required. The spacings most commonly used for rafters are 16 and 24 inches on center, but 12, 20, 32, and 36 inches on centers are also used. Rafter sizes of 2×6, 2×8, 2×10, and 2×12 are most common. The lengths of rafters should be carefully taken from the drawings or worked out by the estimator if the drawings are not to scale. Be certain to add any required overhangs to the lengths of the rafters. The number of rafters for a pitched roof can be determined in the same manner as the number of joists. The principle of pitch versus slope should be understood to reduce mistakes. Figure 13-55 shows the difference between pitch and slope, while Fig. 13-56 shows the length of rafter required for varying pitches and slopes.

Rafters, from the contract documents, determine:

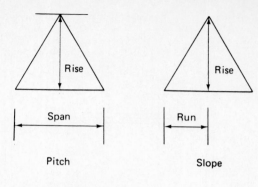

Pitch = $\dfrac{\text{Rise}}{\text{Span}}$

Slope = $\dfrac{\text{Rise}}{\text{Run}}$

Example: Span 40'-0'', Rise 10'-0''

Pitch = $\dfrac{10}{40}$ = $\dfrac{1}{4}$ Pitch

Slope = $\dfrac{10}{20}$ = $\dfrac{6}{12}$ = 6 inches per foot

Figure 13-55   Pitch and Slope

Pitch of roof	Slope of roof	For length of rafter multiply length of run by
$\frac{1}{12}$	2 in 12	1.015
$\frac{1}{8}$	3 in 12	1.03
$\frac{1}{6}$	4 in 12	1.055
$\frac{5}{24}$	5 in 12	1.083
$\frac{1}{4}$	6 in 12	1.12

Note: Run measurement should include
any required overhang.

Example:
For a 24'-0'' span with a 2'-0'' overhang on each
side, $\frac{1}{8}$ pitch, what length rafter is required?

$\dfrac{\text{Span \& Overhang}}{2}$ = $\dfrac{24 + 4}{2}$ = Run = 14'-0''

Rafter length = 14'-0'' x 1.03 = 14.42' = 14'- $5\frac{1}{32}$'

Figure 13-56   Rafter Length

1. size, spacing, and grade of framing required
2. the number of rafters required (divide spacing into length and add one). If the spacing is the same as the ceiling and floor joists the same amount will be required
3. the length of each rafter. (Don't forget to allow for slope (or pitch) and overhang).

This information is compiled on workup sheets.

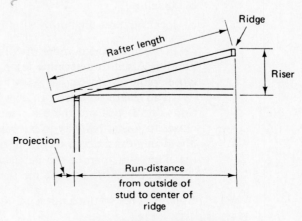

Figure 13-57   Rafters

Roof Assembly
Rafters: 2 × 8, 16″ o.c., #2 KD (52′ of roof)

	NO.	LENGTH	B.F.
Rafters	80	14′	1,493

$$
\begin{aligned}
\text{Run} &= 12'\text{-}0'' \\
\text{Projection} &= \underline{1'\text{-}6''} \\
\text{Horizontal distance} &= 13'\text{-}6''
\end{aligned}
$$

Slope: 3 in. in 12. Use 1.03 factor

$$13'\text{-}6'' \times 1.03 = 13.5 \times 1.03 = 13.905'$$

Use 14′ rafters.

HEADERS AND TRIMMERS

Headers and trimmers are required for any openings (such as chimneys) just as in floor and ceiling joists. For a complete discussion of headers and trimmers refer to Sec. 13-3, "Floor Framing."

COLLAR TIES

These are used to keep the rafters from spreading (Fig. 13–58), and most codes and specifications require them to be a maximum of 5 ft. apart or every third rafter whichever is less. A check of the contract documents will determine:

a. required size (usually 1 × 6 or 2 × 4)
b. spacing
c. location (how high up).

Determine the number of collar ties required by dividing the total length of the building (used to determine the number of rafters) by the spacing and add one to close up the last space.

The length required can be a little harder to determine exactly. Most specifications don't spell out the exact location of the collar ties (up high near ridge, low closer to joists, halfway in between), but the typical installation has the collar ties about one-third down from the ridge with a length of 5 ft. to 8 ft. depending on the slope and span of the particular installation.

The collar tie takeoff for the residence is compiled on the workup sheets.

Collar ties (every third rafter): 2 × 4

	NO.	LENGTH	B.F.
Collar ties	14	6′	56

78 rafters = 39 rows of rafters

$$\frac{39}{3} = 13 + 1 \text{ extra} = 14 \text{ collar ties}$$

RIDGE

The *ridge board* (Fig. 13–58) is taken off next. The contract documents must be checked for the size ridge board required. Quite often no mention of ridge board

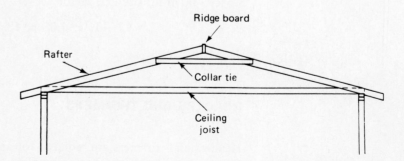

Figure 13-58   Ridge Board

size is made anywhere on the contract documents and this would have to be checked with the architect-engineer (or whoever has authority for the work). Generally, the ridge is 1-inch thick and one size wider than the rafter. In this case, a 2 × 8 rafter would be used with a 1 × 10 ridge board. Other times a 2-inch thickness may be required.

The length of ridge board required will be the length of the building plus any side overhang.

The ridge board information should be gathered and noted on the workup sheet.

Ridge Board: 1 × 10

	L.F.	B.F.
Ridge board	52	43.4

LOOKOUTS

Lookouts are often required when a soffit is boxed-in as shown in Fig. 13-50. The number of lookouts is found by dividing the total lineal feet of boxed-in soffit by the spacing and adding one extra lookout for *each* side of the house it is required on. (If required, front and back add 2 extra lookouts.) The length of the lookout is found by reviewing the sections in details. The detail in Fig. 13-59A indicates the lookout would be the width of the overhang minus the actual thickness of the fascia board and the supporting pieces against the wall. In Fig. 13-59B the lookout is supported by being nailed to the stud wall and its length is equal to the overhang plus stud wall minus the actual thickness of the fascia board.

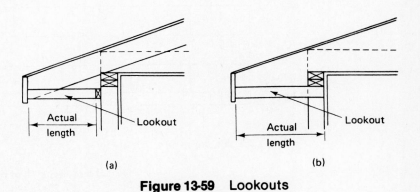

Figure 13-59   Lookouts

The lookouts for this residence have been compiled on the workup sheet.

Lookouts: 2 × 4, 16″ o.c.

	NO.	LENGTH	B.F.
Lookouts	80	1′-4″	71

FINISH MATERIALS

Sheathing for the roof is taken off next. A review of the contract documents should provide the plywood information needed:

1. thickness or identification index (maximum spacing of supports)
2. veneer grades
3. species grade
4. any special installation requirements.

This information should be gathered and noted on the workup sheet.

The plywood is most accurately estimated by making a sketch of the area to be covered and planning the plywood sheet layout. For roof sheathing be certain that the rafter length that is required is used and that it takes into consideration any roof overhang and the slope (or pitch) of the roof. The roof sheathing for the residence is then compiled on the workup sheets.

Each side requires (from Fig. 13-60)

19.5 sheets (for top 3 rows) × 2 = 39
7 sheets for bottom row = 7

46 sheets will take care of both sides

4 × 8 sheets = 46 sheets = 1,472 s.f.

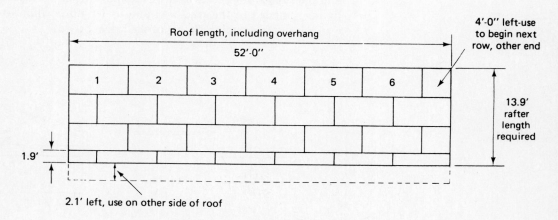

Plywood sheet layout, $\frac{1}{2}$ of roof area

Figure 13-60   Plywood Sheet Layout

Since the carpenters usually install the roofing felt (which protects the plywood until the roofers come), the roofing felt is taken off here or under roofing depending on the estimator's preference. Since felt, shingles and built-up roofing materials are estimated in Chapter 17 that is where it will be estimated for this residence.

The fascia and soffit are estimated in Sec. 13–9.

13-7 TRUSSES

Trusses are available in a variety of shapes (Fig. 12-17) and almost any size desired. The wood and combination wood and metal trusses commonly used in residential construction may be used for floor and roof construction. They offer the advantage of coming to the project prebuilt, ready to install without measuring and cutting, reducing the amount of time to frame the roof.

The specifications and drawings must be checked to determine:

1. type of truss
2. if any one particular manufacturer is specified
3. spacing (which may be on the drawings)
4. type of attachment to building frame required
5. erection requirements.

The estimate must determine the number, type, and size required. When more than one size or shape is required, it should be estimated and listed separately.

13-8 INSULATION

Insulation in light frame construction may be placed between the framing members (studs or joists) or nailed to the rough sheathing. It is used in the exterior walls and the ceiling of most buildings.

Insulation placed between the framing members may be pumped in or laid in rolls or in sheets, while loose insulation is sometimes placed in the ceiling. Roll insulation is available in widths of 11, 15, 19, and 23 inches to fit snugly between the spacing of the framing materials. Sheets of the same widths and shorter lengths are also available. Rolls and sheets are available unfaced, faced on one side, faced on both sides, and foil-faced. The insulating materials may be of glass fiber or mineral fiber. Nailing flanges, which project about 2 inches on each side, lap over the framing members and allow easy nailing or stapling. To determine the number of lineal feet required, the square footage of wall area to be insulated is easily deter-

Stud spacing inches	Insulation width	To determine L.F. of insulation required multiply S.F. of area by	OR	L.F. of insulation required per 100 S.F. of wall area
12	11	1.0		100
16	15	0.75		75
20	19	0.60		60
24	23	0.50		50

Figure 13-61 Insulation Requirements

insulation

mined by multiplying the distance around the exterior of the building, by the height to which the insulation must be carried (often the gable ends of a building may not be insulated). Add any insulation required on interior walls for the gross area. This gross area should be divided by the factor given in Fig. 13-61. If the studs are spaced 16 inches on center, the 15-inch wide insulation plus the width of the stud equals 16″ so a batt 1′–0″ long will cover an area of 1.33 square feet.

Ceiling insulation may be placed in the joists or in the rafters. The area to be covered should be calculated and if the insulation is placed in the joists, the length of the building is multiplied by the width. For rafters (for gabled roofs) the methods shown in Chapter 17 (for pitched roofs) should be used; i.e., the length of the building times the rafter length from the ridge to plate times two (for both sides). Divide the area to be covered by the factor given in Fig. 13-61 to calculate the lineal footage of a given width roll.

Ceiling insulation may also be poured between the joists; a material such as vermiculite is most commonly used. Such materials are available in bags and may be easily leveled to any desired thickness. The cubic feet of material required must be calculated.

Insulation board that is nailed to the sheathing or framing is estimated by the square foot. Sheets in various thicknesses and sizes are available in wood, mineral, and cane fibers. The area to be covered must be calculated and the number of sheets determined.

For all insulation add 5 percent waste when net areas are used. Net areas will give the most accurate take-off. Check the contract documents to determine the type of insulation, thickness, or R value required and any required methods of fastening.

When compiling the quantities, keep each different thickness or type of insulation separate.

Floor.   3½″, R-11, 1,200 s.f. required.

> 1 roll 15″ wide × 56′ long = 75 s.f.
> 1,200 s.f. ÷ 75 s.f. = 16 rolls required

Ceiling.   6″, R-19, 1,200 s.f. required.

> 1 roll 15″ wide × 32′ long = 43 s.f.
> 1,200 s.f. ÷ 43 = 27.9 = 28 rolls required

Walls.   3½″, R-11

> 1 roll 7′–8″ high, 148 l.f. = 1,135 s.f. required.
>
> 15″ wide × 56′ long = 75 s.f.
>
> 1,135 s.f. ÷ 75 s.f. = 15.13 = 16 rolls required

13-9 TRIM

The trim may be exterior or interior. Exterior items include moldings, fascias, cornices, and corner boards. Interior trim may include base, moldings, and chair rails. Other trim items may be shown on the sections and details, or may be included in the specifications. Trim is taken off by the lineal foot required and it usually requires a finish (paint, varnish, etc.), although some types are available prefinished (particularly for use with prefinished plywood panels).

The exterior trim for the residence is limited to the fascia board and soffit.

Fascia board. 1 × 10, Fir, 160 l.f. required—order 170 l.f.

$$52 + 52 + 13.9 + 13.9 + 13.9 + 13.9 = 159.6 \text{ or } 160 \text{ l.f.}$$

Soffit, plywood. ½″ thick, A–C, exterior.

$$1'-6'' \text{ wide, } 104 \text{ l.f.} = 156 \text{ s.f.} = 4.87 \text{ or } 5-4 \times 8 \text{ sheets}$$

Baseboard. 4″ ranch mold, 423 l.f. required—order 440 l.f. Exterior wall = 148 l.f.; interior wall = 149 l.f. times 2 (both sides) minus doors (25 l.f.).

13-10 LABOR

Labor may be calculated at the end of each portion of the rough framing or it may be done for all of the rough framing at once. Many estimators will use a square foot figure for the rough framing based on the cost of past work and taking into consideration the difficulty of work involved. A job such as the small residence being estimated would be considered very simple to frame and would receive the lowest square foot cost and the cost would go up as the building was more involved. Many builders use framing subcontractors for this type work and the sub may price the job by the square foot or as a lump sum. All of these methods provide the easiest approach to the estimator.

When estimators use their own work forces and want to estimate the time involved, they will usually use records from past jobs, depending on how organized they are. The labor would be estimated for the framing by using the appropriate portion of the table for each portion of the work to be done from Fig. 13-62.

Light framing	Unit	Work hours
Sills	100 l.f.	2.0 to 4.5
Joists, floor and ceiling	MBFM	16.0 to 24.0
Walls, interior, exterior (including plates)	MBFM	18.0 to 30.0
Rafters,		
gable roof	MBFM	18.5 to 30.0
hip roof	MBFM	22.0 to 35.0
Cross bridging, wood	100 sets	4.0 to 6.0
metal	100 sets	3.0 to 5.0
Plywood, floor	100 s.f.	1.0 to 2.0
wall	100 s.f.	1.2 to 2.5
roof	100 s.f.	1.4 to 2.8
Trim,		
fascia	100 l.f.	3.5 to 5.0
soffit	100 l.f.	2.0 to 3.5
baseboard	100 l.f.	1.5 to 2.5
molding	100 l.f.	2.0 to 4.0

Figure 13-62 Work Hours Required for Framing

13-11 CHECKLIST

Studs
Wood joists
Rafters
Sheathing
Insulation
Trim
Bracing
Bridging
Plates
Sills

REVIEW QUESTIONS

1. What unit of measure is used for lumber?
2. Determine the fbm for the following order:

$$190/2 \times 4\text{---} 8'\text{--}0''$$
$$1{,}120/2 \times 4\text{---}12'\text{--}0''$$
$$475/2 \times 6\text{---}16'\text{--}0''$$
$$475/2 \times 12\text{---}16'\text{--}0''$$
$$18/2 \times 8\text{---}18'\text{--}0''$$

3. How would you determine the number of studs required for a project?
4. How are the number of joists required determined?
5. What is the difference between *pitch* and *slope,* as the terms pertain to roofing?
6. Determine the length of rafter required for the following conditions if the run is 16'–0":

1/12 pitch
1/6 pitch
3 in 12 slope
4 in 12 slope

7. What unit of measure is used for plywood when it is used for sheathing? How is plywood waste kept to a minimum?

8. What unit of measure is used for roll batt insulation and what type of information should be noted on the workup sheets?

9. What unit of measure is used for wood trim? If it requires a finish, where would this information be noted?

10. What information is required to determine the amount of sill and plates required for a project?

11. Determine the wood framing materials required for the residence in Appendix H.

14
DRYWALL AND WETWALL CONSTRUCTION

14-1 GENERAL

While drywall construction utilizes wallboards and wetwall utilizes plaster and stucco, many components of the two systems of construction are interchangeable. Both require that supporting construction be applied under them, and the same types are used for both. Many of the fasteners, attachments, and accessories are the same or very similar.

All supporting systems and furring should be installed in accordance with the specifications, and the manufacturer's recommendations for spacing, accessories, and installation should be consulted. If the specifications by the architect-engineer are stricter than the manufacturer's, it is those specifications that must be followed. If the project specifications are less stringent than those of the manufacturer, it is advisable to call the architect-engineer's office to be certain what must be bid.

Many manufacturers will not guarantee the performance of their materials on the job unless those materials are installed in accordance with their recommendations.

14-2 SUPPORTING CONSTRUCTION

The wallboard can be applied directly to wood, metal, concrete, or masonry that is capable of supporting the design loads and provides a firm, level, plumb, and true base.

Wood and metal supporting construction often consists of self-supporting framing members including wall studs, ceiling joists, and roof trusses. Wood and metal furring members such as wood strips and metal channels are used over the supporting construction to plumb and align the framing, concrete, or masonry.

Concrete and masonry. These often have wallboard applied to them. When used, exterior and below grade, furring should be applied over the concrete or masonry to protect the wallboard from damage due to moisture in the wall; this is not required for interior walls. Furring may also be required to plumb and align the walls.

The actual thickness of the wall should be checked so that the mechanical and electrical equipment will fit within the wall thickness allowed. Any recessed items such as fire extinguishers and medicine cabinets should be carefully considered.

Wood studs. The most common sizes used are 2 × 4 and 2 × 3, but larger sizes may be required on any particular project: spacing may vary from 12 to 24 inches on center, again depending on job requirements. Openings must be framed around and back-up members should be provided at all corners. The most common method of attachment of the wallboard to the wood studs is by nailing, but screws and adhesives are also used. Many estimators take off the wood studs under rough carpentry, particularly if they are load-bearing. The number of board feet required can be determined from Fig. 13-1, using the length of partition, in conjunction with Fig. 13-2, which takes into consideration the various spacings. Special care should be taken with staggered and double walls so that the proper amount of material is estimated.

Wood joists. The joists themselves are estimated under rough carpentry (Chapter 13). When you intend to apply the wallboard directly to the joists, the bottom faces of the joists should be aligned in a level plane. Joists with a slight crown should be installed with the crown up, and if slightly crooked or bowed joists are used, it may be necessary to straighten and level the surface with the use of nailing stringers or furring strips. The wallboard may be applied by nailing or screwing.

Wood trusses. When used for the direct application of wallboard, trusses often require cross-furring to provide a level surface for attachment.

Stringers attached at third points will also help align the bottom chord of wood trusses and a built-in chamber is suggested to compensate for future deflection. Since the trusses are made up of relatively small members spanning large distances, they have a tendency to be more difficult to align and level for the application of wallboard.

Metal studs. The metal studs (Fig. 14-1) most commonly used are made of 25-gauge, cold-formed steel, electrogalvanized to resist corrosion. Most metal studs have notches at each end and knockouts located about 24 inches on center to facilitate pipe and conduit installation. The size of the knockout, not the size of the stud, will determine the maximum size of pipe, or other material that can be passed through. Often, when large pipes, ducts, or other items must pass vertically or horizontally in the wall, double stud walls are used, spaced the required distance apart. Studs are generally available in thicknesses of 1⅝, 2½, 3½, 4, and 6 inches. The metal runners used are also 25-gauge steel and sized to complement the studs.

A variety of systems has been developed by the manufacturers to meet various requirements of attachment, sound control, and fire resistance. Many of the systems have been designed for ease in erection and yet are still demountable for revising room arrangements. Carefully determine exactly what is required for a particular project before beginning the take-off.

The wallboard may be attached by use of nails, screws, and in certain applica-

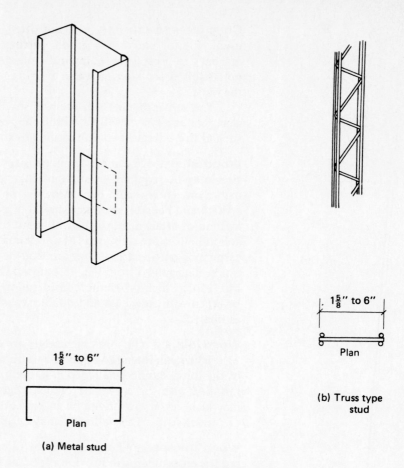

1⅝" to 6"

Plan

(a) Metal stud

1⅝" to 6"

Plan

(b) Truss type
stud

Figure 14-1　Metal Studs

tions, an adhesive. Different shapes of studs are available to accommodate either the screws or the nails.

Metal studs and runners are sold by the pound or by the lineal foot. Once the lineal footage of studs has been determined, it is easy to calculate the weight required. The lineal footage of each different type of wall must be determined. The walls must be separated according to thickness or type of stud, backing board and wallboard as well as according to any variations in ceiling height, application techniques, and stud spacing.

Once a listing of wall lengths and heights has been made, the amount of pieces and board feet may be determined by using Figs. 13–2 and 13–3. Once again, special care should be taken with staggered and double walls so that the proper amount of material may be estimated.

If the studs and runners are sold by the pound, the weight per lineal foot should be determined from the manufacturer's brochures and multiplied by the number of lineal feet to determine the total weight.

Open-web joists.　The joists themselves are estimated under structural steel (Chapter 12). A joist, however, may be used as a base for wallboard. Since the bottom chords of the joists are seldom well aligned and the spacing between joists is

drywall and wetwall construction

often excessive, the most common methods of attachment are with the use of furring and with a suspension system. Each of these methods is discussed in this chapter.

Metal furring.   Metal furring (Fig. 14–2) is used with all types of supporting construction. It is particularly advantageous where sound control or noncombustible assemblies are required. Various types of channels are available. Cold-rolled channels, used for drywall or wetwall construction, are made of 16-gauge steel, ¾ to 2 inches wide and available in lengths of up to 20 feet. These channels must be wire-tied to supporting construction and they are used primarily as a supporting grid for the lighter drywall channels to which the wallboard may be screw attached.

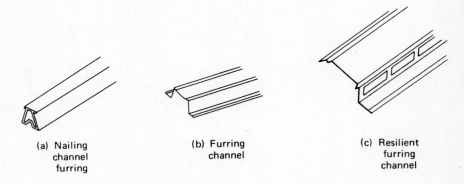

(a) Nailing channel furring

(b) Furring channel

(c) Resilient furring channel

Figure 14-2   Metal Furring

Drywall channels are 25-gauge electrogalvanized steel and are designed for screw attachment of wallboard; nailable channels are also available. The channels may be used in conjunction with the cold-rolled channels or installed over the wood, steel, masonry, or concrete supporting the construction. These drywall channels may be plain or resilient. The resilient channels are often used over wood and metal framing to improve sound isolation and to help isolate the wallboard from structural movement.

Metal furring is sold by the lineal foot. Determine the size and types of furring required, the square footage to be covered, and then the lineal footage of each type. Note the type and spacing of fasteners also. Labor and equipment will depend on the type of supporting construction, height and length of walls, shape of walls (straight or irregular), and fastening.

Wood furring.   Strips are often used with wood frame, masonry, and concrete to provide a suitably plumb, true, or properly spaced supporting construction. These furring strips may be 1 × 2-inches spaced 16-inches on center, or 2 × 2-inches spaced 24-inches on center. Occasionally, larger strips are used to meet special requirements. They may be attached to masonry and concrete with cut nails, threaded concrete nails, and powder or air-actuated fasteners.

When the spacing of the framing is too great for the intended wallboard thickness, cross-furring is applied perpendicular to the framing members. If the wallboard is to be nailed to the cross-furring, the furring should be a minimum of 2 × 2 inches in order to provide sufficient stiffness to eliminate excessive hammer rebound. Thinner furring (1 × 2 or 1 × 3-inches) is often used for screw and adhesive attached wallboard.

The furring is attached by nailing with the spacing of the nails 16- or 24-inches on center. The estimator will have to determine the lineal footage of furring required, the nailing requirements, equipment, and work hours. The labor will vary depending on the height of wall or ceiling, whether straight or irregular walls are present, and the type of framing to which it is being attached.

The commercial building in Appendix A requires truss-type steel studs for interior partitions. The required stud information is compiled on the workup sheet.

EXAMPLE 14-1 — Steel studs (truss type): 16 inches on center (1.33 feet) 3¼ inches wide—all wall except behind plumbing fixtures.

137 l.f. of 3¼ inch wall

$$\frac{\text{length of wall}}{\text{stud spacing}} + 2 \text{ per opening} + 2 \text{ per corner} + 1 \text{ at each intersection}$$

$$\frac{137}{1.33} + 2(4) + 2(1) + 1(8) = 103 + 8 + 2 + 8 = 121 \text{ pieces}$$

8'-0" high total l.f. of 3¼-inch studs 121 (8) = 968 l.f.
Top and bottom runners—137 l.f. × 2 = 274 l.f.
                                               1,242
5% waste = + 62 l.f.
                                               1,304 l.f.

Steel studs (truss type): 16 inches on center (1.33') 6 inches wide—on wall behind plumbing fixtures

$$\frac{25 \text{ l.f.}}{1.33} + 1(2) = 19 + 2 = 21 \text{ pieces } 8'\text{-}0'' \text{ long}$$

Total l.f. of 6-inch studs 21(8) = 168 l.f.
Waste—2 extra studs = 16 l.f.
                                               184 l.f.
Top and bottom runners—25 l.f. × 2 = 50 l.f.
                                               234 l.f.

14-3 SUSPENDED CEILING SYSTEMS

When the plaster, wallboard, or tiles cannot be placed directly on the supporting construction, the wallboard is suspended below the structural system. This may be required if the supporting construction is not properly aligned and true, or if lower ceiling heights are required.

There is a large variety of systems available for use in drywall construction, but basically they can be broken down into two classes (Fig. 14-3): 1) exposed grid system, and 2) concealed grid system. Within each group many different shapes of pieces are used to secure the plaster wallboard or tile, but basically, the systems consist of hangers, main tees (runners) and cross tees (hangers), and furring channels.

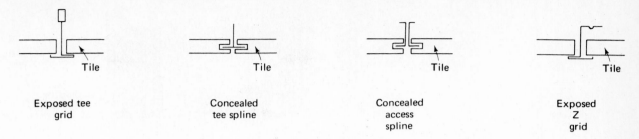

| Exposed tee grid | Concealed tee spline | Concealed access spline | Exposed Z grid |

Figure 14-3 Typical Grid Systems

No matter which type is used, there are accessories, such as wall moldings, splines, and angles that must be considered. For wetwall construction a lath of some type is required.

The suspension system and wallboard may also be used to provide recessed lighting, acoustical control (by varying the type of wallboard and panel), fire ratings and air distribution (special tile and suspension system).

The supporting construction for ceilings generally utilizes wood joists and trusses, steel joists and suspended ceilings, and concrete and masonry, sometimes in conjunction with the various types of furring materials available.

Drywall construction is generally estimated by the square foot, square yard, and square, each estimator using whichever seems most comfortable. The most common approach to estimating drywall partitions is to take the lineal footage of each different type and thickness of wall from the plan and list them on the work-up sheet. Walls that are exactly the same should be grouped together, any variation in the construction of the wall will require that it be considered separately. If the ceiling heights vary throughout the project, the lengths of walls of each height must also be kept separately.

Once you have made a listing of wall lengths and heights, multiply the length times height and determine square footage of partition. With this information the amounts of material required may be estimated. Deduct all openings from the square footage and add 8 to 10 percent for waste.

14-4 TYPES OF ASSEMBLIES

Basically, drywall construction may be broken down into two basic types of construction, single-ply and multi-ply. *Single-ply construction* consists of a single layer of wallboard on each side of the construction, while *multi-ply construction* uses two or more layers of wallboard and often different types of board in the various layers (Fig. 14-4). The multi-ply construction may be semisolid or solid, or may have various combinations of materials. Analyze what is required carefully so that the takeoff and pricing may be as complete and accurate as possible.

Various types of demountable and reusable assemblies are also available. Take the time to analyze the assembly specified, break it down into each piece required, and study its pieces and how they are assembled. Then an estimate can be made. As questions arise about an unfamiliar assembly, don't hesitate to call the supplier or manufacturer for clarification.

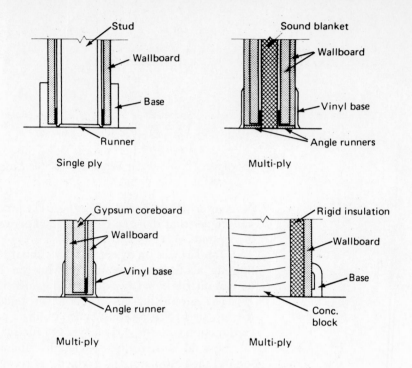

Figure 14-4 Typical Drywall Assemblies

14-5 WALLBOARD TYPES

There are various types of wallboards available for use in drywall construction; among them are gypsum wallboard, tiles, wood panels, and other miscellaneous types. The various types and the special requirements of each are included in this section.

Gypsum wallboard. This type is composed of a gypsum core encased in a heavy manila-finished paper on the face side and a strong liner paper on the back side. It is available in a 4-foot width and lengths of 6, 8, 10, and 12 feet, with thicknesses of ¼, ⅜, ½, and ⅝ inches.

The suspension system itself is available in steel with an electro-zinc coating as well as prepainted and aluminum, and with plain, anodized, or baked enamel finishes. Special shapes—for example, steel-shaped like a wood beam that is left exposed in the room—are also available.

The suspension system may be hung from the supporting construction with 9- or 10-gauge hanger wire spaced about 48 inches on center or it may be attached by use of furring strips and clips.

From the specifications and drawings, the type of system can be determined; the pieces required must be listed and a complete breakdown of the number of lineal feet of each piece is required. Check the drawings for a reflected ceiling plan that will show the layout for the rooms, since this would save considerable time in the estimate. Take note of the size of tile to be used and how the entire system will be attached to the supporting construction. Later this information should be broken

drywall and wetwall construction

down into the average amount of material required per square to serve as a reference for future estimates.

14-6 DRYWALL

Drywall construction consists of wallboard over supporting construction; backing board may also be required.

The types of materials used for this construction will depend on the requirements of the job with regard to appearance, sound control, fire ratings, strength requirements, and cost. Materials may be easily interchanged to meet all requirements.

The supporting construction for partitions generally utilizes wood or steel studs, but often the wallboards are applied over concrete and masonry.

Each component of the drywall construction assembly must be taken off and estimated separately. Plain gypsum wallboard will require a finish of some type, such as painting or wallpaper.

Gypsum wallboard with a rugged vinyl film factory laminated to the panel is also available. The vinyl-finished panel is generally used in conjunction with adhesive fastenings and matching vinyl covered trim; it is also available on fire-resistant gypsum wallboard where fire-rated construction is desired or required.

Fire-resistant gypsum wallboard is available generally in ½- and ⅝-inch thicknesses. These panels have cores containing special mineral materials, and can be used in assemblies that provide up to two-hour ratings in walls, and three-hour ratings in ceilings and columns.

Other commonly used gypsum wallboards include insulating panels (aluminum foil on the back), water-resistant panels (for use in damp areas; they have special paper and core materials), and backing board, which may be used as a base for multi-ply construction and acoustical tile application, and which may be specially formulated for fire-resistant base for acoustical tile application.

Tiles. Tiles used in drywall construction are most commonly available in gypsum, wood and mineral fibers, asbestos, plastic, and metal. Thicknesses vary from 3/16 to 2 13/16 inches, but the most common thicknesses are ½, ⅝, and ¾ inch, with the tile sizes being 12 × 12 inches, 12 × 24 inches, 12 × 36 inches, 16 × 16 inches, 16 × 32 inches, 24 × 24 inches, and 24 × 48 inches. Not all sizes are available in each type and each manufacturer must be checked to determine what he has.

In addition to the different materials, the tiles come in a variety of surface patterns and finishes. They may be acoustical or nonacoustical, and may possess varying edge conditions, light reflection, sound absorption, flame resistance, and flame spread. Tile is even available with small slots that can be opened to provide ventilation in the space below. Plastic louvers and translucent panels are also available.

Prices may be quoted by the piece, square foot, square yard, or particular size package. A square foot take-off will provide the estimating information required.

Tile may be applied to various supporting constructions and is often used in conjunction with suspended ceilings.

Wood panels. Available in many different wood veneers and a variety of finishes, panels are usually 4 feet wide with lengths varying from 6 to 16 feet. The panels may be constructed of a solid piece of wood or laminated plywood with cores of veneer, flakeboard, or lumber. Either hardwood or softwood may be used.

Read the specifications carefully to determine exactly what type of paneling is required. While inexpensive paneling is available at $6 to $15 per sheet and moderately priced panels from $25 to $50 per sheet, there is paneling that sells for $150 per sheet and more, when special face patterns are required. Never guess at a price for materials. Always get a written quote from suppliers and manufacturers.

Paneling is taken off by the square foot, square yard, panels required, or square. Make note of the fastening device and trim to be used.

Miscellaneous panels. There are many types of panels used on walls and ceilings. The majority of them will be priced by the square foot. Among the types available are vinyl-coated plywood, plastic-coated plywood, hardboards, and metal-coated panels. Each separate type has its own method of fastening, accessories, and requirements. Always check with the manufacturer for installation recommendations.

Specifications. Check the type, thickness, sheet size, and method of attachment required. The spacing of fasteners must also be determined either from the specifications, or if it is not given there, from the manufacturer's recommendations. Make a list of accessories, the material they are made of, and the finish required.

Estimating. All wallboard should be taken off by the square foot with the estimator double-checking for panel layout of the job. The first step is to determine the lineal footage of each type of wall, carefully separating any wall with different sizes or types of material, fasteners, or any other variations. Remember that there are two sides to most walls, each side requiring a finish.

After a complete listing has been made, the square footage of wallboard may be determined. Keep in mind the varieties that may be encountered and keep each separate.

Equipment required may simply be wood or metal horses, planks, platforms, and scaffolding as well as small electric tools and staplers. On projects with high ceilings, scaffolds on wheels are often used so the workers can work more conveniently and the scaffold may be easily moved from place to place.

Labor for drywall construction will vary depending on the type of wallboard, trim, fasteners, whether the walls are straight or jogged, height of walls or ceiling, and the presence of other construction underway at the time. Many subcontractors are available with skilled workers, specially trained for this type of work.

14-7 COLUMN FIREPROOFING

Columns may be fireproofed by using drywall construction consisting of layers of fire-resistant gypsum wallboards held in place by a combination of wire, steel studs, screws, and metal angles. Up to a three-hour fire rating may be obtained by using gypsum board and a four-hour fire rating is available when the gypsum board is used in conjunction with gypsum tile (usually 2 or 3 inches thick). In order to receive the fire ratings, all materials must be installed in accordance with Underwriters Laboratories designs that have been approved. A complete takeoff of materials is required. No adhesives may be used.

14-8 ACCESSORIES

Accessories for the application and installation of drywall construction include mechanical fasteners and adhesives, tape and compound for joints, fastener treatment, and trim to protect exposed edges and exterior corners, as well as baseplates and edge moldings.

Mechanical fasteners. Clips and staples may be used to attach the base ply in multi-ply construction. The clip spacing may vary from 16 to 24 inches on center and may also vary depending on the support spacing. Staples should be 16-gauge, galvanized wire with a minimum of a 7/16-inch wide crown with legs having divergent points. Staples should be selected to provide a minimum of ⅝-inch penetration into the supporting structure, they are spaced about 7 inches on center for ceilings and 8 inches on center for walls.

Nails used to fasten wallboard, may be bright, coated, or chemically treated; the shanks may be smooth or annularly threaded with a nailhead that is generally flat or slightly concave. The annularly threaded nails are most commonly used since they provide more withdrawal resistance, require less penetration, and minimize nailpopping. For a fire rating it is usually required to have 1 inch or more of penetration and in this case, the smooth shank nails are most often used. Nails with small heads should never be used for attaching wallboard to the supporting construction. The spacing of nails generally varies from 6 to 8 inches on center, depending on the size and type of nail and the type wallboard being used. Nails are bought by the pound. The approximate weight of the nails that would be required per thousand square feet (MSF) of gypsum board varies between 5 and 7 pounds.

Screws may be used to fasten to both wood and metal supporting construction and furring strips. In commercial work, the drywall screws have virtually eliminated the use of nails. Typically, these screws have self-drilling, self-tapping threads with flat Phillips recessed heads for use with a power screwdriver. The drywall screws are usually spaced about 12 inches on center except when a fire rating is required when the spacing is usually 8 inches on center at vertical joists. There are three types of drywall screws: one for fastening to wood, one for sheet metal, and one for gypsum board. The approximate number of screws required per 1,000 square feet of gypsum board based on vertical board application and spacing of 12 inches on center is 1,000. If 8-inch on center spacing is used on the vertical joints, about 1,200 screws are required. If the boards are applied horizontally with screws spaced 12 inches on center, only about 820 screws are required.

Adhesives may be used to attach single-ply wallboard directly to the framing, concrete, or masonry, or to laminate the wallboard to a base layer. The base layer may be gypsum board, sound-deadening board, or rigid-foam insulation. Often, the adhesives are used in conjunction with screws and nails that provide either temporary or permanent supplemental support. Basically, the three classes of adhesives used are stud adhesives, laminating adhesives, and contact adhesives. There are also various modifications within each class. Information regarding exact adhesives required should be obtained from the specifications of the project and cross-checked with the manufacturer. Determine the special preparation, application, and equipment requirements from the manufacturer. Information concerning coverage per gallon and curing requirements should also be obtained.

Trim.   A wide variety of trims is available in wood and metal for use on drywall construction. The trim is generally used to provide maximum protection and neat, finished edges throughout the building. The wood trim is available unfinished and prefinished in an endless selection of sizes, shapes, and costs. The metal trim is available in an almost equal amount of sizes and shapes with finishes ranging from plain steel, galvanized steel, prefinished painted to trim with permanently bonded finishes which match the wallboard; even aluminum molding, plain and anodized, is available. Most trim is sold by the lineal foot, so the takeoff should also be made in lineal feet. Also to be determined is the manner in which the moldings are to be attached to the construction.

Joint tape and compounds.   Joint tape and compounds are employed when a gypsum wallboard is used, and it is necessary to reinforce and conceal the joints between wallboard panels and to cover the fastener heads. These items provide a smooth, continuous surface in interior wall and ceilings.

The tape used for joint reinforcement is usually a fiber tape designed with chamfered edges feathered thin and with a cross-fiber design.

Joint compounds are classified as follows: 1) embedding compound, used to embed and bond the joint tape; 2) topping compound, used for finishing over the embedding compound (it provides final smoothing and leveling over fasteners and joints); and 3) all-purpose compound, which combines the features of both of the other two, providing embedding and bonding of the joint tape and giving a final smooth finish. The compounds are available premixed by the manufacturer or in a powdered form to be job-mixed.

The amount of tape and compound required for any particular job will vary depending on the number of panels used with the least number of joints and the method of fastening specified. To finish 1,000 square feet of surface area, about 380 lineal feet of tape and 50 pounds of powder joint compound (or 5 gallons of ready-mixed compound for the average job) will be required.

Blankets.   Various types of blankets are used in conjunction with the drywall construction. The blankets are most commonly placed in the center of the construction, between studs, or on top of the suspended ceiling assembly. The two basic types of blankets are heat-insulating and sound-control. The heat-insulating blankets are used to help control heat loss (winter) and heat gain (summer), while the sound-control blankets are used to improve Sound Transmission Classification (STC) ratings of the assembly being used. Both types are available in a variety of thicknesses and widths. Once the square footage of wall or ceiling that requires the blanket has been determined, the amount of blanket required is virtually the same, but note the stud spacing so that the proper width blanket will be ordered. Check the specifications to determine any special requirements for the blankets, such as aluminum foil on one or both sides, paper on one or both sides, and method of attachment. The most common method of attachment is with staples.

EXAMPLE 14-2 — Commercial Building: For the building being estimated, 1-inch-thick rigid insulation on the exterior walls and ⅜-inch-thick gypsum board are required on the interior of the concrete block walls. Determine the square footage of the rigid insulation and gypsum board required for the project. The lengths of wall used to calculate the block may be used to help determine the area of wall to be covered (Appendix A).

Gross area of wall 397 (8 feet high) = 3176 s.f.
Deduct window and door openings − 304 s.f. (Sec. 10–5)
Net area to be covered 2872 s.f.

Rigid insulation: ½″ thick, 4 × 8-foot sheets.
Allow 3% waste:

$$2872\,(1.03) = \frac{2959}{32} = 92.5 \text{ sheets}$$

Order 95 to 100 sheets.
Gypsum board: 3/8″ thick, 4 × 8-foot sheets.
Allow 5% waste:

$$2872\,(1.05) = \frac{3017 \text{ s.f.}}{32} = 94.4 \text{ sheets}$$

Order 95 to 100 sheets.

EXAMPLE 14-3 — Metal trim is required around all openings and where the gypsum board abuts the ceiling:
Doors:

2 @ 21.5 l.f. = 43 l.f.
1 @ 24 l.f. = 24 l.f.

Windows:

2 @ 28 l.f. = 56 l.f.
4 @ 20 l.f. = 80 l.f.
Gypsum board abuts ceiling = 397 l.f.
600 l.f.
+ waste 50 l.f.
650 l.f. of ⅜″ trim

Taping joints is also required. Enough tape must be provided to cover the vertical joints between the gypsum board sheets. According to the information given in this article, 1,000 square feet of surface area requires 380 lineal feet of tape and 5 gallons of ready-mix compound.

3,000 s.f. = surface area
1,140 l.f. of tape

15 gallons of ready-mix compound are required.

EXAMPLE 14-4 — Residential Building (Appendix G):
Exterior walls: gypsum board, ½″ × 4′ × 8′
Ceiling height: 8′-0″
148 l.f. (from Sec. 13-4)

$$\frac{148 \times 8}{32} = 37 \text{ sheets} = 1184 \text{ s.f.}$$

Interior walls: ½″ gypsum board each side, 4 × 8
Ceiling height: 8′-0″
149 l.f. (from Sec. 13-4)

$$\frac{149 \times 8 \times 2}{32} = 75 \text{ sheets} = 2400 \text{ s.f.}$$

Ceilings: ½″ gypsum board, 4 × 8

$$24′ \times 50′ = 1{,}200 \text{ s.f.} = 38 \text{ sheets}$$

$$
\begin{aligned}
\text{Sheets required} = \quad &37 \text{ (exterior walls)} \\
+ \quad &75 \text{ (interior walls)} \\
+ \quad &\underline{38} \text{ (ceiling)} \\
&150 \text{ sheets}
\end{aligned}
$$

5% waste   <u>9</u>

Total 159 sheets required

Metal trim:

$$
\begin{aligned}
\text{Corners—2 @ 8′} &= \quad 16 \text{ l.f.} \\
\text{At ceiling—148 l.f.} \times 2 \text{ sides} &= \underline{296 \text{ l.f.}} \\
&\quad\; 312 \text{ l.f.}
\end{aligned}
$$

Labor. Subcontractors who specialize in drywall installations generally do this type of work and they may price it on a unit basis (per square foot) or a lump sum. The time required for the installation is shown in Fig. 14–5.

The hourly wages used are based on local labor and union conditions.

Drywall	Work hours per 100 s.f.
Gypsum board	
nailed to studs:	
walls	1.0 to 2.2
ceiling	1.5 to 2.8
jointing	1.0 to 1.8
glued:	
walls	0.8 to 2.4
ceiling	1.5 to 3.0
jointing	0.5 to 1.2
Ceilings over 8′-0″ high	add 10% to 15%
Screwed to metal studs	add 10%
Rigid insulation, glued to wall	1.2 to 3.0

Figure 14-5 Work Hours Required for Drywall Installation

EXAMPLE 14-5 — Commercial Building:
Gypsum board walls, glued: 2872 s.f.
Rate of work:

Drywall hanger, 1.0 work hour per 100 s.f.
Taper (jointing), 0.8 work hour per 100 s.f.
     Total 1.8 work hours per 100 s.f.

drywall and wetwall construction

Labor: Hanger and taper, $18.42 per hour

Time:

$$28.72 \times 1.8 = 51.7 \text{ work hours}$$

Cost:

$$51.7 \times 18.42 = \$952.31$$

EXAMPLE 14-6 — Residential Building: Gypsum board walls, nailed to studs:

$$1,184' + 2,384' = 3568 \text{ s.f.}$$

Rate of work:

Hanger, 1.2 work hours per 100 s.f.
Taper, 1.4 work hours per 100 s.f.
Total 2.6 work hours per 100 s.f.

Labor: Hanger and taper, $9.84 per hour

Time:

$$35.68 \times 2.6 = 92.8 \text{ work hours}$$

Cost:

$$92.8 \times \$9.84 = \$913.15$$

Ceiling: 1200 s.f.
Rate of work:

Hanger, 1.8 work hours per 100 s.f.
Taper, 1.6 work hours per 100 s.f.
Total 3.4 work hours per 100 s.f.

Labor: Hanger and taper, $9.84 per hour

Time:

$$12 \times 3.4 = 40.8 \text{ work hours}$$

Cost:

$$40.8 \times \$9.84 = \$401.47$$

14-9 WETWALL CONSTRUCTION

Wetwall construction consists of supporting construction, lath, and plaster. The exact types and methods of assembly used for this construction will depend on the requirements of the particular job regarding appearance, sound control, fire ratings, strength requirements, and cost.

The supporting construction may be wood, steel, concrete, gypsum, tile, masonry, or lath. Certain types of lath used with the plaster are self-supporting. The plaster itself may be two- or three-coat with a variety of materials available for each coat.

Proper use of plasters and bases provides the secure bond necessary to develop the required strength. A mechanical bond is formed when the plaster is pressed through the holes of the lath or mesh and forms keys on the back side. A suction, or chemical bond, is formed when the plaster is applied over masonry and gypsum bases, with the tiny needle-like plaster crystals penetrating into the surface of the base. Both mechanical and suction bonds are developed with perforated gypsum lath.

14-10 PLASTER

Plaster is a material that in its plastic state can be troweled to form and when set provides a hard covering for interior surfaces, such as walls and ceilings. Plaster is the final step in wetwall construction (although other finishes may be applied over it). Together with the supporting construction and some type of lath, the plaster will complete the assembly. The type and thickness of plaster used will depend on the type of supporting construction, the lath, and the intended use. Plaster is available in one-coat, two-coat, and three-coat work, and plaster is generally classed according to the number of coats required. The last and final coat applied is called the *finish coat,* while the coat, or combination of coats, applied before the finish coat is referred to as the *base coat.*

Base coats. Base-coat plasters provide a plastic working material that conforms to the required design and serves as a base over which the finish coats are applied. Base coats are available mill-mixed and job-mixed. Mill-mixed base coats are available with an aggregate added to the gypsum at the mill. Aggregates used include wood fibers, sand, perlite, and vermiculite. For high-moisture conditions a portland cement and lime plaster base coat is available.

Three-coat plaster must be used on metal lath. The first coat (scratch coat) must be of a thickness sufficient to form keys on the back of the lath, fill it in completely, and cover the front of the lath. The thickness may vary from $\frac{1}{8}$ to $\frac{1}{4}$ inch. The second coat (brown coat) ranges from $\frac{1}{4}$ to $\frac{3}{8}$ inch thick and the finish coat $\frac{1}{16}$ to $\frac{1}{8}$ inch thick.

Two-coat plaster may be used over gypsum lath and masonry. The first coat is the base coat (scratch or brown), and the second coat is the finish coat. Base coats range from $\frac{1}{4}$ to $\frac{3}{8}$ inch in thickness and the finish coat $\frac{1}{16}$ to $\frac{1}{8}$ inch. Perforated gypsum lath will require enough material to form the mechanical keys on the back of the sheets.

Finish coats. Finish coats serve as leveling coats and provide either a base for decorations or the required resistance to abrasion. Several types of gypsum-finish plasters are available, including those that require the addition of only water and those that blend gypsum, lime, and water; or gypsum, lime, sand, and water. The finish coat used must be compatible with the base coat. Finishing materials may be classified as prepared finishes, smooth trowel finishes, and sand float finishes. Finish coat thicknesses range from $\frac{1}{16}$ to $\frac{1}{8}$ inch thick.

Specialty finish coats are also available. One such specialty coat is radiant heat plaster for use with electric cable ceilings. It is a high-density plaster that allows higher operating temperature for the heating system as it provides more efficient heat transmission and greater resistance to heat deterioration. Applied in two coats— the first to embed the cable, the second a finish coat over the top—its total thickness is about $\frac{1}{4}$ to 3/8 inch. It is usually mill-prepared and requires only the addition of water.

One-coat plaster is a thin-coat interior product used over large sheets of gypsum plaster lath in conjunction with a glass fiber tape to finish the joints. The plaster coat is $\frac{1}{16}$ to $\frac{3}{32}$ inch thick.

Keene cement plaster is used where a greater resistance to moisture and surface abrasions is required and is available in smooth and sand-float finish. It is a dead burned gypsum mixed with lime putty and is difficult to apply unless sand is added to the mixture, but with sand as an additive, it is then less resistant to abrasion.

Acoustical plasters, which absorb sound, are also available. Depending on the type used, they may be troweled or machine-sprayed onto the wall. Trowel applications are usually stippled, floated, or darbied to a finish. Some plasters may even be tinted various colors. Thicknesses range from $\frac{3}{8}$ to $\frac{1}{2}$ inch.

Special plasters for ornamental plastering work, such as moldings and cornices, are also available. Molding and casting plasters are most commonly used for such work.

Stucco. This material is used in its plastic state. It can be troweled to form and when set, it provides a hard covering for exterior walls and surfaces of a building or structure. Stucco is generally manufactured with portland cement as its base ingredient and with clean sand and sometimes lime added. Generally applied as three-coat work the base coats are mixed about one part portland cement to three parts clean sand. If lime is added, no more than six to eight pounds per 100 pounds of portland cement should be used in the mix. The lime tends to allow the mix to spread more easily. The finish coat is usually mixed 1:2 (cement to sand), and no coat should be less than $\frac{1}{4}$-inch thick.

Stucco is usually applied to galvanized metal lath that is furred out slightly from the wall, but can also be applied directly to masonry. Flashing is often required and must be included in the estimate.

Various special finishes may also be required and will affect the cost accordingly. Finishes may be stippled, broomed, pebbled, swirled or in other designs.

Specifications. The specifications should state exactly what type of plaster is required and where. The types often will vary throughout the project and each must be kept separately. The number of coats, thickness of each coat, materials used, and the proportions of the mix must all be noted. Check also the type of finish required and what trueness of the finish coat is required. Check the room finish schedule on the drawings also since often, finishes and the finish coat required are spelled out in the schedule. Determine also the accessories, grounds, trim, and anything else that may be required for a complete job.

Estimating. Gypsum plasters are usually packed in 100-pound sacks and priced by the ton. The estimator makes the wetwall takeoff in square feet, converts it to square yards, and must then consider the number of coats, thickness of coats, mixes to be used, and the thickness and type of lath required. The amounts of materials required may be determined with the use of Fig. 14–6. This table gives the cubic feet of plaster required per 100 square yards of surface. Depending on the type of plaster being used, the approximate quantities of materials can be determined. The mix design varies from project to project and must be carefully checked. Figure 14–7 shows some typical quantities of materials that may be required. Many projects have mixes, designed for a particular use, included in the specifications. Read them

Per $\frac{1}{16}''$ thickness of plaster allow	Perforated plaster board lath — add	Metal lath Add
5.2 C.F.	4.0 C.F.	8.0 C.F.

Example: For 100 sq. yd. of wall area
a $\frac{1}{4}''$ thickness over metal lath will require:

5.2(4) + 8 = 20.8 + 8 = 28.8 C.F.
of plaster per 100 sq. yd.

Figure 14-6   Approximate Plaster Quantities

carefully and use the specified mix to determine the quantities of materials required. Once the quantities of materials have been determined, the cost for materials may be determined and the cost per square yard (or per 100 square yards) calculated.

Mix	Maximum amount of aggregate, in cu. ft. per 100 lbs of gypsum plaster	Volume obtained from mix shown (C.F.)
100:2	2	2
100:$2\frac{1}{2}$	$2\frac{1}{2}$	$2\frac{1}{2}$
100:3	3	3

Aggregate weights vary.   Sand 95–110 lb. per C.F.
Perlite 40–50 lb. per C.F.
Vermiculite 40–50 lb. per C.F.

Figure 14-7   Plaster Materials

Labor time and costs for plastering are subject to variations in materials, finishes, local customs, type of job, and heights and shapes of walls and ceilings. The ability of workers to perform this type of work varies considerably from area to area. In many areas skilled plasterers are few; this means that the labor cost will be high and there may be a problem with the quality of the work done. It is advisable to contact the local unions and subcontractors to determine the availability of skilled workers. In most cases one helper will be required to work with two plasterers.

If the plastering is bid on a unit basis, be certain that you understand how the yardage will be computed, since the methods of measuring vary in different localities. The yardage may be taken as the gross area, the net area, or the gross area minus the openings that are over a certain size. In addition, curved and irregular work may be charged and counted extra; it will not be done for the same costs as the flatwork.

Equipment required includes a small power mixer, planks, scaffolds, mixing tools, mixing boxes, and miscellaneous hand tools. Machine-applied plaster will require special equipment and accessories, depending on the type used.

Labor.   Subcontractors who specialize in wetwall installations do this type of work and they may price it on a unit basis (per square yard) or lump sum.

EXAMPLE 14-7 — All interior walls are wetwall construction requiring 3¼-inch truss-type steel stud, the lath is diamond mesh, and the plaster thickness is ¾ inch. The wall behind the plumbing fixtures requires a 6-inch stud. Determine the area (in square yards) of wall requiring plaster (Appendix A).

l.f. of interior walls:

$$35 + 28 + 54 + 10 + 10 + 25 = 162 \text{ l.f. @ 8 feet} = 1296 \text{ s.f.}$$

Less door openings: $\qquad\qquad 4 @ 3'\text{-}0'' \times 7'\text{-}0'' = -84 \text{ s.f.}$

$$\text{Net area to be plastered} = 1212 \text{ s.f.}$$
$$\text{Multiplied by 2 sides of wall to be plastered} \quad \underline{\times 2}$$
$$2424 \text{ s.f.}$$

$$\text{Square yards of plaster} = \frac{2424 \text{ s.f.}}{9 \text{ s.f.}} = 270 \text{ square yards}$$

14-11 LATH

Lath is used as a base and the plaster is bonded to the lath. Types of lath include gypsum tile, gypsum plaster, board, metal, and wood.

The type lath required will be specified and will vary, depending on the requirements of the project. Read the specifications carefully and note on your workup sheet the type of lath required and where it is required. It is not unusual for more than one type to be used on any one job.

Gypsum plaster lath.   In sheet form, this provides a rigid base for the application of gypsum plasters. Special gypsum cores are faced with multilayered laminated paper. The different types available are plain gypsum lath, perforated gypsum lath, fire-resistant, insulating, and radiant-heat lath. Depending on the supporting construction, the lath may be nailed, stapled, or glued. The type of spacing of the attachments depends on the type of construction and the thickness of the lath. Gypsum lath may be attached to the supporting construction by use of nails, screws, staples, or clips.

Plain gypsum lath is available in thicknesses of ⅜ and ½ inch with a face size of 16 × 48 inches. The ⅜-inch thickness is also available in 16 × 96-inch sizes. (A 16⅜6-inch width is available in certain areas only.) When the plaster is applied to this base, a chemical bond holds the base to the gypsum lath. Gypsum lath 24 × 144 inches is also available in certain areas.

Perforated gypsum lath is available ⅜-inch thick with a face size of 16 × 48 inches. Holes ¾ inch in diameter are punched in the lath, spaced 16 inches on center. The perforated lath permits higher fire ratings since the plaster is held on by mechanical as well as chemical bonding.

Fire-resistant gypsum lath has a specially formulated core of special mineral materials. It has no holes, but provides additional resistance to fire exposure. It is available ⅜-inch thick with a face size of 16 × 48 inches.

Insulating gypsum lath is plain gypsum lath with aluminum foil laminated to the back face. It serves as a plaster base, an insulator against heat and cold, and a vapor barrier. It is available in ⅜- and ½-inch thicknesses with a face size of 16 × 48 inches.

Radiant-heat lath is a large gypsum lath for use with plaster in electric cable ceilings. It has improved the heat emission of the electric cables and has increased their resistance to heat deterioration. It is available 48 inches wide, and in ½- and ⅜-inch thickness and lengths of 8 to 12 feet. This type of lath is used with plaster that is formulated for use with electric cable heating systems.

Estimating. The gypsum lath is sold by the sheet or 1,000 square feet, and the estimator will calculate the number of square feet required (the square yards of plaster times nine equals square feet), and divide by the number of square feet in a sheet. Note the type and thickness required. Depending on the number of jogs and openings, about 6 percent should be allowed for waste. The materials used for attachment must be estimated and a list of accessories made.

Metal lath. Metal lath is sheet steel that has been slit and expanded to form a multitude of small mesh openings. Ordinary expanded metal lath (such as diamond mesh, flat rib lath) is used in conjunction with other supporting construction. There are also metal laths that are self-supporting (such as 3/8- inch rib lath), requiring no supporting construction.

Metal lath is available painted, galvanized, or asphaltum-dipped; sheet sizes are generally 24 × 96 inches (packed 16 square yards per bundle) or 27 × 96 inches (20 square yards per bundle). Basically, the three types of metal lath available for wetwall construction are diamond, flat-rib lath and ⅜-inch rib lath. Variations in the designs are available through different manufacturers.

The metal lath should be lapped not less than ½ inch at the sides and 1 inch at the ends. The sheets should be secured to the supports at a maximum of 6 inches on center. The metal lath is secured to the steel studs or channels by use of 18-gauge tie wires about 6 inches on center. For attachment to wood supporting construction nails with a large head (about ½ inch) should be used.

Diamond lath. This is an all-purpose lath that is ideal as a plaster base, as a reinforcement for walls and ceilings, and as fireproofing of steel columns and beams. It is easily cut, bent, and formed for curved surfaces. It is available in weights of 2.5 and 3.4 pounds per square yard; both sizes are available in copper alloy steel either painted or asphaltum-coated. Galvanized diamond lath is available only in 3.4 pounds per square yard.

Flat rib lath is a ⅛-inch lath with "flat ribs," which make a stiff type of lath. This increased stiffness generally permits wider spacing between supports than diamond lath and the design of the mesh allows the saving of plaster. The main longitudinal ribs are spaced 1½ inches apart with the mesh set at an angle to the plane of the sheet. Available in copper alloy and steel in weights of 2.75 and 3.4 pounds per square yard, and in galvanized steel in a weight of 3.4 pounds per square yard, it is used with wood or steel supporting construction on walls and ceilings, and for fireproofing.

The ⅜-inch rib lath combines a small mesh with heavy reinforcing ribs. The ribs are ⅜-inch deep, 4½ inches on center. Used as a plaster base, it may be employed in studless wall construction and in suspended and attached ceilings. The rib lath permits wider spacing of supports than flat rib and diamond lath. This type is also used as a combination form and reinforcement for concrete floor and roof slabs. Copper alloy steel lath is available in 3.4 and 4.0 pounds per square yard, while the galvanized is available in 3.4 pounds per square yard.

Specifications. The type of lath, its weight, and finish must be checked. The spacing of the supporting construction will affect the amount of material and labor required to attach the lath. The type and spacing of attachment devices should be checked as well as a list of accessories.

Estimating. The metal lath is taken off by the square yard in the same manner as plaster. It is usually quoted at a cost per 100 square yards with the weight and finish noted. For plain surfaces add 6 to 10 percent for waste and lapping; for beams, pilasters, and columns add 12 to 18 percent. When furring is required, it is estimated separately from the lath. Determine what accessories will be required and the quantity of each.

Wood lath. Although largely displaced by the other types of laths available, on occasion wood lath is encountered. The most commonly used size is ⅜-inch thick, 1⅜ inches wide (actual) and 48 inches long spaced ⅜ inch apart. Therefore, one lath takes up the space of 1¾ × 48 inches. The wood lath would be taken off by the square yard, with about 14.3 pieces required per square yard. They are attached by nails or staples.

Gypsum tile. This is a precast, kiln-dried tile used for nonload-bearing construction and fireproofing columns. Thicknesses available are 2 inches (solid), 3 inches (solid or hollow), and 4 and 6 inches (hollow). The 2-inch tile is used only for fireproofing, not for partitions. A face size of 12 × 30 inches (2.5 square feet) is available. Used as a plaster base, it provides excellent fire and sound resistance. Gypsum tile may be taken off as part of wetwall construction or under masonry.

Specifications. Determine the type and thickness required; make a list of all clips and accessories; and decide how the gypsum tile is to be installed.

Estimating. The number of units required must be determined. If the square feet, squares, or square yards have been determined, their area can easily be converted to the number of units required. Also note the thickness required. Resilient clips are sometimes used also. Make a list of accessories and the amount of each required.

The lath required for the building will be included in the subcontractor's bid, but the estimator should check the subcontractor's proposal to be certain that it calls for the same lath as the contract documents. The lath takeoff for the commercial building is shown here.

EXAMPLE 14-8— Determine the amount of lath required for the plaster in Sec. 14-10. The quantity of lath required will equal the amount of plaster in Sec. 14-10. Since only one type of lath was used, no further breakdown in the figures is necessary.

From Sec. 14-10, there are 270 square yards to be covered.

Metal lath, 16 square yards per bundle: 17 bundles provides 272 square yards (including waste).

14-12 ACCESSORIES

The accessories available for use with wetwall construction include various types of corner beads, control and expansion joints, screeds, partition terminals, casing beads, and a variety of metal trim to provide neat edged, cased openings. Metal ceiling and floor runners are also available as are metal bases. Resilient channels may also be used. These accessories are sold by the lineal foot, so the estimator makes the take-off accordingly.

A complete selection of steel clips, nails, staples, and self-drilling screws is available to provide positive attachment of the lath. Special attachment devices are available for each particular wetwall assembly. The estimator will have to determine the number of clips or screws required on the project. The specifications will state the type of attachment required and may also give fastener spacing. The manufacturer's recommended fastener spacing may also be checked to help determine the number required.

Accessories required should be included in the subcontractor's bid but the estimator should check the subcontractor proposal against the contract documents to be certain they are the same size, thickness of metal, and finish. The accessories take-off for the commercial building is shown here.

EXAMPLE 14-9 — Accessories required for the wetwall construction include corner beads for outside corners, cornerite for interior corners, casing beads as a plaster stop around door openings and floors and ceiling runners to hold the metal stud.

Corner beads: 2 corners @ 8 feet each = 16 l.f.

Cornerite: 10 corners @ 8 feet each = 80 l.f.

Casing beads: 4 doors @ 21.5 feet each = 86 l.f.

Floor and ceiling runners (from Example 14-1):

162 l.f. of wall, runners, floor and ceiling—162 l.f. (2) = 324 l.f.

Attach mesh to steel studs with No. 18 gauge tie wire 6 inches on center.

Waste must be considered for each item. The amounts required above are generally small and an allowance of 5 percent is sufficient.

14-13 CHECKLIST

Wetwall:

    lath, metal
    furring
    studs
    channels
    lath, gypsum
    gypsum block
    corner beads
    accessories

number of coats

type of plaster

tie wire

molding

stucco

Drywall:

studs

wallboards

furring

channels

tape

paste

adhesives

staples

clips

nails

screws

REVIEW QUESTIONS

1. Determine the amounts of materials required for the drywall and/or wetwall for the building in Appendix E. Use workup sheets and sketches.

2. What is the difference between wetwall and drywall?

3. Why should the walls with various heights, thickness, and finishes be listed separately?

4. What procedure would be used to estimate the number of steel studs required?

5. What procedure would be used to estimate the amount of runners required for the steel studs?

6. List the types of lath used for wetwall and the unit of measure for each.

7. What advantages are there in using subcontractors for this portion of the work?

8. When considering the use of a subcontractor for any given portion of work, what disadvantages must be considered?

9. Determine the amounts of materials required for the drywall for the residence in Appendix H.

15

WINDOWS, CURTAIN WALLS, DOORS, AND HARDWARE

15-1 WINDOW AND CURTAIN WALL FRAMES

Window and curtain wall frames may be made of wood, steel, aluminum, bronze, stainless steel, and plastic. Each material has its particular types of installation and finishes, but from the estimator's point there are two basic types of window, stock and custom-made.

Stock windows are more readily available and, to the estimator, more easily priced as to the cost per unit. The estimator can count up the number of units required and list the accessories to work up his material cost.

Custom-made frames cannot be accurately estimated by the estimator. Approximate figures can be worked up based on the square footage and type of window, but exact figures can be obtained only from the manufacturer. In this case the estimator will call either the local supplier or manufacturer's representative to be certain that they are bidding the job. Often, copies of the drawings and specifications are sent to them, which they may use to prepare a proposal.

When checking proposals for the windows on a project, make a note whether the glass or other glazing is included, where delivery will be made, and whether installation is done by the supplier, a subcontractor, or the general contractor. Be certain the proposal includes all the accessories that may be required, including mullions, screens, sills, etc., and that the material being bid is in conformance with the specifications.

If the contractor is going to install the windows, check to see if they will be delivered preassembled or if they must be assembled on the job. The job may be bid by the square foot, or lineal foot of frame, but the most common method is to bid in a single lump sum.

Shop drawings should always be required for windows and curtain wall frames since even stock sizes vary slightly in terms of masonry and rough openings required for their proper installation. Custom-made windows always require shop drawings so that the manufactured sizes will be coordinated with the actual job conditions.

If you, the estimator, decide to do a complete takeoff of materials required, you should: (1) determine the lineal footage of each different shape required of each type of material; (2) determine the type and thickness of glazing required; and (3) calculate the sheet sizes required.

Wood frames.   More than 60 percent of all stock window frames are made from ponderosa pine. Other commonly used species are southern pine and Douglas fir. Custom-made frames may be of any of the above species or of some of the more exotic woods, such as redwood and walnut. Finishes may be plain, primed, preservative-treated, and even of wood that is shielded with vinyl.

The same general estimating procedure is followed here as in other frames since these frames may be stock or custom-made. Shop drawings should be made and carefully checked so that all items required are covered by the proposal. If painting is required, be certain it is covered in that portion of the specifications.

EXAMPLE 15-1 — The wood windows required for the residence have been taken off and compiled here.

4 double-hung wood windows, 32/36 (2′-8″ × 3′-0″)

2 d.h. wood, 36/44

Labor.   The time required to install the windows is taken from the table in Fig. 15-1. The hourly wages used are based on local labor and union conditions.

Windows	Work hours per unit
Wood:	
in wood frame	2.0 to 3.5
in masonry	2.5 to 5.0
Metal:	
in wood frame	2.0 to 3.5
in masonry	2.0 to 4.0
Over 12 s.f.	add 20%
Bow and bay windows	2.5 to 4.0

Figure 15-1   Work Hours Required for Window Installation

EXAMPLE 15-2:

Wood windows: 4 double-hung wood windows, 42/52 (3′-6″ × 4′-4″): 2 double-hung wood windows, 42/52 (3′-6″ × 4′-4″) with mullions

Rate of work: 2.5 work hours per window

Labor: Carpenter, $11.87 per hour

Time:

$$6 \times 2.5 = 15 \text{ work hours}$$

Cost:

$$15 \times \$11.87 = \$178.05$$

Aluminum windows and curtain wall. The shapes are made by extrusion and roll forming. Finishes include mill finish (natural silvery sheen), anodizing (protective film; it may be clear or a variety of color tones), paint, lacquer, and colored or opaque coatings. Obviously the finish required must be noted, since it will affect the price of the materials. The specifications will require certain thicknesses of metal throughout the construction of the frames.

Care on the job is important, since exposed aluminum surfaces may be subject to splashing with plaster, mortar, or concrete masonry cleaning solutions that often cause corrosion of the aluminum. Exposed aluminum should be protected during construction with either a clear lacquer or strippable coating. Concealed aluminum that is in contact with concrete, masonry, or absorbent material (such as wood or insulation) and may become intermittently wet must be protected permanently by coating either the aluminum or adjacent materials. Coatings commonly used include bituminous paint and zinc-chromate primer.

The costs of such protection must be included in and are part of the cost of the project. If the frames are to be installed by a subcontractor, check to see if he has included the required protection in his proposal; if not, the estimator will have to make an allowance for it. Also check to see who will put the protection on. If a strippable coating is used on exposed aluminum, note who must remove the coating, if the subcontractor has included this in his price, or if the general contractor will have to take care of it. Any piece of work that must be done costs money; check to see where it is included.

Before the job is bid, the estimator should check the details showing how the windows are to be installed. They may have to be installed as the construction progresses (where the window is built into the wall), or they may be slipped into openings after the building has been constructed. The method is important since there is always a time lag for approvals, and manufacture (of custom units) and coordination will be important.

EXAMPLE 15-3 — The aluminum windows required for the commercial building have been taken off and compiled below.

> 4 aluminum horizontal sliding windows, ins. glass, screens, 4′ × 6′
>
> 24 l.f. exterior sill (aluminum)
>
> 24 l.f. interior sill (aluminum)

On large projects the window supplier sometimes bids on installation also. On smaller projects such as this commercial building, the contractor will usually install the windows. The time required for window installation is taken from the table in Fig. 15–1. The hourly wages are based on local labor and union conditions.

EXAMPLE 15-4:

Aluminum windows: 4 required
Rate of work: 2.8 work hours per window
Labor: $11.87 per hour
Time:

$$4 \times 2.8 = 11.2 \text{ work hours}$$

Cost:

$$11.2 \times \$11.87 = \$132.94$$

The aluminum curtain walls are almost always installed by a subcontractor, usually the curtain wall supplier. The time required for curtain wall installation is shown in Fig. 15-2.

Metal curtain walls	s.f. per work hour
Up to 1 story high	8–18
1 to 3 stories	6–12
over 3 stories	6–10

Includes glazing.
The larger the openings for glass
or panels the more work which
can be done per work hour.

Figure 15-2 Work Hours Required for Curtain Wall Installation

EXAMPLE 15-5 — Commercial Building:

Two 6′ × 8′ windows (96 s.f. of window area) requiring:

56 l.f. of 2″ × 4″ aluminum tubing
96 s.f. of ¼″ insulating glass
56 l.f. of glazing bead (aluminum)

Exterior sills: 16 l.f. required
Interior sills: 16 l.f. required

15-2 ACCESSORIES

The items that may be required for a complete job include glass and glazing, screening, weatherstripping, hardware, grilles, mullions, sills, and stools. The specifications and details of the project must be checked to find out what accessories must be included, so make a list of each accessory and what restrictions there are for each.

Glass. Glass is discussed in Sec. 15-14. At this point note whether or not it is required, what thickness, type, and quality are required, and if it is to be a part of the unit or installed at the job site. The square footage of each type of glass must be known as well.

Glazing. *Glazing* is the setting of whatever material is installed in the frames. Most times the material is glass, but porcelain, other metal panels, exposed aggregate plywood, plastic laminates, and even stone veneer and precast concrete have been set in the frames. Many stock frames are glazed at the factory, but it is not uncommon that stock windows, especially steel frames, be glazed on the job. Custom-made frames are almost always glazed on the job. Check with the subcontractor and supplier to determine who glazes them. Glazing costs will depend on the total amount

accessories

involved, size and quality of the material to be glazed, and the type of glazing method used. If the frame is designed so that the material can be installed from inside the building, no scaffolding will be required and work will proceed faster. On wood frames glazing compounds are usually employed while on metal frames either glazing compounds or neoprene gaskets are used.

Screens. Screening mesh may be painted or galvanized steel, plastic-coated glass fiber, aluminum, and bronze. If specified, be certain the screens are included in the proposals received. If the screens are not bid by the supplier or subcontractor whose bid is used for frames, prices for screens must be obtained elsewhere. If so, a word of caution: Be certain that the screens and frames are compatible, and that a method of attaching the screen to the frame is determined. The various sizes required will have to be noted so an accurate price for the screening can be arrived at.

Hardware. Most types of frames require hardware. It may consist only of locking and operating hardware with the material of which the frame is made. Since the hardware is almost always sent unattached, it must be applied at the job site. Various types of locking devices, handles, hinges, and cylinder locks on sliding doors are often needed. Materials used may be stainless steel, zinc-plated steel, aluminum, or bronze, depending on the type of hardware and where it is being used.

Weatherstripping. Most specifications require some type of weatherstripping, and many stock frames come with weatherstripping factory installed. The only thing to be careful of in that case is that the factory-installed weatherstripping is the type specified. Some of the more common types of weatherstripping are vinyl, poly-propylene-woven pile, neoprene, metal flanges and clips, polyvinyl, and adhesive-backed foam. It is usually sold by the lineal foot for the rigid type and by the roll for the flexible. Do the takeoff in lineal feet, find out how it is sold, and work up a cost.

Mullions. *Mullions* are the vertical bars that connect adjoining sections of frames. The mullions may be of the same material as the frame, or a different material, color, or finish. Mullions may be small "T"-shaped sections that are barely noticeable or large elaborately designed shapes used to accent and decorate. The mullions should be taken off by the lineal footage with a note as to thicknesses, finish, and color.

Sills. The *sill* is the bottom member of the frame. The member on the exterior of the building, just below the bottom member of the frame, is another sill. This exterior sill serves to direct water away from the window itself. These exterior sills may be made of stone, brick, precast concrete, tile, metal, or wood. The details of the frame and its installation must be studied to determine the type of material, its size, and how it fits in. Basically, there are two types of sills, the *slip sill* and the *engaged sill*. The slip sill is slightly smaller than the opening for the frame and can be slipped in place after the construction of the walls is complete or either just before or just after the frames have been installed (depending on the exact design). The engaged sill must be installed as the walls go up, since it is wider than the opening and extends into the wall construction. Sills are taken off by the lineal foot, and the takeoff should be accompanied by notes and sketches that show exactly what is required.

Stools. The interior member at the bottom of the frame (sill) is called the *stool*. The stool may be made of stone, brick, precast concrete, terrazzo, tile, metal, and wood. It may be of the slip or engaged variety. The quantity should be taken off in lineal feet, with notes and sketches showing material size and installation requirements.

Flashing.   Flashing may be required at the head and sill of the frame. Check the specifications and details to determine if it is required and the type required. Usually, the flashing is installed when the building is being constructed, but check the installation details to see how it is to be installed. Check also to see who buys the flashing and who is supposed to install it—don't neglect these seemingly small items, since they total up to a good deal of time and expense.

Lintels.   The horizontal supporting member over the opening is called a *lintel*. It may be wood, concrete, steel, or block with steel reinforcing. The lintels are *not* installed as parts of the windows, but rather are installed as the building progresses. However, this is the time to double check that they have been included, both as material and installation costs.

15-3 DOORS

Doors are generally classified as interior or exterior, although exterior doors are often used in interior spaces. The list of materials of which doors are made includes wood, aluminum, steel, glass, stainless steel, bronze, copper, plastics, and hardboard. Doors are also grouped according to the mode of their operation. Some different types of operation are illustrated in Fig. 15–3. Accessories required include glazing, grilles and louvers, weatherstripping (for sound, light, and weather), molding, trim, mullions, transoms, and more. The frames and hardware must also be included.

There are many specially constructed doors that serve a particular need: some examples are fire-rated, sound-reduction, and lead-lined doors. The frames may be of as many different materials as the doors themselves, and the doors sometimes come "prehung" in the frame. The doors may be prefinished at the factory or job-finished.

Wood doors.   Wood doors are basically available in two types, solid- and hollow-core. Solid-core doors may have a core of wood block or low-density fibers and are generally available in thicknesses of 1⅜ and 1¾ inches for fiber-core and 1¾ inches through 2½ inches in the wood block core. Widths of 5′-0″ are available in both types with a height of 12′-0″ as maximum for woodblock core; the fiber-core door is available in 6′-8″, 7′-0″, and 8′-0″ heights, but different manufacturers offer different sizes. In checking the specifications, note the type of core required, any other special requirements such as the number of plies of construction, and the type face veneer.

Hollow-core doors are any doors in which the cores are not solid. The core may have interlocking wood grid strips, ribs, struts, or corrugated honeycomb. The specifications should spell out exactly what type is required. Thicknesses usually range from 1⅜ inches to 1¾ inches with widths of 1′-0″ to 4′-0″ and heights to 8′-0″.

The face veneer required in the specifications should be carefully checked as there is a considerable difference in prices. Refer to Fig. 15–4 for a quick reference cost guide of various face veneers.

From the specifications determine whether the doors are to be prefinished at

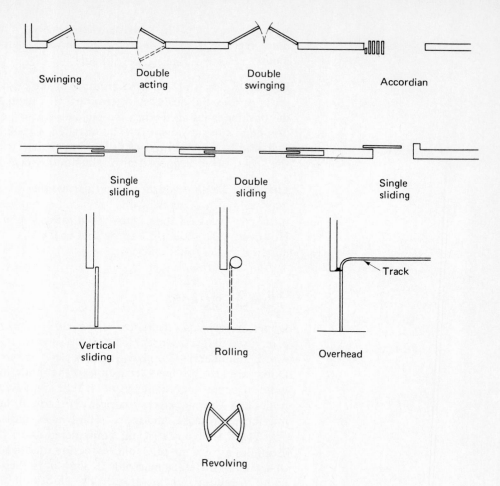

Swinging Double acting Double swinging Accordian

Single sliding Double sliding Single sliding

Vertical sliding Rolling Overhead Track

Revolving

Figure 15-3 Typical Door Operations

the factory or job-finished. Decide who will finish the doors if they must be job-finished and what type of finish is required. Doors that are prefinished at the factory may require some touch-up on the job, and damaged doors will have to be replaced.

No matter what type of door is required, the items to be checked are the same. Read the specifications carefully.

Metal doors.   The materials may be aluminum, steel, stainless steel, bronze, or copper. The doors may be metal frames with large glazed areas, hollow metal, tin-clad, or a variety of other designs. The important thing is to read the specifications, study the drawings, and bid what is required. Often, special doors, frames, and hardware are specified and the material custom made. Finishes may range from lacquered and anodized aluminum and primed or prefinished steel to the natural and satin-finished bronze. Bid what is specified.

Door swings.   The swing of the door is important to the proper coordination of door, frame, and hardware. The names are often confused so they are shown, in

Each veneer has an index value comparing it with the price of
Rotary cut premium grade birch. Flucuation in veneer costs will
cause variations in the index values.

Veneer	Index value
Rotary sound ph. mahongany	72
Rotary sound natural birch	92
Rotary good natural birch	98
Rotary premium grade birch	100
Rotary premium red oak	100
Rotary premium select red birch	116
Plain sliced premium red oak	133
Plain sliced premium african mahogany	150
Plain sliced premium natural birch	165
Plain sliced premium cherry	170
Plain sliced premium walnut	180
Sliced quartered premium walnut	250
Plain sliced premium teak	280
Laminated plastic faces (premachined)	250

Figure 15-4   Veneer Cost Comparisons

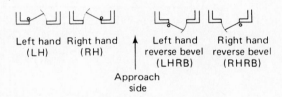

Left hand (LH)   Right hand (RH)   Approach side   Left hand reverse bevel (LHRB)   Right hand reverse bevel (RHRB)

Figure 15-5   Door Hand and Swing

Fig. 15–5 to act as a reminder. Learn the door swings; when checking door, frame, and hardware schedules, you must check the swing as well.

15-4 FIRE-RATED DOORS

Fire-rated doors are produced under the factory inspection and labeling service program of Underwriters Laboratories, Inc. (U.L.) and can be identified by the labels on the door. The label states the rating given to that particular door. The ratings given are also related to the temperatures at which the doors are rated. Doors labeled in this manner commonly are called "labeled doors." The letters used, hourly ratings, temperatures, and common locations, are given in Fig. 15–6. Fire doors may be of hollow metal, sheet metal, composite wood with incombustible cores, and other composite constructions. Special hardware will be required and the frame may also have to be labeled.

fire-rated doors

Underwriters' label classification	Heat transmission rating	Most commonly recommended locations in building
A 3 hr. situations	Temp. rise 30 minutes 250° max.	Openings in fire walls. Areas of high hazard contents. Curtain and division walls.
A* 3 hr. situations		Openings in fire walls. Areas of high hazard contents. Curtain and division walls.
B 1½ hr. situations	Temp. rise 30 minutes 250° max.	Openings in stairwells, elevator, shafts, vertical shafts.
B* 1½ hr. situations		Openings in stairwells, elevator shafts, vertical shafts.
C* ¾ hr. situations		Openings in room partitions and corridors
D* 1½ hr. situation		Openings in exterior walls where exposure to fire is severe.
E* ¾ hr. situations		Openings in exterior walls where exposure to fire is moderate.

*Not rated for heat transmission.

Figure 15-6 Fire Door Requirements

Fire-rated doors may be conventionally operated or be horizontal or vertical slide doors that close when the fusible link releases.

15-5 ACOUSTICAL DOORS

Acoustical doors are doors specially designed and constructed for use in all situations where sound control is desired. The doors are generally metal or wood with a variety of cores used. Since a wide range of sound control is offered by the doors available, use caution in selecting the right door to meet the specifications. Where high sound control is required, double doors may be desirable. Acoustical doors alone will not solve the problem of noise transmission around the door. An automatic drop seal for the bottom of the door and adjustable door stops (or other types of sound control devices) are used at the door jamb and head where the door meets the frame.

Often, the seals and devices are sold separately from the door. In these cases, determine the lineal footage of the seal needed and include its cost and labor charges in the estimate.

15–6 OVERHEAD DOORS

Overhead doors are available in all sizes. They are most commonly made of wood and steel although aluminum and stainless steel are sometimes used. The overhead doors are first designated by the type of operation of the door: rolling, sectional, canopy, and others. The estimator must determine the size of door, size of opening, type and style of door, finishes required, door hanger type, installation details, and whether it is hand-, electric- or chain-operated. The overhead door is priced as a unit with all required hardware. Often, the supplier will act as a subcontractor and install the doors.

Miscellaneous doors.   There is a wide variety of specialized doors, each designed to fill a particular need, among them are rolling grilles, which roll horizontally or vertically to provide protection or control traffic. Other specialty doors are revolving, dumbwaiter, rubber shock-absorbing doors, cooler and freezer doors as well as blast and bullet-proof doors. The estimating procedures are basically the same, regardless of the type of door required.

15–7 FOLDING DOORS AND PARTITIONS

Folding (accordian) doors and *partitions* offer the advantage of increased flexibility and more efficient use of the floorspace in a building. They are available in fiberglass, vinyl, and wood, or in combinations of these materials. They can be made to form a radius, have concealed pockets and overhead track; a variety of hardware is also available. Depending on the type of construction, the maximum opening height ranges from 8 to 21 feet and the width from 8 to 30 feet. The doors may be of steel construction with a covering, rigid fiberboard panels, laminated wood, and solid wood panels. Specially constructed dividers are also available with higher sound-control ratings. The exact type required will be found in the specifications as will the hardware requirements. Both the specifications and drawings must be checked for installation details and the opening sizes required.

15–8 PREFITTING AND MACHINING (PREMACHINING)

The doors may be machined at the factory to receive various types of hardware. Factory machining can prepare the door to receive cylindrical, tubular, mortise, unit, and rim locks. Other hardware such as finger pulls, door closers, flush catchers, and hinges (butts) are also provided for. Bevels are put on the doors and any special requirements taken care of.

Premachining is popular since it cuts job labor costs to a minimum, but coordination is important since the work is done at the factory from the hardware manufacturer's templates. This means that approved shop drawings, hardware and door schedules, and the hardware manufacturer's templates must be supplied to the door manufacturer.

From the estimator's point of view, premachining takes an item that is difficult to estimate and simplifies it considerably. Except where skilled door hangers are available, premachined doors offer cost control with maximum results. For this reason, they are being used more and more frequently.

15-9 PREFINISHING

Prefinishing of doors is the process of applying the desired finish on the door at the factory instead of finishing the work on the job. Doors that are premachined are often prefinished as well. Various coatings are available, including varnishes, lacquer, vinyl, and polyester films (for wood doors). Pigments and tints are sometimes added to achieve the desired visual effect. Metal doors may be prefinished with baked-on enamel or vinyl-clad finishes. Prefinishing can save considerable job-finishing time and generally yields a better result than job-finishing. Doors which are prefinished should also be premachined so that they will not have to be "worked on" on the job. The prefinished door must be handled carefully and installed at the end of the job so that the finish will not be damaged; it is often difficult to repair a damaged finish. Care during handling and storage is also requisite.

Door takeoff.

EXAMPLE 15-6 — Commercial Building:
Doors:

2—3/0 × 6/8, 1¾ in. thick, solid, prefinished

1—7/0 × 10/0, primed, 4-panel garage door

4—3/0 × 6/8, 1¾ thick, solid, prefinished

EXAMPLE 15-7 — Residential Building:
Doors, Interior:

1—1/6 × 6/8, hollow core, wood, prefinished
2—2/4 × 6/8, h.c., wood, prefinished
8—2/6 × 6/8, h.c., wood, prefinished
1—5/0 × 6/8, bifold, louvered, wood, prefinished

Doors, Exterior:

1—2/8 × 6/8, solid core, wood, prefinished

1—2/8 × 6/8, screen/storm, wood,

1—3/0 × 6/8, solid core, wood, prefinished

Labor.   The time required to install the doors is taken from the tables in Fig. 15–7. The hourly wages are based on local labor and union conditions.

EXAMPLE 15-8 — Commercial Building:
Doors:
  6 solid core, metal frames
  1 garage door, overhead 7/0 × 10/0
Rate of work per unit:
  2.5 work hours per unit—solid core doors, metal frames

Doors and frames	Work hours per unit
Residential, wood	
prehung	2.0 to 3.5
pocket	1.0 to 2.5
not prehung	3.0 to 5.0
Overhead	4.0 to 6.0
heavy duty	add 20%
Commercial	
aluminum entrance,	
per door	4.0 to 6.0
wood	3.0 to 5.0
metal, prefitted	1.0 to 2.5

Figure 15-7 Work Hours Required for Door Installation

4.5 work hours per unit—overhead garage door
Labor: Carpenter, $21.36 per hour
Time:

$$6 \times 2.5 = 15 \text{ work hours}$$
$$\underline{1 \times 4.5 = 4.5 \text{ work hours}}$$
$$19.5 \text{ work hours}$$

Cost:

$$19.5 \times \$21.36 = \$416.52$$

EXAMPLE 15-9 — Residential Building:
Doors:
   14 single
   1 bi-fold
Rate of work per unit:
   2.2 work hours per unit—single
   3.5 work hours per unit—bi-fold
Labor: Carpenter, $11.87 per hour
Time:

$$14 \times 2.2 = 30.8 \text{ work hours}$$
$$\underline{1 \times 3.5 = 3.5 \text{ work hours}}$$
$$34.3 \text{ work hours}$$

Cost:

$$34.3 \times \$11.87 = \$407.14$$

15-10 DOOR FRAMES

The door frames are made of the same type of materials as the doors. In commercial work, the two most common types are steel and aluminum. The steel frames (also called *door bucks*) are available in 14-, 16-, and 18-gauge steel and come with a prime coat of paint on them. Steel frames are available knocked-down (KD), set-up and spot-welded (SUS), or set-up and arc-welded (SUA). Many different styles and shapes are available, and the installation of the frame in the wall varies considerably also. Usually steel frames are installed during the building of the surrounding construction. For this reason, it is important that the door frames be ordered quickly; slow delivery will hinder the progress of the job. Check also how the frames are anchored to the surrounding construction.

The sides of the door frame are called the *jambs* and the horizontal pieces at the top are called the *heads.* Features available on the frames include a head 4 inches wide (usually about 2 inches), lead lining for X-ray frames, anchors for existing walls, base anchors, a sound-retardent strip placed against the stop, weatherstripping, and various anchors. Fire-rated frames are usually required with fire-rated doors.

When frames are ordered, the size and type of door, the hardware to be used, and the swing of the door all must be known. Standard frames may be acceptable on some jobs, but often special frames must be made.

Factory finishes for steel frames are prime-coated and must be finished on the job. Aluminum, bronze, brass, and stainless steel have factory-applied finishes, but have to be protected from damage on the job site. Wood frames are available primed and prefinished.

EXAMPLE 15-10 — Determine the number of frames required for the commercial building. From the door schedule it is noted there are two exterior doors and four interior doors, all doors 3/0 × 7/0 (Appendix A).

16-gauge steel frames, masonry construction: 2 required, 3/0 × 7/0

16-gauge steel frames, wetwall construction: 4 required, 3/0 × 7/0

When steel frames are used in masonry construction, it is critical that the frames be installed as the masonry is laid. This means that the frames will be required very early in the construction process. Since door frames for most commercial buildings must be specially made from approved shop drawings and coordinated with the doors and hardware used, the estimator must be keenly aware of this item and how it may affect the flow of progress on the project.

15-11 HARDWARE

The hardware required on a project is broken down into two categories: rough and finished. *Rough hardware* comprises the bolts, screws, nails, small anchors, and any other miscellaneous fasteners. This type of hardware is not included in the hardware schedule, but it is often required for installation of the doors and frames. *Finished hardware* is the hardware that is exposed in the finished building and includes items such as hinges (butts), hinge pins, door-closing devices, locks and latches, locking bolts, kickplates, and other miscellaneous articles. Special hardware is required on exit doors.

Finished hardware for doors is either completely scheduled in the specifications, or a cash allowance is made for the purchase of the hardware. If a cash allowance is made, this amount is included in the estimate (plus sales taxes, etc.) only for the purchase of materials. When hardware is completely scheduled in the specification or on the drawings, this schedule should be sent to a hardware supplier for a price. Only on small projects will the estimator work up a price on hardware unless the firm is experienced in this type of estimating. The cost for installation of finished hardware will vary depending on what type and how much hardware is required on each door, and whether the door has been premachined or not.

EXAMPLE 15-11 — For the commercial and residential buildings on which we have been working, a hardware allowance is made in the outline specifications:

Commercial building (Appendix A):

Hardware allowance (from specs), Materials only $250

Residential building *(Appendix G):*

Hardware allowance, Materials only $150

15-12 ACCESSORIES

Items that may be required to complete the job include weatherstripping, sound control, light control, and saddles. The specifications and details will spell out what is required and a list containing each item must be made. The takeoff for accessories should be made in lineal feet or the number of each size piece, for example: "five saddles, 3'-0" long."

Weatherstripping for the jambs and head may be metal springs, interlocking shapes, felt or sponge, neoprene in a metal frame, and woven pile. At the bottom of the door (sill), the weatherstripping may be part of the saddle, attached to the door, or both. It is available in the same basic types as used for jambs and heads, but the attachment may be different. Metals used for weatherstripping may be aluminum, bronze, or stainless steel.

Saddles are most commonly wood, aluminum, or bronze. Various shapes, heights, and widths may be specified. Sound control and light control usually employ felt or sponge neoprene in an aluminum, stainless steel, or bronze housing. The sill protection usually is automatic, closing at the sill, while at the jambs and head it is usually adjustable.

15-13 CHECKLIST FOR DOORS AND FRAMES

1. sizes required and the number required
2. frame and core types specified
3. face veneer specified (wood and veneer doors)
4. prefinished or job-finished (if so, specify the finish)
5. prehung or job-hung (if so, who will do it)
6. special requirements—
   a. louvers
   b. windows
   c. fire rating
   d. lead-lining
   e. sound control

7. type, size, style of frame, and the number of each required
8. method of attachment of the frame to the surrounding construction
9. finish required on the frame and who will apply it
10. hardware—what types are required?
11. hardware—who will install it?
12. accessories—types required and time to install them.

Everything takes time and costs money; someone must do it, so it must be included in the estimate.

15-14 GLASS

Glass is the most common material to be glazed into the frames for windows, curtain walls, storefronts, and doors. The most commonly used types are plate glass, clear window glass, wire glass, and patterned glass. Clear window glass is available in thicknesses of 0.085 to 2.30 inches thick, and the maximum size varies with the thickness and type. Generally available as single- and double-strength, heavy sheet and picture glass with various qualities is available in each classification. Clear window glass has a characteristic surface wave that is more apparent in the larger sizes.

Plate glass is available in thicknesses of $\frac{3}{16}$ to $1\frac{1}{4}$ inches and as polished plate glass, heavy-duty polished plate, rough, or polished plate. The more common types are regular, grey, bronze, heat-absorbing, and tempered.

Wire glass is available with patterned and polished finishes, and various designs of the wire itself. The most common thickness is $\frac{1}{4}$ inch and it is also available in colors. This type of glass is used where fire-retardent and safety characteristics (breakage) may be required.

Patterned glass, used primarily for decoration, is available primarily in $\frac{7}{32}$- and $\frac{1}{8}$-inch thicknesses. Pattern glass provides a degree of privacy, yet allows diffused light into the space.

Other types of glass available include a structural-strength glass shaped like a channel, tempered, sound control, laminated, insulating, heat- and glare-reducing, colored, and bullet-resisting.

The frame may be single-glazed, of one sheet of glass, or double-glazed, with two sheets of glass, which provides increased sound and heat insulation. If the specifications call for double-glazed, then twice as many square feet of glass will be required.

Glass is estimated by the square foot with sizes taken from the working drawings. Since different frames may require various types of glass throughout the project, special care must be taken to keep each type separate. Also to be carefully checked is which frames need glazing, since many windows and doors come with the glazing work already completed. The type of setting blocks and glazing compound required should be noted as well.

REVIEW QUESTIONS

1. Determine the window and/or curtain wall and door materials required for the building in Appendix E. Use workup sheets and sketches.

2. What accessories should be checked for when taking off windows and curtain wall?

3. Define glazing. Why must the estimator determine who will perform the required glazing?

4. What information is required to price a door?

5. What advantages are there in prefitting and prefinishing doors?

6. Determine the door and hardware requirements for the building in Appendix E. Use workup sheets.

7. Why should the type finish required on the door and door frames be noted on the workup sheet?

8. Describe briefly the ways hardware may be handled on a project.

9. What precautions must an estimator make when using an allowance, from the specifications, in the estimate?

10. What is the unit of measure for glass and why should the various types and sizes required be listed separately?

11. Determine the window and door materials required for the residence in Appendix H.

16

FINISHES

16-1 FLOORING

Flooring may be made of wood, resilient tile or sheets, carpeting, clay and ceramic tiles, stone, and terrazzo. Each type has its own requirements as to types of installation, depending upon job conditions, subfloor requirements, methods of installation, and moisture conditions.

16-2 WOOD FLOORING

The basic wood flooring types are strip, plank, and block. The most widely used wood for flooring is oak. Other popular species are maple, southern pine and Douglas fir—with beech, ash, cherry, cedar, mahogany, walnut, and teak also available. The flooring is available unfinished or factory-finished.

Strip flooring is flooring up to 3¼-inches wide and comes in various lengths. *Plank flooring* is from 3¼ inches to 8″ wide with various thicknesses and lengths. The most common thickness is 25/32 inch, but other thicknesses are available. It may be tongue-and-grooved, square-edged, or splined. Flooring may be installed with nails, screws, or mastic. When a mastic is used, the flooring used should have a mastic recess so that excess mastic will not be forced to the face of the flooring. Nailed wood flooring should be blind nailed (concealed); nail just above the tongue with the nail at a 45° angle. Waste on strip flooring may range from 15 to 40 percent, depending on the size of the flooring used. This estimate is based on laying the flooring straight in a rectangular room, without any pattern involved.

Strip and plank flooring may be sold by either the square foot or board foot measure. The estimator should estimate the square footage required on the drawings,

Size of wood Flooring (actual)	To change sq. ft. to board feet, multiply the sq. ft. amount by
$\frac{25}{32}'' \times 1\frac{1}{2}''$	1.55
$\frac{25}{32}'' \times 2\frac{1}{4}''$	1.383
$\frac{25}{32}'' \times 3\frac{1}{4}''$	1.29
$\frac{3}{8}'' \times 1\frac{1}{2}''$	1.383
$\frac{1}{2}'' \times 2''$	1.30

Values allow 5% waste.

FIGURE 16-1   Wood Flooring, Board Measure

noting the size of flooring required, type of installation, and then, if required, should convert square feet to board feet, (Fig. 16–1) and not forget to add waste.

Block flooring is available as parquet (pattern) floors, which consist of individual strips of wood or larger units that may be installed in decorative geometric patterns. Block sizes range from 6 × 6 inches to 30 × 30 inches, thickness from $\frac{5}{16}$ to ¾ inch. They are available tongue-and-grooved or square edged. Construction of the block varies considerably, it may be pieces of strip flooring held with metal or wood splines in the lower surface; *laminated blocks,* cross-laminated plies of wood; or the *slat blocks,* which are slats of hardwood assembled in basic squares and factory assembled into various designs. Block flooring is estimated by the square foot with an allowance added for waste (2 to 5 percent). The type of flooring, pattern required, and method of installation must be noted.

The wood flooring may be unfinished or factory finished. Unfinished floors must be sanded with a sanding machine on the job and then finished with a penetrating sealer, which leaves virtually no film on the surface, or with a heavy solids type finish, which provides a high luster and protective film. The penetrating sealer will usually require a coat of wax also. The sanding of the floors will require from three to five passes with the machine. On especially fine work, hand sanding may be required. The labor required will vary, depending on the size of the space and the number of sanding operations required. The surface finish may require two or three coats to complete the finishing process. Factory-finished wood flooring requires no finishing on the job, but care must be taken during and after installation to avoid damaging the finish.

In connection with the wood floor, various types of supporting systems may be used. Among the more common are treated wood sleepers, combination of ⅜-inch hot asphalt fill and treated sleepers, steel splines, and cork underlayment.

16-3 RESILIENT FLOORING

Resilient flooring tiles may be made of asphalt, vinyl-asbestos, vinyl, rubber, cork, and linoleum. Resilient sheets are available in vinyl and linoleum. The flooring may be placed over wood or concrete subfloors by use of the appropriate adhesive. The location of the subfloor, below grade, on grade, or suspended above grade, will affect the selection of resilient flooring since moisture will adversely affect some of them. All types may be used on suspended wood subfloors and concrete also as long as it is sufficiently cured. Where moisture is present below grade and on grade, all of the materials may be used *except* linoleum, cork, and rag felt-backed vinyl.

Tile sizes range from 9 × 9 inches to 12 × 12 inches, except for rubber tile, which is available up to 36 × 36 inches and vinyl accent strips, which come in various sizes. Thicknesses range from 0.050 to ⅛ inch except for rubber and cork tiles, which are available in greater thicknesses. Sheet sizes most commonly used are 6 and 12 feet wide with 4'-6" also available.

The subfloor may require an underlayment on it to provide a smooth, level, hard, clean surface for the placement of the tile or sheets. Over wood subfloors, panel underlayments of plywood, hardboard, or particleboard may be used, while the concrete subfloors generally receive an underlayment of mastic. The panel underlayment may be nailed or stapled to the subfloor. The mastic underlayments may be latex, asphalt, polyvinyl-acetate resins, or portland and gypsum cements.

Adhesives used for the installation of resilient flooring may be troweled on with a notched trowel or brushed on. Since there are so many types of adhesives, it is important that the proper adhesive be selected for each application. Check the project specifications and the manufacturer's recommendations to be certain they are compatible.

The wide range of colors and design variations are in part responsible for the wide use of resilient flooring. For the estimator, this means taking care to bid the color, design, thickness (gauge), size, and finish that is specified.

Accessories include wall base, stair tread, stair nosings, thresholds, feature strips, and reducing strips. The color and design variations are more limited in the accessories than in the flooring materials.

Specifications. From the specifications the color, design, gauge (thickness), size of tile or sheet, adhesives, subfloor preparation, and any particular pattern requirements may be found. Fancy patterns may be shown on the drawings and in most cases will have a higher percentage of waste.

Note which areas will require the various types of resilient flooring since it is seldom that one type, color, size, design, etc., will be used throughout the project. The wall base may vary in height from area to area, this can be determined from the specifications, room finish schedule and details.

Estimating. Resilient tile is estimated by the square foot: the actual square footage of the surface to be covered plus an allowance for waste. The allowance will depend on the area and shape of the room. When designs and patterns are made of tile or a combination of feature strips and tile, a sketch of the floor should be drawn up and an itemized breakdown of required materials made. The cost of laying tile will vary with the size and shape of the floor, size of the tile, type of subfloor and underlayment, and the design. Allowable waste percentages are shown in Fig. 16–2.

Feature strips must be taken off in lineal feet and, if they are to be used as part of the floor pattern, the square footage of feature strips must be subtracted from the floor area of tile or sheets required.

For floors that need sheet flooring, the estimator must do a rough layout of the floors involved to determine the widths required, where cutting of the roll will occur,

Area (sq. ft.)	Percent waste
Up to 75	10–12
75–150	7–10
150–300	6–7
300–1,000	4–6
1,000–5,000	3–4
5,000 and up	2–3

Figure 16-2 Approximate Waste for Resilient Tile

and how to keep waste to a minimum. Waste can amount to 30 and 40 percent if the flooring is not well laid out or if small amounts of different types are required.

Each area requiring different sizes, designs, patterns, types of adhesives, or anything else that may be different must be kept separately if the differences will affect the cost of material and amount of labor required for installation (including subfloor preparation).

Wall base (also referred to as *cove base*) is taken off by the lineal foot. It is available in vinyl and rubber, with heights of 2½, 4, and 6 inches and in lengths of 42 inches, 50, and 100 feet. Corners are preformed so the number of interior and exterior corners must be noted.

Adhesives are estimated by the number of gallons required to install the flooring. To determine the number of gallons, divide the total square footage of flooring by the coverage of the adhesive per gallon (in square feet). The coverage usually ranges from 150 to 200 square feet per gallon, but will vary depending on exact type used and the subfloor conditions.

EXAMPLE 16–1 — Commercial Building: Resilient tile is required on the floors in all the rooms. A 4-inch high base is also required throughout the building. To determine the square feet of floor area to be covered, refer to Sec. 9–2 (Appendix A):

Net s.f. of concrete slab	= 7,172 s.f.
Area of floor taken up by (Sec. 4–10)	
interior walls (137 ft.) (4¾ in.)	
= (137 ft.) (0.40 ft.) = 55 s.f.	
(25 ft.) (7½ in.)	
= (25 ft.) (0.62 ft.) = 16 s.f.	
	− 71 s.f.
Net s.f. of floor to be covered	7,101 s.f.
Waste	+ 110 s.f.
Resilient flooring required	7,211 s.f.

Resilient flooring shall be ⅛ inch thick, 12 × 12 inches, vinyl-asbestos, marbleized design, light colors.

Adhesives required to cover 7,100 square feet, with an average coverage of 150 square feet per gallon.

$$\frac{7,100}{150} = 47.3 \text{ gallons} = 50 \text{ gallons required}$$

Vinyl base, ⅛ by 4 inches, is required throughout the project. To determine the lineal feet of base required, the footage of wall must be calculated.

From Sec. 14–6, l.f. of exterior wall = 397 l.f.
From Sec. 14–10, l.f. of interior wall = 162 (2 sides) = 324 l.f.
= 721 l.f.

Less deductions at doors:

6 @ 3′-0″ = 18 feet
1 @ 10′-0″ = 10 feet
− 28 l.f.
693 net l.f.
Waste = + 37 l.f.
730 l.f. base required

Adhesive required to apply base, about 250 square feet; order a 5-gallon pail.

EXAMPLE 16–2 — Residential Building:
Flooring: Resilient tile, ⅛″ thick, 9 × 9 in. vinyl

Bath 1: 8′-0″ × 5′-0″ = 40 s.f.
Bath 2: 5′-2″ × 4′-6″ = 24 s.f.
Kitchen: 3′-4″ × 8′-0″ = 27 s.f.
5′-6″ × 6′-0″ = 33 s.f.
Total 124 s.f.

Baseboard: Vinyl, ⅛ by 4 in.

Bath 1: 13′-8″ l.f.
Bath 2: 17′-0″ l.f.
Kitchen: None
Total 30′-8″ l.f.

Adhesive: coverage is 150 s.f. per gallon, 250 l.f. per gallon. Order 2 gallons of adhesive.

Labor. Subcontractors who specialize in resilient floors installation generally do this type of work, and they may price it on a unit basis (per square foot) or on a lump sum. The time required for resilient floor installation is shown in Fig. 16–3.

Tile	Work hours per 100 s.f.
Resilient squares	
9″ × 9″	1.5 to 2.5
12″ × 12″	1.0 to 2.2
Seamless sheets	0.8 to 2.4
Add for felt underlay	add 10%
Less than 500 s.f.	add 15%

Figure 16-3 Work Hours Required for Resilient Tile Installation

16-4 CARPETING

Carpeting is selected and specified by the type of construction and the type of pile fibers. The types of construction (how they are made) are tufted, woven, and knitted; punched and flocked have become available more recently. In comparing carpeting of similar construction factors such as pile yard weight, pile thickness, and the number of tufts per square inch are evaluated in different carpets. Carpeting is generally available in widths of 9, 12, 15, and 18 feet, but not all types come in all widths.

Pile fibers used include wool, nylon, acrylics, modacrylics, and polypropylene for long-term use. Acetate, rayon, and polyester fibers are also used. The type of pile used will depend on the type of use intended and the service required.

The installed performance of the carpet is not dependent on any single factor, but on all of the variables involved in the construction of the carpet and the pile characteristics.

The cushion, over which the carpeting is installed, may be manufactured of animal hair, rubberized fibers, or cellular rubber. The cushion increases the resilience and durability of the installation. The cushion may be bonded to the underside of the carpet, but it is more common to have separate cushions. The type of cushion used will depend on the intended usage of the space, and a variety of designs is available for each type of material. The various cushions within each group are rated by weight in ounces per square yard. The heavier the cushion, the better and more expensive it will be. Cushioning is generally available in widths of 27, 36, and 54 inches; 6, 9, and 12 feet.

Specifications. The specifications should state the type of carpeting required, the pile yarn weight, thickness, number of tufts per inch, construction, backing, rows, and other factors relating to the manufacture of the carpet. Many specifications will state a particular product "or equal," which means the estimator will either use the product mentioned or ask other suppliers (or manufacturers) to price a carpet which is equal in quality. In the latter case, the estimator should compare the construction specifications of the carpeting to be certain that the one chosen is "equal." Different types of carpeting may be used throughout the project. Take note of what types are used and where they are used.

The cushion type required, its material, design, and weight, must be noted. If variations in the type of cushion required throughout the project are evident, they should be noted.

Estimating.   Carpeting is estimated by the square yard with special attention given to the layout of the space for the most economical use of the materials. Waste and excess material may be large unless there is sufficient planning. Each space requiring different types of carpeting, cushion, or color must be figured separately. If the specifications call for the color to be selected by the architect-engineer at a later date, it may be necessary to call and try to determine how many different colors may be required. In this manner, a more accurate estimate of waste may be made.

Certain types of carpeting can be bought by the roll only, and it may be necessary to purchase an entire roll for a small space. In this case waste may be high, since the cost of the entire roll must be charged to the project.

The cushion required is also taken off in square yards, with the type of material, design, and weight noted. Since cushions are available in a wider range of widths, it may be possible to reduce the amount of waste and excess material.

EXAMPLE 16-3 — Residential Building:

	LENGTH OF 12′-WIDE CARPET REQUIRED
Bedroom 1	= 8′
Bedroom 2	= 8′-2″
Bedroom 3	= 11′-5″
Bedroom 4	= 10′-0″
Living-Dining room	= 21′-6″
	+ 8′-6″
Hall (to D.R.) 3′ × 30′ = 90 s.f.	+ 9′-0″
Closets	= 6′-0″
	82′-7″ of 12′-wide carpeting = 990.96 s.f.

Padding:

$$82.58′ \times 12′ = 990.96 \text{ s.f.}$$

Labor.   Subcontractors who specialize in carpet installations do this type of work, and they may price it on a unit basis (per square yard) or lump sum. The time required for the installation is shown in Fig. 16-4.

Carpet	Squares per work hour
Carpet and pad, wall to wall	8 to 20
Carpet, pad backing, wall to wall	10 to 22
Deduct for gluing to concrete slab	10%
Less than 10 squares	add 15-20%

Figure 16-4   Work Hours Required for Carpet Installation

finishes

16-5 TILE

Tile may be used on floors and walls. The tile used for floors is usually ceramic or quarry tile, while the tile used for walls and wainscots may be ceramic, plastic, or metal.

Ceramic tile is available in exterior or interior grades, glazed or unglazed. Individual tile size may range from $\frac{3}{8} \times \frac{3}{8}$ inch to $8\frac{1}{2} \times 4\frac{1}{4}$ inches. Tile may come in individual pieces or sheets of from $\frac{1}{2}$ to $2\frac{1}{4}$ square feet per sheet. Tile mounted in sheets will be much less expensive to install than unmounted tile. Ceramic tiles come in various shapes and a wide range of sizes and colors. The tile may be installed by use of portland cement mortar, dry-set mortar, organic adhesives, and epoxy mortars. The portland cement mortar is used where leveling or slopes are required in the subfloor; the thickness of this mortar ranges from $\frac{3}{4}$ to $1\frac{1}{4}$ inches, and it requires damp curing. The mortar will receive a coat of neat grout cement coating and the tile will be installed over the neat cement. The other methods are primarily thin-set ($\frac{1}{16}$ to $\frac{1}{4}$ inches) the one-coat operations. After the tile has been installed the joints must be grouted. The grouts may be portland cement-based, epoxies, resins, and latex.

Specifications. The type of tile (material) should be determined for each space for which it is specified. The type of tile and the finish often vary considerably throughout the job. The specifications usually give a group that will be selected from and a price range (e.g., American Olean, price range A). The groupings and price ranges vary among manufacturers, so care must be used in the use of specifications written in this manner. Other specifications will spell out precisely what is required in each area; this makes it easier to make an accurate estimate. Each area requiring different types, sizes or shapes of tile must be taken off separately.

The methods of installation must be noted and, if the methods vary throughout the job, they must be kept separately; the type of grout required in each area must also be noted.

The types of trimmers are also included in the specifications. The amount of trimmers required is kept separate from the rest of the tile takeoff since it is more expensive. Note exactly what is required, since some trimmer shapes are much more expensive than others. Bid what is specified, and if the specifications are not clear, the architect-engineer should be called.

Estimating. Floor and wall tiles are estimated by the square foot. Each area must be kept separate, according to the size and type being used. It is common to have one type of tile on the floor and a different type on the walls. The different colors also vary in cost, even if the size of the tile is the same, so caution is advised. The trim pieces should be taken off by the lineal feet of each type required. There is a large variety of sizes and shapes at varying costs, so check the specifications carefully and bid what is required. If portland cement mortar is used as a base, it is installed by the tile contractor. This requires the purchase of cement, sand, and sometimes wire mesh. Tile available in sheets is much more quickly installed than individual tiles. Adhesives are sold by the gallon or sack and approximate coverage is obtained from the manufacturer. The amount of grout used depends on the size of the tile.

When estimating wall tile, note the size of the room, number of internal and external corners, height of wainscot, and types of trim. Small rooms require more labor than large ones. A tile setter can set more tile in a large room than in several smaller rooms in a given time period.

Accessories are also available and, if specified, should be included in the estimate. The type and style are in the specifications. They may include soap holders, tumbler holders, toothbrush holders, grab rails, paper holders, towel bars and posts, door stops, hooks, shelf supports, or combinations of these. These accessories are sold individually, so the number required of each type must be taken off. The accessories may be recessed, flush, or flanged, and this must also be noted.

16-6 PAINTING

The variables that affect the cost of painting include the material painted, the shape and location of the surface painted, the type of paint used, and the number of coats required. Each of the variables must be considered, and the takeoff must list the different conditions separately.

Although painting is one of the items commonly subcontracted out, the estimator should still take off the quantities so the subcontractor's proposal can be checked. In taking off the quantities, the square feet of surface is taken off the drawings, and all surfaces that have different variables must be listed separately. With this information the amount of materials can be determined by use of the manufacturer's information on coverage per gallon.

The following methods for taking off the painting areas are suggested, with interior and exterior work listed separately.

Interior.

Walls—actual area in square feet
Ceiling—actual area in square feet
Floor—actual area in square feet
Trim—lineal footage (note width); amount of door and window trim
Stairs—square footage
Windows—size and number of each type, square feet
Doors—size and number of each type, square feet
Baseboard radiation covers—lineal feet (note height)
Columns, beams—square feet

Exterior.

Siding—actual area in square feet
Trim—lineal footage (note width)
Doors—square feet of each type
Windows—square feet of each type
Masonry—square footage (deduct openings over 50 square feet)

The specifications should list the type of coating, number and type of coats, and finish required on the various surfaces throughout the project. Interiors receive different treatment than exteriors; different material surfaces require different applications and coatings—all of this should be in the specifications. Paints may be applied by brush, roller, or spray gun, and the method to be used is also included in the specifications.

Sometimes the specifications call for prefinished and factory-finished materials to be job-finished also. Except for possible touch-ups, this is usually due to an oversight in the architect-engineer's office and the estimator should seek clarification from them. The most common items to be factory-finished are doors, floorings, windows, baseboard, radiation covers, and grilles. The estimator should keep a sharp eye out to see that each item of work is figured only once.

Structural-steel work often requires painting also. It often comes to the job primed, with only touch-up of the prime coat required. Sometimes, it is delivered unprimed with the priming done on the job. Touch-up painting is impossible to figure accurately and depends on the type of structural system being used, but an average of 5 to 10 percent of the area is usually calculated as the touch-up required.

The structural steel work is taken off by the tonnage of steel required with the types and sizes of the various members required. It must be noted which type of steel is to be painted: steel joists, rectangular or round tubes, H sections, or any other type. The shape of the member will influence the cost considerably and the square footage to be painted per ton of steel may vary from 150 for large members to 500 square feet for trusses and other light framing methods.

EXAMPLE 16–4 — Commercial Building: The interior walls in our example require two coats of latex-base paint, and the ceiling, inside the building, requires one coat of masonry paint. Determine the square feet of area that must be covered for each item (Appendix A).
Walls:

$$\begin{array}{ll}
\text{from Sec. 14-6} & \text{2,862 s.f. (exterior)} \\
\text{from Sec. 14-10} & \underline{\text{2,424 s.f. (interior)}} \\
\text{Net wall area} & \text{5,296 s.f. requiring two coats of paint}
\end{array}$$

Ceiling: Use the same figures for the square footage as for the concrete slab in Sec. 9-2. A ceiling area of 7,172 square feet is to be painted with one coat of masonry paint.

The metal door frames require two coats of semi-gloss enamel paint. Determine the lineal feet of frame to be painted.
Trim:

$$\begin{array}{lll}
\text{6 doors @ 17 l.f.} & = & \text{102 l.f.} \\
\text{1 door \ @ 24 l.f.} & = & \underline{\text{24 l.f.}} \\
& & \text{126 l.f. of door frame to be painted}
\end{array}$$

All exposed surfaces of metal door frames shall receive one coat of semi-gloss enamel paint.

EXAMPLE 16–5 — Residential Building:
Interior walls and ceiling: Primer plus one coat
Exterior wood: Stain one coat
Trim (doors and windows): Primer plus one coat
Interior walls:

$$1,184 + 2,384 + 1,200 = 4,768 \text{ s.f.}$$

Exterior plywood:

$$148 \times 8 = 1{,}184 + \text{gables} = 1{,}184 + 72 = 1{,}256 \text{ s.f.}$$

Trim:

$$14 \text{ doors—average } 17 \text{ l.f.} = 238 \text{ l.f.}$$
$$4 \text{ windows—32/36} = 46 \text{ l.f.}$$
$$2 \text{ windows—36/44} = 27 \text{ l.f.}$$
$$\text{Baseboard (interior)} = 148 + 149 + 149 = 446 - 30 \text{ res.} = 416 \text{ l.f.}$$
$$\text{Fascia} = 160 \text{ l.f.}$$
$$\text{Soffit} = 156 \text{ s.f.}$$

Labor. Subcontractors who specialize in painting and staining often do this type of work and they almost always price it on a lump-sum basis. The time required for painting and staining is shown in Fig. 16–5.

Painting, brushes	s.f. or l.f. per hour
Interior	
primer and 1 coat	200–260 s.f.
primer and 2 coats	150–200 s.f.
stain, 2 coats	150–200 s.f.
Trim	
primer and 1 coat	120–180 l.f.
primer and 2 coats	100–160 l.f.
stain, 2 coats	100–160 l.f.
Exterior	
primer and 1 coat	160–220 s.f.
primer and 2 coats	120–180 s.f.
stain, 2 coats	200–280 s.f.

Figure 16-5 Work Hours Required for Painting and Staining

16–7 CHECKLIST

Floors:

    type of material

    type of fastener

    spikes

    nails

    adhesives

    screws

    finish

    thickness

    size, shape

    accent strips

    pattern

    cushion

base

corners

Painting:

filler

primer

paint, type

number of coats

shellac

varnish

stain

check specifications for all areas requiring paint

REVIEW QUESTIONS

1. When block flooring is estimated, what unit of measure is used and what information should be noted on the workup sheets?

2. When resilient flooring is estimated, what unit of measure is used and what information should be noted on the workup sheets?

3. Determine the amount of flooring required for the building in Appendix E. Use workup sheets.

4. What is the unit of measure used in estimating carpeting? What procedure should be followed to minimize waste?

5. What unit of measure is used in estimating ceramic tile and what type of information should be noted on the workup sheets?

6. Assume the building in Appendix E requires ceramic tile floors and a ceramic tile wainscot 4' high in rooms 103 and 104. Determine the s.f. of wall and floor tile required.

7. Determine the interior areas of the building in Appendix E to be painted. Use workup sheets.

8. Determine the exterior areas of the building to receive a finish. Use workup sheets.

9. What type of finish is required on the interior doors on the building? (Appendix E)

10. Assume the interior exposed concrete floors require 2 coats of concrete sealer (Appendix E). What area must be covered and, based on an average coverage of 75 s.f. per gallon per coat, how many gallons of sealer are required?

11. Determine the flooring materials required for the residence in Appendix H.

12. Determine the exterior and interior areas of the residence in Appendix H to be painted.

17
ROOFING

17-1 ROOFING

Roofing is considered to include all the material that actually covers the roof deck. It includes any felt (or papers) as well as bituminous materials that may be placed over the deck, it is most commonly installed by the roofer and included in that portion of the work. Flashing required on the roof, at wall intersections, joints around protrusions through the roof, and the like is also included in the roofing takeoff. protrusions through the roof, and the like is also included in the roofing take-off. Miscellaneous items such as fiber blocking, cant strips, curbs, and expansion joints should also be included. The sections and details, as well as the specifications, must be carefully studied to determine what is required and how it must be installed.

17-2 ROOF AREAS

In estimating the square footage of area to be roofed you must consider the shape of the roof (Fig. 17-1). When measuring a flat roof with no overhang or with parapet walls, take the measurements of the building from the outside of the walls. This method is used for parapet walls since it allows for the turning up of the roofing on the sides of the walls. If the roof projects beyond the building walls, the dimensions used must be the overall outside dimensions of the roof, and must include the overhang. The drawings should be carefully checked to determine the roof line around the building, as well as where and how much overhang there may be. The floor plans, wall sections, and details should be checked to determine the amount of overhang and type of finishing required at the overhang. Openings of less than 30 square feet should not be deducted from the area being roofed.

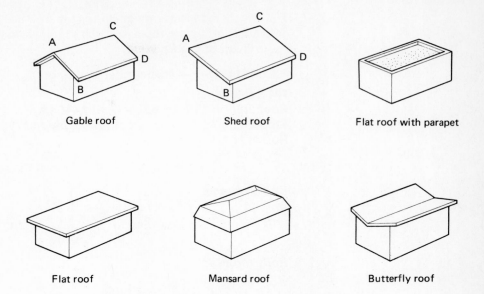

Figure 17-1 Typical Roof Shapes

To determine the area of a gable roof, multiply the length of the ridge (A to C) by the length of the rafter (A to B) by two (for the total roof surface).

The area of a shed roof is the length of the ridge (A to C) times the length of rafter (A to B). The area of a regular hip roof is equal to the area of a gable roof that has the same span, pitch, and length. The area of a hip roof may be estimated just as the area of a gable roof is: the length of roof times the rafter length times two.

The length of rafter is easily determined from the span of the roof, the overhang, and slope. Refer to Fig. 13–56 for information required to determine the lengths of rafters for varying pitches and slopes.

17-3 SHINGLES

Asphalt shingles.   Available in a variety of colors, styles, and exposures, strip asphalt shingles are 12 and 15 inches wide, and 36 inches long. They are packed in bundles that contain enough shingles to cover 25.33 or 50 square feet of roof area. The *exposure* (amount of shingle exposed to the weather) generally is 4, 4.5, or 5 inches. Individual shingles 12 to 16 inches long are sometimes also used.

Asphalt shingles may be specified by the weight per square, which may vary from 180 to 350 pounds. Shingles may be fire-rated and wind-resistant, depending on the type specified.

In determining the area to be covered, always allow one extra course of shingles at the eaves since the first course must always be doubled. Hips and ridges are taken off by the lineal foot and considered 1′-0″ wide to determine the square footage of shingles required. Waste averages 5 to 8 percent. Galvanized, large-headed nails, ⅞ to 1¾ inches long, are used on asphalt shingles. From 1½ to 3 pounds of nails are required per square.

Asphalt shingles are generally placed over an underlayment of building paper or roofing felt. The felt is specified by the type of material and weight per square.

The felt should have a minimum top lap of 2 inches and end lap of 4 inches, so to determine the square footage of felt required, multiply the roof area by a lap and waste factor of 5 to 8 percent.

EXAMPLE 17-1 — Residential Building:

15 lb. felt paper
Roof area = 13.9′ × 52′ × 2 = 1,446 s.f.
Slope: 3 in. 12, 2 layers felt required, lapped 19″ and exposed 17″
Roll of felt:

$$36″ \text{ wide} \times 144′ \text{ long} = 432 \text{ s.f.}$$

Roll coverage:

$$\text{s.f. in roll} \times \frac{\text{Exposure}}{\text{Width}} = 432 \times \frac{17}{36} = 204 \text{ s.f.}$$

Rolls required:

$$\frac{1,446}{204} = 7.08 \text{ (order 8 rolls)}$$

Shingles: 235 lbs., 1,446 s.f. = 14.46 squares, plus double starter row

$$52 \text{ l.f. of eave} \times 2 \text{ eaves} = 104 \text{ l.f.}$$

Shingles: 12″ × 36″ (3′-0″)

104 l.f. ÷ 3 l.f. = 35 shingles extra for starter row (.35 squares)

```
  14.46
+   .44
  14.90 squares
```

Order 15.25 squares (this allows for waste).
Ridge: Use ridge shingles, with 5″ exposure

$$52 \text{ l.f. of ridge} = 624 \text{ in.} = 624″ \div 5 = 125 \text{ pieces}$$

Labor. Subcontractors who specialize in roofing will do this work and they may price it on a unit basis (per square) or a lump sum. The time required to install shingles is shown in Fig. 17-2.

Asbestos shingles. Also referred to as *mineral fiber roof shingles,* asbestos shingles are available in hexagonal and rectangular shapes as well as in strip shingles of 14 by 30 inches. Bundle sizes range from 20 to 33⅓ square feet each, with the weight per square between 250 and 500 pounds. Roof slopes as low as 3 in 12 may be used, provided the proper underlayment is employed.

Asbestos shingles are estimated in the same manner as asphalt shingles. Measurements are taken in the same sequence and with the same general approach. A starter course will be required (measure the length of the eaves), and the hips and eaves must also be measured as extra material will be required at these points.

Asbestos shingles may be laid in a variety of patterns and shingle layouts; the

Roofing materials	Work hours per square
Shingles,	
asphalt, asbestos, (strip)	0.8 to 4.0
asphalt, asbestos, (single)	1.5 to 6.0
wood (single)	3.0 to 5.0
metal (single)	3.5 to 6.0
heavyweight asphalt and asbestos	add 50%
Tile	
clay	3.0 to 5.0
metal	3.5 to 6.0
Built-up	
2 ply	0.8 to 1.4
3 ply	1.0 to 1.6
4 ply	1.2 to 2.0
aggregate surface	0.3 to 0.5

Figure 17-2 Work Hours Required for Shingle Installation

type of design and pattern selected will greatly affect the cost of labor for the project. Underlayment and nails also must be included in the cost: allow 2 to 4 pounds of nails per square. Shingle waste ranges from 8 to 15 percent and underlayment waste, from 5 to 8 percent.

Wood shingles. Available in various woods, the best of which are cypress, cedar, and redwood, wood shingles come hand-split (rough texture) or sawed, the hand-split variety commonly referred to as *shakes*. Lengths are 16, 18, and 24 to as long as 32 inches, while random widths of 4 to 12 inches are common. Wood shingles are usually sold by the square based on a 10-inch weather exposure. A double-starter course is usually required and in some installations roofing felt is also needed. Valleys will require some type of flashing, while hips and ridges require extra material to cover the joint, and thus extra long nails become requisite. Nails should be corrosion-resistant; they are 1¾ to 2 inches long and from 2 to 4 pounds per square.

17-4 BUILT-UP ROOFING

Built-up roofing consists of layers of overlapping roof felt with each layer set in a mopping of hot tar or asphalt. Such roofing is usually designated by the number of plies of felt that are used, for example, a three-ply roof has four coats of bituminous material (tar or asphalt) and three layers of felt. While built-up roofing is used primarily for flat and near flat roofs it can be used on inclines of as great as 9 inches per foot, providing certain special bituminous products are used.

The specifications must be carefully read since there is no one type of system applicable for all situations. There may be vapor barrier requirements, varying amounts of lapping of felt, and different weights and types of felts and bituminous materials may be used. The deck that is to receive the roofing must be considered, as must the service it is intended to sustain.

Many specifications require a bond on the roof. This bond is furnished by the material manufacturer and supplier through the roofing contractor. It generally guarantees the watertightness of the installation for a period of years, usually 10, 15, or 20. During this period of time the manufacturer will make repairs on the roof that become necessary due to the normal wear and tear of the elements to maintain the roof in a watertight condition. Each manufacturer has its own specifications or limitations in regard to these bonds, and the roofing manufacturer's representative should be called in for consultation. Special requirements may pertain to approved roofers, approved flashing, and deck inspection; photographs may also be necessary. If it appears that the installation may not be approved for bonding, the architect-engineer should be so informed during the bid period. The manufacturer's representative should call on the architect-engineer and explain what the situation is so it can be worked out before the proposals are due.

The takeoff follows the same general procedure as any other roofing:

1. The number of squares to be covered must be noted.

2. Base sheets—a base sheet or vapor barrier, when required, note the number of plies, lap (usually 4 inches), weight per ply per square.

3. Felt—determine the number of plies, weight of felt per square, and type of bituminous impregnation (tar or asphalt).

4. Bituminous material—note the type, number of coats, and the pounds per coat per square (some specifications call for an extra heavy top coat or pour).

5. Aggregate surface—(if required) the type and size of slag, gravel, or other aggregate, and the pounds required per square should all be noted.

6. Insulation—(if required) note type, thickness and any special requirements (refer also to Sec. 17-12).

7. Flashing—note type of material, thickness, width, and lineal feet required.

8. Trim—determine the type of material, size, shape and color, method of attachment, and the lineal footage required.

9. Miscellaneous materials—blocking, cant strips, curbs, roofing cements, nails and fasteners, caulking, and taping are some examples.

Bituminous materials are used to cement the layers of felt into a continuous skin over the entire roof deck. Types of bituminous material used are coal tar pitch and asphalts. The specifications should indicate the amount of bituminous material to be used for each mopping so that each ply of felt is fully cemented to the next, and in no instance should felt be allowed to touch felt. The mopping between felts averages 25 to 30 pounds, while the top pour (poured, not mopped), which is often used, may be from 65 to 75 pounds per square. It is this last pour into which any required aggregate material will become embedded.

Felts are available in 15- and 30-pound weights, 36-inch widths, and 432- or 216-square foot rolls. With a 2-inch lap a 432-square foot roll will cover 400 square feet, while the 216-square foot roll will cover 200 square feet. In built-up roofing a starter strip 12 to 18 inches wide is applied; then over that one a strip 36 inches wide is placed. The felts that are subsequently laid overlap the preceding felts by 19, 24⅔, or 33 inches, depending on the number of plies required. Special applications sometimes require other layouts of felts. The specifications should be carefully checked to determine exactly what is required as to weight, starter courses, lapping, and plies. Waste averages about 8 to 10 percent of the required felt and this allows for the material used for starter courses.

Aggregate surfaces such as slag or gravel are often embedded in the extra-heavy top pouring on a built-up roof. This aggregate acts to protect the membrane against the elements, such as hail, sleet, snow, and driving rain. It also provides weight against wind uplift. To insure embedment in the bituminous material, the aggregate should be applied while the bituminous material is hot. The amount of aggregate used varies from 250 to 500 pounds per square. The amount required should be listed in the specifications. The amount of aggregate required is estimated by the ton.

Since various types of materials may be used as aggregates, the specifications must be checked for the size, type, and gradation requirements. It is not unusual for aggregates such as marble chips to be specified, and this type of aggregate will result in much higher material costs than gravel or slag. Always read the specifications thoroughly.

The joints between certain types of roof deck materials, such as precast concrete, gypsum, and wood fiber, may be required to have caulking and taping of all joints. This application of flashing cement and a 6-inch-wide felt strip will minimize the deleterious effects on the roofing membrane of uneven joints, movement of the units, and moisture transmission through the joint; it also reduces the possibility of bitumen seepage into the building. The lineal footage of joints to be covered must be determined for pricing purposes.

EXAMPLE 17–2 — Commercial Building: The roofing required is a three-ply, built-up roof with a 20-year bond. Rigid insulation, 1 inch thick, cant strips, and aluminum gravel stop are also required. Determine the quantities of each material.

Roofing: Same net area as the precast slabs in Sec. 12–2.

$$7{,}890 \text{ s.f.} = 78.9 \text{ squares} = 79 \text{ squares of three-ply roofing}$$
$$\text{with a 20-year bond}$$

Insulation: Same net area as roofing

$$
\begin{array}{r}
79 \text{ squares} \\
\text{Waste} + \underline{ 1 \text{ square}} \\
80 \text{ squares (8000 s.f.) of 1-inch thick rigid insulation}
\end{array}
$$

Cant strip: Required around the perimeter of the roof decking, refer to App. A for dimensions.

$$
\begin{array}{r}
127 + 67 + 127 + 67 + 10 + 10 = 408 \text{ l.f.} \\
\text{Waste} \quad + \underline{ 12 \text{ l.f.}} \\
420 \text{ l.f. of } 1\tfrac{1}{2}'' \times 12 \text{ inch cant strip}
\end{array}
$$

Gravel stop: Required around the perimeter of the roof, 9 inches high, 0.032 inch thick, aluminum; (from above) about 420 lineal feet are required.

Labor. Subcontractors who specialize in roofing will submit unit (per square) and lump-sum bids. The time required to install built-up roofing is shown in Fig. 17–2.

17-5 CORRUGATED SHEETS (INCLUDING RIBBED, V-BEAM)

Corrugated sheets may be made of aluminum, galvanized steel, and various combinations of materials such as zinc-coated steel. They are available in various sizes, thicknesses, and corrugation shapes and sizes, in a wide range of finishes. Common steel gauges of from 12 to 29 are available with lengths of 5 to 12 feet and widths of 22 to 36 inches. Aluminum thicknesses of 0.024, 0.032, 0.040, and 0.050 inch are most commonly found, with sheet sizes in widths of 35 and 48⅓ inches and lengths of 3 through 30 feet.

Estimate the corrugated sheets by the square; note the type of fastening required. Corrugated sheets require that corrugated closures seal them at the ends. The enclosures may be metal or rubber and are estimated by the lineal foot.

The amount of end and side lap must also be considered when estimating the quantities involved. There is such a variety of allowances that it is not possible to include them here. Given the information, the supplier or manufacturer's representative will supply the required information. Fasteners should be noncorrosive and may consist of self-taping screws, weldable studs, cleats, and clips. The specifications and manufacturer's recommendations should be checked to determine spacing and any other special requirements; only then can the number of fasteners be estimated. Flashing and required trim should be estimated by the lineal foot.

17-6 METAL ROOFING

Materials used in metal roofing include steel (painted, tinned, and galvanized), copper, and terne. Basically, in this portion we are considering the flat and sheet roofing. The basic approach to estimating this type is the same as that used for other types of roofing: first the number of squares to be covered must be determined; then consideration must be given to available sheet sizes (often 14 × 20 inches and 20 × 28 inches for tin) and quantity in a box. Items that require special attention are the fastening methods, and the type of ridge and seam treatment.

Copper roofing comes in various sizes; the weights range from 14 to 20 ounces per square foot, and the copper selected is designated by its weight (16 ounce copper). The copper may have various types of joining methods; among the most popular are standing seam and flat seam, but many other types may be used. Study the drawings and specifications carefully, since the type of seam will affect the coverage of a copper sheet. When you use a standing seam, 2¾ inches in width is lost from the sheet size (with regard to actual coverage) to make the seam. This means then when standing seams are used, 12 percent must be added to the area being covered to allow for the forming of the seam; if end seams are required, an allowance must also be added for them. Each different roofing condition must be planned so that there will be a proper allowance for seams, laps, and waste.

17-7 SLATE

The slate is available in widths of 6 to 16 inches and lengths of 12 to 26 inches. Not all sizes are readily available and it may be necessary to check the manufacturer's current inventory. The basic colors are blue-black, gray, and green. Slate may be purchased smooth or rough textured. Slate shingles are usually priced by the amount required to cover a square if the manufacturer's recommended exposure is used. If the exposure is different from the manufacturer's, the formula shown in Sec. 17–9 may be used. To the squares required for the roof area, add one square foot per foot of length for hips and rafters. Slate thicknesses and corresponding weights are shown in Fig. 17–3, assuming a 3-inch lap. Each shingle is fastened with two large-head, solid copper nails, 1¼ or 1½ inches long. The felt required under the slate shingles must also be included in the estimate. Waste for slate shingles varies from 8 to 20 percent, depending on the shape, number of irregularities, and number of intersections on the roof.

Thickness of slate (inches)	$\frac{3}{16}$	$\frac{1}{4}$	$\frac{3}{8}$	$\frac{1}{2}$	$\frac{3}{4}$	1
Weight per square (3″ lap)	700–800	900–1000	1300–1400	1700–1800	2500–2800	3400–3600

Figure 17-3 Approximate Weight of Slate per Square

17–8 TILE

The materials used for roofing tiles are cement, metals, and clay. To estimate the quantities required, the roof areas are obtained in the manner described earlier. The number, or lineal footage, of all special pieces and shapes required for ridges, hips, hip starters, and terminals, and any other special pieces must be carefully taken off. If felt is used under the tile, its cost must also be included in the estimate of the job, and furring strips must be installed for certain applications.

As unfamiliar installations arise, call the manufacturer, the manufacturer's representative, or a local dealer. They will review the project with you, help you in arriving at material takeoffs, and perhaps even suggest a local subcontractor with experience in installing the tile.

17–9 SHEETS, TILE, AND SHINGLES FORMULA

To determine the number of sheets, tiles, or shingles required to cover a square (100 square feet) of roof area for any required lap or seam, the following formula may be used.

$$N = \frac{\text{(one square)}}{(W - S)} \frac{\text{(square inches)}}{(L - E)} = \frac{100\,(144)}{(W - S)(L - E)}$$

N = Number of sheets (tiles or shingles)
W = Width of sheet (inches)
L = Length of sheet (inches)
S = Side lap or seam lap requirement (for some types there may be none)
E = End lap or seam requirement

To determine the square feet of any given roofing sheet, tile, or shingles required to cover one square, the following formula may be used.

$$A = \frac{100\ WL}{(W - S)\ (L - E)}$$

A = square feet of material required per square

If the weight per square must be determined, multiply A times the weight of the material per square foot.

17-10 LIQUID ROOFING

Liquid roofing materials were developed primarily for free-form roofs. Composition varies among the manufacturers, but the application sequences are similar. First, primer is applied liberally over the entire surface; then all major imperfections are caulked, joints taped, and flashing applied at all intersecting surfaces. Next, three coats of liquid roofing are applied; depending on the deck and slope, they may either both be base and finish coats or only the finish coat. Finish coats are available in a variety of colors. Equipment is simple consisting of only rollers, hand tools, brushes, and joint-tape dispenser. Labor will be affected primarily by the shape of the roof, ease of moving about, etc. On some buildings it is necessary to erect scaffolding in order to apply the roofing; in such cases the cost obviously will go up accordingly. Before bidding this type of roofing, discuss the project with the manufacturer's representative so that the latest technical advice may be incorporated in your bid.

17-11 FLASHING

Flashing is used to help keep water from getting under the roof covering and from entering the building wherever the roof surface meets a vertical wall. It usually consists of strips of metal or fabric shaped and bent to a particular form. Depending on the type of flashing required, it is estimated by the piece, lineal footage with the width noted, or square feet. Materials commonly used include copper, asphalt, tinned, painted, plastic, rubber, composition, and combinations of these materials. Bid the gauge of thickness and width specified. Expansion joints are estimated by the lineal foot and it may be necessary that curbs be built-up and the joint cover either prefabricated or job-assembled. Particular attention to the details on the drawings is required so that the installation is understood.

17-12 INSULATION

Insulation included as a part of roofing is of the type that is installed on top of the deck material. This type of insulation is rigid and is in a sheet (or panel). Rigid insulation may be made of urethane, fiberboards, or perlite and is generally available in lengths of 3 to 12 feet in 1-foot increments and widths of 12, 16, 24, and 48 inches. Thicknesses of ½, ¾, 1, 1½, 2, 2½, 3, and 4 inches are available. Insulation is estimated by the square with a waste allowance of 5 percent, provided there is proper planning and utilization of the various sizes.

When including the insulation, keep in mind that its installation will require extra materials, either in the line of additional sheathing paper, moisture barriers, moppings, or a combination of these. Also, the specifications often require two layers of insulation, usually with staggered joints. This requires twice the square footage of insulation (to make up two layers), an extra mopping of bituminous material, and extra labor. Read the specifications carefully—never bid something you do not fully understand.

NOTE: Some roofing manufacturers will not bond the performance of the roof unless the insulation meets their specifications. This item should be checked and if the manufacturer, or his representative, sees any problem, the architect-engineer should be notified so that the problem may be cleared up during the bid period.

Refer to Sec. 17–4 for an example of roof insulation takeoff.

17-13 ROOFING TRIM

Trim such as gravel stops, fascia, coping, ridge strips, gutters, downspouts, and soffits, is taken off by the lineal foot. All special pieces used in conjunction with the trim (e.g., elbows, shoes, ridge ends, cutoffs, corners and brackets), are estimated by the number of pieces required.

The usual wide variety of materials and finishes is available in trim. Not all trim is standard stock; much of it must be specially formed and fabricated. This will add considerably to the cost of the materials and to the delivery time—so check this out and never assume that the trim needed is standard stock. (See the example at the end of Sec. 17–4.)

17-14 LABOR

The labor cost of roofing will depend on the hourly output and hourly wages of the workers. The output will be governed by the incline of the roof, size, irregularities in the plan, openings (skylights, etc), and the elevation of the roof above the ground, since the higher the roof, the higher the materials have to be hoisted to the work area.

Costs can be controlled by the use of work gangs that are familiar with the type of work and experienced in working together. For this reason much of the roofing done on projects is handled by a specialty contractor with equipment and trained personnel.

17-15 EQUIPMENT

Equipment requirements vary considerably, depending on the type of roofing used and the particular job being estimated. Most jobs will require hand tools for the workers, ladders, some scaffolding, and some type of hoist—regardless of the type of roofing being applied. Specialty equipment for each particular installation may also be required. Built-up roofing may demand that mops, buggies, and heaters (to heat the bituminous materials) be employed; some firms have either rotary or stationary felt layers. Metal roofing requires shears, bending tools, and soldering outfits.

Equipment costs are estimated either by the square or by the job, with the cost including such items as transportation, set-up, depreciation, and replacement of miscellaneous items.

17-16 CHECKLIST

Paper

Felt

Composition (roll)

Composition (built-up)

Tile (clay, metal, concrete)

Shingles (wood, asphalt, asbestos, slate)

Metal (copper, aluminum, corrugated, steel)

Insulation

Base

Solder

Flashing

Ridges

Valleys

Fasteners

Trim

Battens

Blocking (curbs)

Cant strips

REVIEW QUESTIONS

1. What is the unit of measure used for shingle roofing?
2. Determine the amount of insulation required on the building (Appendix E). Use workup sheets.
3. Determine the amount of cant strips and other accessories required on the building.
4. Determine the amount of area to be covered with built-up roofing (Appendix E).
5. Determine the amount of slag required for the building (Appendix E).
6. What unit of measure is used for metal flashing? for fabric flashing?
7. Determine the amount of flashing required for the building (Appendix E).
8. What is the unit of measure for roofing trim such as gravel stops and fascias? What information about the items should be noted on the workup sheet?
9. What are the advantages in using a subcontractor for the roofing?
10. Determine the amount of roofing material required for the residence in Appendix H.

18
ELECTRICAL

18-1 ELECTRICAL WORK

Under single contracts the electrical work is the responsibility of a single prime contractor. In most cases this means that electrical contractors will submit prices on the work to be completed to the prime contractor. The general contractor will include an electrical contractor price plus overhead and profit in his bid price on the project. In this case the prime contractor is directly responsible for the work to the owner and must coordinate all of the parties involved in the project.

With separate contracts the electrical work will be bid directly to the owner by the electrical contractor, and the owner will select the contractor and sign a contract. The responsibility for the electrical work is the electrical contractor's, and while there shall be certain mutual responsibilities and coordination according to the General Conditions, Article 6, the general contractor's responsibilities are not as great as they would be under single contracts.

All electrical work must be installed in accordance with the code regulations of a given area. Throughout the United States, the National Electric Code is used extensively. State and local regulations must also be considered. Before beginning the takeoff, the contractor should review the plans and carefully read the specifications. Often, the specifications will require that "all work and installations be in conformance with all applicable national, state, and local codes." This means that contractors are responsible for compliance with the laws; if they are, then they better be familiar with them. The codes contain information regarding wiring design and protection, wiring methods and materials, equipment, and special occupancies as well as other information.

Field experience in construction will be helpful in understanding the problems involved in electrical work and how the electrical aspect should be integrated into the rest of the construction. Without field experience, an understanding of the fundamentals of electrical work, and an ability to read and understand the drawings and specifications, it will be difficult to do a meaningful takeoff on these items.

18-2 SINGLE CONTRACTS

Since an electrical subcontractor will undoubtedly do the electrical work, that contractor will do the bidding. Learning to estimate electrical work would require a course on this one subject; however, by using some common sense and complete files from past jobs, it is possible to obtain approximate estimates for checking whether or not the bids submitted to you are reasonable. With experience and complete files, it is possible to come within a few percent of the low bid.

The wiring is considered the *rough* work and the fixtures are considered the *finish* work. The wiring will usually be concealed in a conduit, which is installed throughout the building as it is erected. The wiring is pulled through the conduit much later in the job. The fixtures are usually the last items to go into the building, often after the interior finish work is complete.

To work up your estimate, go through the plans and specifications in a systematic manner, taking each different item and counting the number of each. Be sure and keep them separate. For example, floor outlets are different from wall outlets. Do not hesitate to check off (lightly) each item as you count it. This will cut down the possibility of estimating the same items twice. Include on your list all outlets, floor plugs, distribution panels, junction boxes, lighting panels, telephone boxes, switches, television receptacles, fixtures, and any other items such as snow-melting mats.

Always determine exactly where the responsibility begins for the wiring. Does it begin at the property line, at the structure, 10 feet from the structure? If a transformer is required, know who pays for it, who installs it, and who provides the base on which it will be set.

Different types of construction affect the installation of rough and finish work. When using steel joists, you are usually afforded ample space through which to run conduits easily. Cast-in-place concrete requires that there be close cooperation between the general contractor and the electrical contractor, since the conduit (as well as sleeves) often must be cast in the concrete and fixture hangers cast in. The use of hollow-core, prestressed, precast concrete causes other problems, such as where to run the conduit and how to hang fixtures properly. The conduit can be run in the holes (or joints) that are in the direction of the span, but you must carefully plan to run them in other directions unless the conduit can be exposed in the room. These problems greatly increase the amount of conduit required as well as the cost of the installation. Since installation is difficult, it becomes also more expensive. Similar problems occur when using precast double tees, except in that case no holes are available in the spanning direction. The problem can be alleviated to some extent by pouring a 2½- to 3-inch concrete floor over the slabs to run the conduit in.

One last point as to the estimate: Do not guess the price of fixtures. Prices vary considerably. What seems to you an inexpensive fixture may be very expensive. Never trust your own guess—check.

Light fixture manufacturers often prefer not to give anyone but the actual electrical subcontractor a firm price; however, if you are insistent, they will cooperate. This is one reason that it is important to be on a friendly basis with as many people as you can. Others often provide the key to success in whatever you are doing.

The selection of an electrical subcontractor should not be based on price alone, although price is an important consideration. Other factors, such as the speed with which the subcontractors complete their work and the cooperation they show in dealing with the prime contractor and other subcontractors are very important. Nothing causes hard feelings faster than subcontractors who are uncooperative. Since prime contractors are responsible to the owners for all work, it is in their own best interest to consider all factors when selecting an electrical subcontractor.

The major areas of coordination required between the electrical contractor and the general contractor are outlined in Sec. 18-4, while Fig. 18-1 shows typical electrical symbols and abbreviations.

18-3 SEPARATE CONTRACTS

The takeoff and bidding are done by the electrical contractor. This does not mean the estimator for general construction should not review the drawings and specifications for this work. Often, when projects are being bid under separate contracts, the contractors for each phase receive the drawings only for that phase (or portion of work) on which they will be bidding. (The bidders for general construction may receive no mechanicals.) In this case a trip to the plan room or to the office of the mechanical contractor who is bidding the project should be made so that the drawings and specifications may be investigated.

Even under separate contracts, there are many areas of mutual responsibility and coordination. Keep in mind that the entire building must fit together and operate as one unit. The major areas of coordination for single and separate contracts are outlined in Sec. 18-4.

18-4 COORDINATION REQUIREMENTS

Listed below are the major areas of coordination required between the electrical and general contractors.

Item	Coordination Requirements
1. Underground utilities	Location, size, excavation by whom, from where?
2. Equipment	Recessed depth, size of access openings, method of feed, supports, by whom, size limitations.
3. Distribution	Outlet locations, materials, method of feed, chases, in walls, under floor, overhead, special considerations.
4. Terminal fixtures and devices	Location, method of support, finish, color, and material.
5. Mounting surfaces	What is mounting surface? Can it work?
6. Specialty equipment	Field provisions, storage.
7. Scheduling	Work to be done? When required? Job to be completed on time-who-why-when?

Coordination of work among the mechanical contractors themselves is also important, since often all three of the mechanical contractors have work to do on a

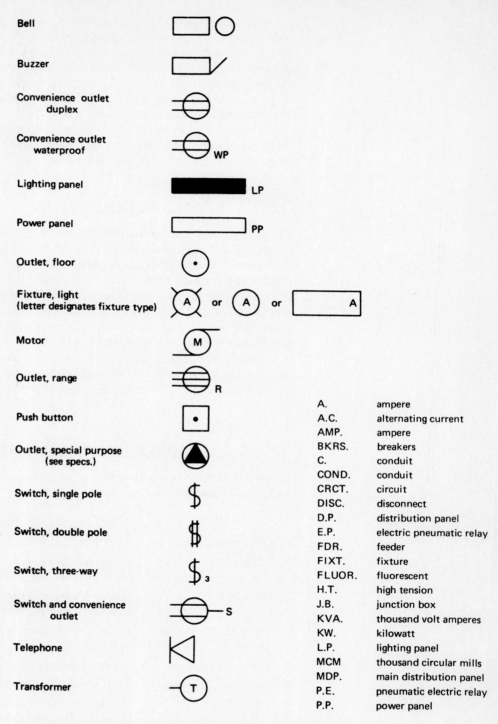

Figure 18-1 Electrical Symbols and Abbreviations Commonly Used

particular piece of equipment. For example, the heating contractor may install the boiler unit in place, the electrical contractor may make all power connections, and the plumbing contractor may connect the water lines. There are many instances of several trades connecting to one item. Coordination and an understanding of the work to be performed by each contractor is important to a smooth-running job.

18-5 CHECKLIST

Rough:

      conduit (sizes and lengths)

      wire (type, sizes, and lengths)

      outlets (floor, wall, overhead)

      switches (2-, 3-, and 4-way)

      panelboards

      breakers (size, no. of each)

      outlets, weatherproof

      control panels

      power requirements

Finish:

      fixtures (floor, wall, ceiling, etc.)

      mounting requirements

      clocks

      time clocks

      bells (buzzers)

      alarm systems

      TV outlets

      heaters

Miscellaneous:

      hook-up for various items from the other trades (motors, boilers, etc.)

REVIEW QUESTIONS

1. What is the difference between single and separate contracts?

2. Why are the mechanical and electrical portions of the work often put out under separate contracts?

3. Why should the estimator review the electrical portions of the job whether it is single or separate contracts?

4. Why is cooperation and coordination so important between the various contractors on a project?

5. How will the various types of construction affect the cost of the electrical work?

19
PLUMBING

19-1 PLUMBING WORK

Plumbing, like electrical work, may be bid under either single or separate contracts. Regardless of the type of contract, the areas of responsibility and coordination still exist, and the general contractor should have a basic understanding of what his responsibilities to the plumbing contractor are.

The plumbing work will have to be installed in accordance with all applicable building codes, federal, state, and local. The code most commonly used in the United States is the American Standard National Plumbing Code. The code contains information relating to water supply, pipe sizes of all types, sewage drainage systems, vents and venting, storm water drainage, and various design data.

Field experience, familiarity with plumbing principles, and an understanding of the drawings and specifications are key points to the successful bidding of this item. Typical symbols and abbreviations are shown in Fig. 19-1.

19-2 SINGLE CONTRACTS

Plumbing is another area of work that will be subbed out to a subcontractor. This sub will prepare an estimate and submit a bid to you. Even so, you should have a general knowledge of plumbing so that you can determine whether the subcontractor's bid is reasonable. Again, by use of common sense and complete files an approximate estimate can be obtained.

Using the working drawings and specifications, prepare a complete list of everything that will be required. First, determine exactly where the responsibility for the plumbing begins. It may begin at the property line, 10 feet outside the structure, at the structure or somewhere else; always check so it will be clear in your mind.

Piping symbols:

Vent — — — — — — — — — —

Cold water —— · —— · —— · ——

Hot water — · · — · · — · · — · ·

Hot water return —— — —— — ——

Gas —— G —— —— G ——

Soil, waste or leader —————————
(above grade)

Soil, waste or leader — — — — — — —
(below grade)

Fixture symbols:

Baths

Water closet (with tank)

Water closet (flush valve)

Shower

Lavatory

DW Dishwasher

SS Service sink

HWH Hot water heater

HWT
HWT Hot water tank

DF Drinking fountain

M Meter

HB Hose bib

C O CO Cleanouts

FD Floor drain

RD Roof drain

A.F.D.	area floor drain
B.W.V.	backwater valve
CODP.	deck plate cleanout
C.W.	cold water
C.W.R.	cold water return
DEG.	degree
D.F.	drinking fountain
D.H.W.	domestic hot water
DR.	drain
D.W.	dishwasher
F.	fahrenheit
FDR.	feeder
FIXT.	fixture
F.D.	floor drain
F.H.	fire hose
F.E.	fire extinguisher unit
H.W.	hot water
H.W.C.	hot water circulating line
H.W.R.	hot water reserve
H.W.S.	hot water supply
H.W.P.	hot water pump
I.D.	inside diameter
LAV.	lavatory
LDR.	leader
O.D.	outside diameter
(R)	roughing only
R.D.	roof drain
S.C.	sill cock
S.S.	service sink
TOIL.	toilet
UR.	urinal
V.	vent
W.C.	water closet
W.H.	wall hydrant

Figure 19-1 Plumbing Symbols and Abbreviations Commonly Used

The takeoff for plumbing includes the costs of water pipes, gas pipes, sewer pipes, drains, soil and vent stacks, soil, waste, vent pipes, and all rough work required for fixtures. Information required includes size, length, pipe material, weight, and fittings (unions, hangers, T's, Y's, elbows, etc.). Also include kinds and sizes of cleanouts, grease traps, plugs, traps, valves, and the like. The trenches required for plumbing pipe, especially from the road to building, may be the responsibility of the plumbing subcontractor or the general contractor. If the general contractor is responsible for trenching, it should be noted as such on your estimate. Pipe materials include cast iron, copper, lead, galvanized, and plastic.

The finish plumbing list should include a list of all items and fixtures furnished and installed by the contractor. Keep all different types and sizes of fixtures separate. In pricing the fixture, be sure you also consider the accessories and fittings required for each fixture. The listing requires care in counting the fixtures on the working drawings and making note of all specification requirements. When listing your prices for the fixture, note whether the fixtures and accessories were priced separately. The make, kind, type, and quality will greatly affect the price of the fixtures and accessories. Price what is in the specification only; do not guess at prices. For example, the Sears-Roebuck tub probably won't meet the specifications and the one specified may easily cost two and three times more, so never use catalogs as references. Items that should be included in this takeoff include lavatory, water closet, shower with stall, shower without stall, bathtub, bathtub with shower, slop sink, water heater, dishwasher, garbage disposal, laundry tubs, and water cooler. Another situation that must be noted is that which occurs when an item is supplied by the owner or others to the plumbing subcontractor for installation. It is important that you know exactly the limits of responsibility in this case, since this is often an area of dispute.

In order to improve your accuracy from job to job, keep a complete file of all costs. Require your subs to give you breakdowns of costs when they submit an application of payment to you. Constantly check one job against another. It is not unusual that, by the use of your file, you will be able to detect not only high bids from your subs, but pick out a low bid and call it to his attention. If you had suspected a subcontractor's bid to be too high, you should not hesitate to call and discuss it. Call if you suspect the bid is too low and a reasonable profit cannot be made. Job performance will probably suffer if the bid is too low. Cooperation and respect are two key ingredients to success. Treat your subs well and they will reciprocate.

19-3 SEPARATE CONTRACTS

The takeoff and bidding is done by the plumbing contractors. The estimator for general construction must still check to determine the areas of mutual responsibility and coordination—they are similar to those required for the electrical work. There is always considerable work done in coordination, wall chases, anchoring supports, sleeves, and many other items outlined in Sec. 19-4.

19-4 COORDINATION REQUIREMENTS

Listed below are the major areas of coordination required between the plumbing and general contractors.

Item	*Coordination Requirements*
1. Underground utilities	Location, size, excavation by whom, from where?
2. Building entrance	Floor sleeves, supports.
3. Mechanical room equipment	Supports required, location, anchors, by whom?
4. Distribution	Wall sleeves, hangers, chases, in wall, roof vents, access doors.
5. Fixtures	Method of support, feed, outlets, built-in, floor drains, vents.
6. Finishes	Factory or field.
7. Specialty equipment	Field provisions, storage.
8. Scheduling	Work to be done, when required? Job to be completed on time—who—why—when?

Once again, coordination among the mechanical contractors is also important in the understanding of who is responsible for what? why? when?

19-5 CHECKLIST

Rough:

> permits
>
> excavation and backfill
>
> water, gas, and sewage lines
>
> required pipes and fittings
>
> cleanouts
>
> valves
>
> tanks
>
> sleeves

Finish:

> water closets
>
> bath tubs
>
> lavatories
>
> drinking fountains
>
> showers
>
> tubs
>
> service sinks
>
> water heater
>
> urinals
>
> washers and dryers
>
> dishwashers

Miscellaneous:

> Hook-up to equipment supplied by owner or other contractors may be required.

REVIEW QUESTIONS

1. How do the various codes affect the installation of the mechanical portions of the project?

2. What type of work is most generally included under plumbing?

3. Why would an estimator call a subcontractor if it is suspected that the subcontractor's bid is too low?

4. Why should the estimator review the plumbing portions of the project whether it is single or separate contracts?

5. How do the various types of construction affect the cost of the plumbing work?

20

HEATING, VENTILATING, AND AIR CONDITIONING

20-1 HVAC WORK (HEATING, VENTILATING, AND AIR CONDITIONING)

As in electrical and plumbing work, this portion of the work may be bid under single or separate contracts. Bidding under separate contracts for HVAC work is similar to bidding under separate contracts for electrical and plumbing work. Understand the mutual responsibilities and coordination required (Sec. 20-4).

Like all work, this portion must be designed and installed in accordance with all federal, state, and local codes. Many different codes may have control, depending on the types of systems used.

Again, the field experience, understanding of design principles, and ability to understand drawings and specifications are key points. There is a tremendous variety of installations that may be used in this portion in dealing with electrical heat, ventilating ceilings, and various methods and fuels used for heating and air conditioning.

20-2 SINGLE CONTRACTS

Subcontractors who specialize in this type of work will be submitting bids to you, so your area of responsibility falls within the preparation of an approximate bid for comparison.

Referring to the working drawings and specifications, prepare a complete list of major items required. Determine exactly where the responsibility for each portion rests. If responsibility boundaries are unclear, request the architect or engineer to clarify them. Never assume that someone will do something unless it is so specified or indicated.

The takeoff includes piping, ductwork, drains, equipment, fixtures, and accessories. Check the specification to see who is responsible for trenching, both from the road to the building and within the building. In your takeoff list all equipment separately as to types and sizes.

You should have a general knowledge of heating, ventilating, and air conditioning in order for you to understand the equipment involved. There are many different systems that could be used on a building. If a system is used with which you are unfamiliar, call in the manufacturer's representative so it may be fully explained to you. In this way you will understand what is required of the general contractor and subcontractor in order to have a successful installation.

Check the specifications carefully to be sure that the materials you (or your subcontractor) bid are those specified. So that your accuracy may improve from job to job, keep a complete file of all costs, breakdown of costs from subs, and check job against job. It is suggested that you go through the specifications item by item to be sure that they are all included in your takeoff.

Another method used for approximate estimates is the use of square feet of building and square feet or radiation prices. Also, prices per btu, cfm, and tons of air conditioning are also sometimes quoted. If used with caution, these pricing procedures may be effectively employed for comparison of prices.

20-3 SEPARATE CONTRACTS

The same general conditions prevail with HVAC work under separate contracts as when the electrical and plumbing are bid as separate contracts. Understand the mutual responsibilities involved and what the coordination requirements are. The work must still be integrated into the building construction. Review the drawings and check that there is space for the ducts, pipes, and other mechanical lines.

20-4 COORDINATION REQUIREMENTS

Listed below are the major areas of coordination required between the HVAC and general contractors.

Item	*Coordination Requirements*
1. Underground utilities	Location, size, excavation by whom, from where?
2. Equipment	Method of support, location, by whom, anchors, access for receiving and installing, size limitations, flues, roof curbs.
3. Piping	Wall sleeves, size limitations, chases, in walls, under floors, floor sleeves, expansion compensators, access doors.
4. Ductwork	Sizes, support, access doors, drops, outlet sizes, chases, outside air louvers, roof curbs, lintels.
5. Terminal equipment	Recesses for CUH, RC, FC, etc., size of radiation, method of concealing, grille fastenings.

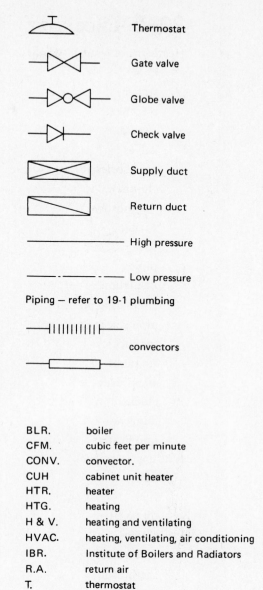

BLR.	boiler
CFM.	cubic feet per minute
CONV.	convector.
CUH	cabinet unit heater
HTR.	heater
HTG.	heating
H & V.	heating and ventilating
HVAC.	heating, ventilating, air conditioning
IBR.	Institute of Boilers and Radiators
R.A.	return air
T.	thermostat
T.R.	top register
U.H.	unit heater
VENT.	ventilate, ventilation

Figure 20-1 HVAC Symbols and Abbreviations Commonly Used

6. Mechanical-Electrical   Who is doing what?
   responsibility
7. Finishes   Field or factory?
8. Scheduling   Who? Why? When?

20-5 CHECKLIST

Heating (hot water):

    boiler

    stoker

    oil tanks

    gauges

    valves

    accessories

    chimney

    thermostats

    wiring

    fuel

    piping

    insulation

    circulating pumps

    piping accessories

    radiators

    fin tubes

    enclosures

    clocks

    hangers

    unit heaters

Heating (warm air):

    boiler

    fuel

    oil tanks

    gauges

    accessories

    thermostats

    wiring

    chimney

    ducts

    diffusers

    fans

    filters

    humidifiers

    dehumidifiers

    insulation

    baffles

Air Conditioning:

central

units

coolant

fans

piping

diffusers

registers

wiring

ducts

filters

humidifiers

dehumidifiers

inlets

returns

fresh air

louvers

thermostats

REVIEW QUESTIONS

1. Under which of the mechanicals would electric heat be most likely placed?
2. How do the various types of construction affect the cost of the heating work?
3. Why are subcontractors hired, under single contracts, to perform the mechanical work?

21
PROFIT

21-1 PROFIT

Profit is not included in a chapter with other topics since it is the last thing considered and must be taken separately. Keep in mind that construction is among the tops in the percentage of business failures. Would you believe that some people even forget to add profit?

First, let us understand that by *profit* we mean the amount of money added to the total estimated cost of the project; this amount of money should be clear profit. All costs relating to the project, including project and office overhead and salaries, are included in the estimated cost of the building.

There are probably more different approaches to determining how much profit should be included than could be listed. Each contractor and estimator seem to have a different approach. A few typical approaches are listed below:

1. Add a percentage of profit to each item as it is estimated, allowing varying amounts for the different items; for example, 8 to 15 percent for concrete work, but only 3 to 5 percent for work subcontracted out.

2. Add a percentage of profit to the total price tabulated for materials, labor, overhead, and equipment. The percent would vary from small jobs to larger jobs (perhaps 20 to 25 percent on a small job and 5 to 10 percent on a larger one), taking into account the accuracy of the takeoff and pricing procedures used in the estimate.

3. Various methods of selecting a figure are employed that will make a bid low while not being too low, by trying to analyze all of the variables and other contractors who are bidding.

4. There are "strategies of bidding" that some contractors (and estimators) apply to bidding. Most of the strategies require bidding experience to be accumulated and competitive patterns from past biddings to be used as patterns for future biddings. This will also lend itself nicely to computer operations. Interested in such a strategy? If so, refer to *Construction Contracting* by Richard H. Clough (Wiley-Interscience), 2nd ed.; 1969, Appendix L.

5. Superstition sometimes plays a part. And why not, since superstition is prevalent throughout our lives? Many contractors and estimators will use only certain numbers to end their bids; for example, some always end with a 7 or take a million-dollar bid all the way down to 50 cents.

My own method is to, first, include all of the costs of the project before profit is considered. Then a review of the documents is made to find if the drawings and specifications were clear, if you understood what you were bidding on, and how accurate a takeoff was made (it should always be as accurate as possible). The other factors to be considered are the architect-engineer, their reputation, and how the work is handled.

After reviewing the factors, you, the contractor, must decide how much money (profit, over and above salary) you want to make on this project. This is the amount that should be added to the cost of construction to give the amount of the bid (after it is adjusted slightly to take into account superstitions and strategy types of bids). Exactly what is done at this point, slightly up or down, is an individual matter, but you should definitely know your competitors, keep track of their past bidding practices, and use those against them whenever possible.

Since profit is added at the end of the estimate, the estimator has a pretty good idea of the risks and problems that may be encountered. Discuss these risks thoroughly with other members of the firm. It is far better to bid what you feel is high enough to cover the risks than to neglect the risks, bid low, and lose money. There is some-times a tendency to "need" or "want" a job so badly that risks are completely ignored. Try to avoid this sort of foolishness—it only invites disaster. If a project entails substantial risk and it is questionable that a profit can be made, consider not even bidding it and let someone else have the heartaches and the loss. Always remember: Construction is a business in which you are supposed to make a fair, reasonable profit.

A

DRAWINGS AND OUTLINE SPECIFICATIONS OF A SMALL COMMERCIAL BUILDING

OUTLINE SPECIFICATIONS—STAR BUILDING

Section 1—Excavation

Excavation shall be carried 2 feet beyond all walls.
Excavations shall be carried to depths shown on the drawings.
All excess earth not needed for fill shall be removed from the premises.

Section 2—Concrete

Concrete shall develop a compressive strength of 2,500 pounds per square inch (psi) at 28 days.
Reinforcing steel shall be A-15 Intermediate Grade.
Reinforcing mesh shall be 6 × 6 10/10 welded wire mesh.
All concrete shall receive a machine-troweled finish.
Vapor barrier shall be 4 mil polyethylene.
Expansion joint filler shall be ½ × 4 inch asphalt impregnated fiberboard.
Concrete lintels shall be precast.

Section 3—Masonry

Concrete masonry units shall be 16 inches in length.
Masonry wall reinforcement shall be truss design and installed every second course.
Brick shall be standard size, red.
All mortar joints ⅜-inch thick.

Section 4—Roof Planks

Prestressed hollow core.
Slabs to be erected by manufacturer.

Section 5—Door Frames, Doors, Windows

Door frames shall be 16-gauge steel.
Doors shall be flush, solid core, seven-ply construction with a rotary birch veneer. Prefinish interior doors with semi-gloss lacquer and exterior doors with polyurethane varnish.
Overhead door shall be 1¾ inches thick, solid and flush design. All required hardware for overhead door shall be furnished.
Hardware for doors (except overhead)—allow $250 for the purchase of the hardware. Hardware to be installed under this contract.

Section 6—Roofing

Roofing shall be three-ply, built-up roof, 20-year bond.
Cant strips shall be asphalt-impregnated fiberboard.
Insulation shall be styrofoam.
Gravel stop shall be 0.032-inch aluminum.

Section 7—Finishes

All exposed concrete shall receive one coat of concrete paint.
All exposed surfaces of gypsum board and plaster wall shall receive two coats of latex paint.
Floors—1/8″ thick, 12 × 12 inches, vinyl-asbestos, marbelized design, light colors.
Base—1/8″ thick, 4 inches high, vinyl.

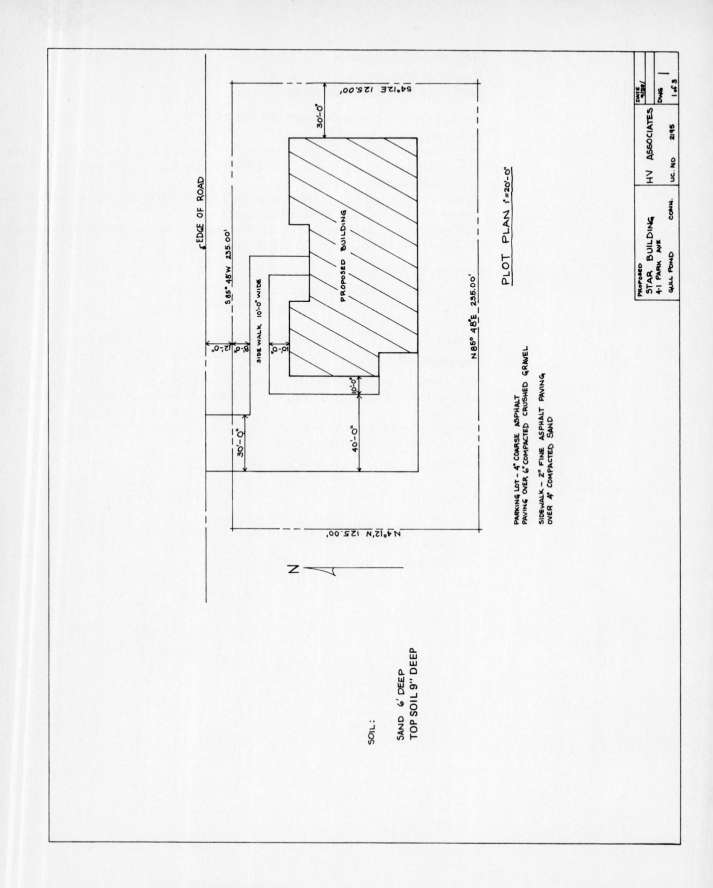

PLOT PLAN 1"=20'-0"

PROPOSED BUILDING

SIDEWALK 10'-0" WIDE

EDGE OF ROAD

S 85° 48' W 235.00'

N 86° 48' E 235.00'

S 4°12'E 125.00'

N 4°12'N 125.00'

30'-0"

30'-0"

40'-0"

12'-0"

8'-0"

10'-0"

10'-0"

N

PARKING LOT - 4" COARSE ASPHALT
PAVING OVER 6" COMPACTED CRUSHED GRAVEL

SIDEWALK - 2" FINE ASPHALT PAVING
OVER 4" COMPACTED SAND

SOIL:

SAND 6' DEEP
TOP SOIL 9" DEEP

PROPOSED
STAR BUILDING
4-1 PARK AVE
GULL POND   CONN.

HV ASSOCIATES

LIC. NO   2195

DATE
9/28/

DWG

1 of 3

appendix a

340

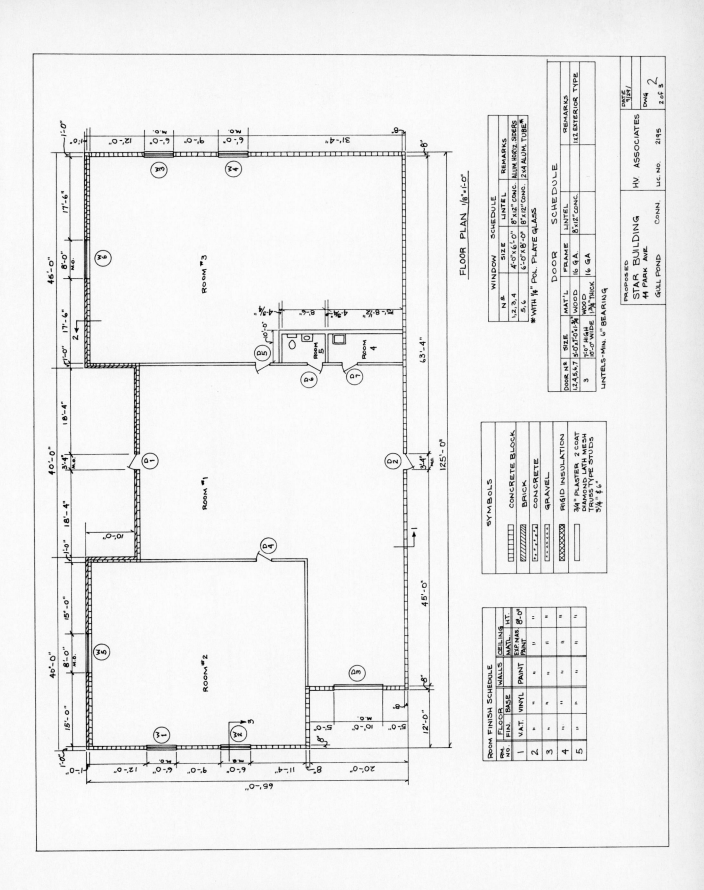

FLOOR PLAN 1/8"=1'-0"

WINDOW SCHEDULE

Nº	SIZE	LINTEL	REMARKS
1,2,3,4	4'-0"x6'-0"	8"x12" CONC.	ALUM. HORIZ. SLIDERS
5,6	6'-0"x8'-0"	8"x12" CONC.	2x4 ALUM. TUBE*

*WITH 1/4" POL. PLATE GLASS

DOOR SCHEDULE

DOOR Nº	SIZE	MAT'L	FRAME	LINTEL	REMARKS
1,2,4,5,6,7	3'-0"x7'-0"x1¾"	WOOD	16 G.A.	8"x12"CONC.	
3	7'-0" HIGH 10'-0" WIDE	WOOD 1¾"THICK	16 GA		1x2 EXTERIOR TYPE

LINTELS-MIN. 6" BEARING

PROPOSED
STAR BUILDING
44 PARK AVE.
GULL POND   CONN.

HV. ASSOCIATES
LIC. NO.   2195

DATE 1/28/
DWG   2 of 3

SYMBOLS

	CONCRETE BLOCK
	BRICK
	CONCRETE
	GRAVEL
	RIGID INSULATION
	3/4" PLASTER 2 COAT DIAMOND LATH MESH TRUSS TYPE STUDS 3/4" & 6"

ROOM FINISH SCHEDULE

RM. NO.	FLOOR FIN.	BASE	WALLS	CEILING MAT'L	HT.
1	V.A.T.	VINYL	PAINT	EXP. MAS. PAINT	8'-0"
2	"	"	"	"	"
3	"	"	"	"	"
4	"	"	"	"	"
5	"	"	"	"	"

ROOM #3
ROOM #1
ROOM #2
Room 5
Room 4

small commercial building   341

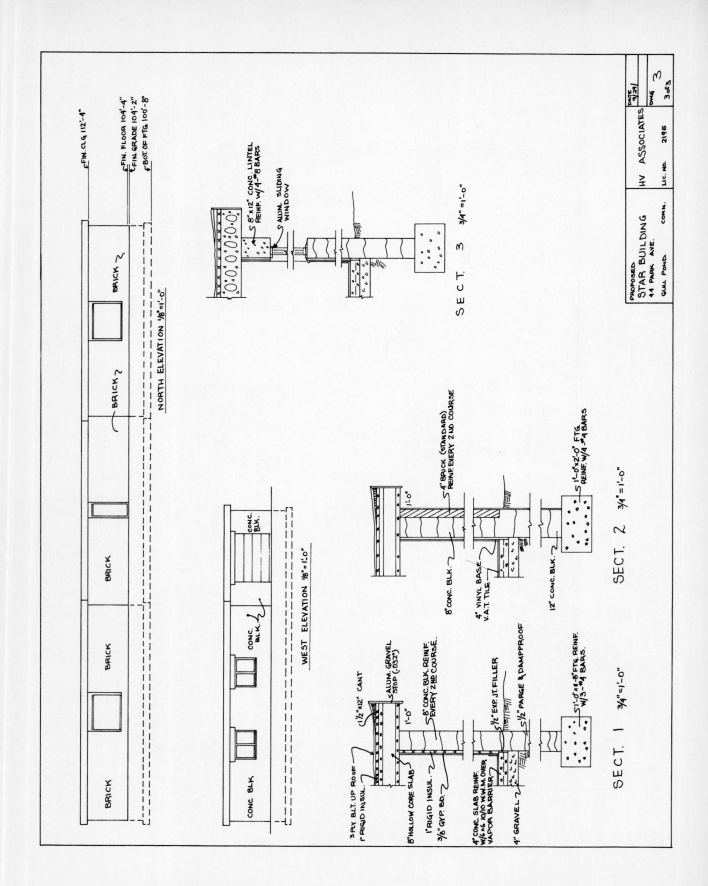

NORTH ELEVATION 1/8"=1'-0"

FIN. CLG 112'-4"
FIN. FLOOR 104'-4"
FIN. GRADE 104'-2"
BOT. OF FTG. 100'-8"

BRICK

BRICK

BRICK

BRICK

BRICK

WEST ELEVATION 1/8"=1'-0"

CONC. BLK.

CONC. BLK.

8"x12" CONC. LINTEL REINF. W/4-#8 BARS

ALUM. SLIDING WINDOW

SECT. 3 3/4"=1'-0"

4" BRICK (STANDARD) REINF. EVERY 2ND COURSE

1'-0"

8" CONC. BLK.

4" VINYL BASE
V.A.T. TILE

12" CONC. BLK.

1'-0"x2'-0" FTG. REINF. W/4-#4 BARS

SECT. 2 3/4"=1'-0"

3 PLY B'LT. UP ROOF
1" RIGID INSUL.

1 1/2"x12" CANT

ALUM. GRAVEL STOP (.032")

8" CONC. BLK. REINF. EVERY 2ND COURSE

1'-0"

8" HOLLOW CORE SLAB

1" RIGID INSUL.
3/8" GYP. BD.

4" CONC. SLAB REINF. W/6x6 10/10 W.W.M. OVER VAPOR BARRIER

4" GRAVEL

1/2" EXP. JT. FILLER

1/2" PARGE & DAMPPROOF

1'-0"x1'-8" FTG. REINF. W/3-#4 BARS.

SECT. 1 3/4"=1'-0"

PROPOSED
STAR BUILDING
44 PARK AVE.
GULL POND   CONN.

HV ASSOCIATES
LIC. NO.   2195

DATE 9/29/
DWG. 3
3 of 3

B
COMMON TERMS USED IN THE BUILDING INDUSTRY

ADDENDA — Statements or drawings that modify the basic contract documents after the latter have been issued to the bidders, but prior to the taking of bids.

ALTERNATES — Proposals required of bidders reflecting amounts to be added to or subtracted from the basic proposal in the event that specific changes in the work are ordered.

ANCHOR BOLTS — Bolts used to anchor structural members to concrete or the foundation.

APPROVED EQUAL OR — The term used to indicate that material or product finally supplied or installed must be equal to that specified and as approved by the architect (or engineer).

AS-BUILT DRAWINGS — Drawings made during the progress of construction, or subsequent thereto, illustrating how various elements of the project were actually installed.

ASTRAGAL — A closure between the two leafs of a double-swing or double-slide door to close the joint. This can also be a piece of molding.

AXIAL — Anything situated around, in the direction of, or along an axis.

BASEPLATE — A plate attached to the base of a column which rests on a concrete or masonry footing.

BAY — The space between column center lines or primary supporting members, lengthwise in a building. Usually the crosswise dimension is considered the *span* or *width module,* and the lengthwise dimension is considered the *bay spacing*.

BEAM — A structural member that is normally subjected to bending loads and is usually a horizontal member carrying vertical loads. (An exception to this is a purlin.) There are three types of beams:

a. *continuous beam:* A beam that has more than two points of support.

b. *cantilevered beam:* A beam that is supported at only one end and is restrained against excessive rotation.

c. *simple beam:* A beam that is freely supported at both ends, theoretically with no restraint.

BEAM AND COLUMN — A primary structural system consisting of a series of beams and columns; usually arranged as a continuous beam supported on several columns with or without continuity that is subjected to both bending and axial forces.

BEAM-BEARING PLATE — Steel plate with attached anchors that is set in top of a masonry wall so that a purlin or a beam can rest on it.

BEARING — The condition that exists whenever one member or component transmits load or stress to another by direct contact in compression.

BENCH MARK — A fixed point used for construction purposes as a reference point in determining the various levels of floor, grade, etc.

BID — Proposal prepared by prospective contractor specifying the charges to be made for doing the work in accordance to the contract documents.

BID BOND — A surety bond guaranteeing that a bidder will sign a contract, if offered, in accordance with his or her proposal.

BID SECURITY — A bid bond, certified check, or other forfeitable security guaranteeing that a bidder will sign a contract, if offered, in accordance with his or her proposal.

BILL OF MATERIALS — A list of items or components used for fabrication, shipping, receiving, and accounting purposes.

BIRD SCREEN — Wire mesh used to prevent birds from entering the building through ventilators or louvers.

BOND — Masonry units interlocked in the face of a wall by overlapping the units in such a manner as to break the continuity of vertical joints.

BONDED ROOF — A roof that carries a printed or written warranty, usually with respect to weathertightness, including repair and/or replacement on a prorated cost basis for a stipulated number of years.

BONUS AND PENALTY CLAUSE — A provision in the proposal form for payment of a bonus for each day the project is completed prior to the time stated, and for a charge against the contractor for each day the project remains uncompleted after the time stipulated.

BRACE RODS — Rods used in roofs and walls to transfer wind loads and/or seismic forces to the foundation (often used to plumb building but not designed to replace erection cables when required).

BRIDGING — The structural member used to give lateral support to the weak plane of a truss, joist, or purlin; provides sufficient stability to support the design loads, sag channels, or sag rods.

BUILT-UP ROOFING — Roofing consisting of layers of rag felt or jute saturated with coal tar pitch, with each layer set in a mopping of hot tar or asphalt; ply designation as to the number of layers.

CAMBER — A permanent curvature designed into a structural member in a direction opposite to the deflection anticipated when loads are applied.

CANOPY — Any overhanging or projecting structure with the extreme end unsupported. It may also be supported at the outer end.

CANTILEVER — A projecting beam supported and restrained only at one end.

CAP PLATE — A horizontal plate located at the top of a column.

CASH ALLOWANCES — Sums that the contractor is required to include in the bid and contract sum for specific purposes.

CAULK — To seal and make weathertight the joints, seams, or voids by filling with a waterproofing compound or material.

CERTIFICATE OF OCCUPANCY — Statement issued by the governing authority granting permission to occupy a project for a specific use.

CERTIFICATE OF PAYMENT — Statement by an architect informing the owner of the amount due a contractor on account of work accomplished and/or materials suitably stored.

CHANGE ORDER — A work order, usually prepared by the architect and signed by the owner or the owner's agent, authorizing a change in the scope of the work and a change in the cost of the project.

CHANNEL — A steel member whose formation is similar to that of a "C" section without return lips; may be used singularly or back-to-back.

CLIP — A plate or angle used to fasten two or more members together.

CLIP ANGLE — An angle used for fastening various members together.

COLLATERAL LOADS — A load, in addition to normal live, wind, or dead loads, intended to cover loads that are either unknown or uncertain (sprinklers, lighting, etc.).

COLUMN — A main structural member used in a vertical position on a building to transfer loads from main roof beams, trusses, or rafters to the foundation.

CONTRACT DOCUMENTS — Working Drawings, Specifications, General Conditions, Supplementary General Conditions, the Owner-Contractor Agreement, and all Addenda (if issued).

CURB — A raised edge on a concrete floor slab.

CURTAIN WALL — Perimeter walls that carry only their own weight and wind load.

DATUM — Any level surface to which elevations are referred (see *Bench Mark*).

DEAD LOAD — The weight of the structure itself, such as floor, roof, framing, and covering members, plus any permanent loads.

DEFLECTION — The displacement of a loaded structural member or system in any direction, measured from its no-load position, after loads have been applied.

DESIGN LOADS — Those loads specified by building codes, state, or city agencies, or owner's or architect's specifications to be used in the design of the structural frame of a building. They are suited to local conditions and building use.

DOOR GUIDE — An angle or channel guide used to stabilize and keep plumb a sliding or rolling door during its operation.

DOWNSPOUT — A hollow section such as a pipe used to carry water from the roof or gutter of a building to the ground or sewer connection.

DRAIN — Any pipe, channel, or trench for which waste water or other liquids are carried off, i.e., to a sewer pipe.

EAVE — The line along the sidewall formed by the intersection of the inside faces of the roof and wall panels; the projecting lower edges of a roof, overhanging the walls of a building.

EQUAL, OR — (See *Approved Equal.*)

ERECTION — The assembly of components to form the completed portion of a job.

EXPANSION JOINT — A connection used to allow for temperature-induced expansion and contraction of material.

FABRICATION — The manufacturing process performed in the plant to convert raw material into finished metal building components. The main operations are cold forming, cutting, punching, welding, cleaning, and painting.

FASCIA — A flat, broad trim projecting from the face of a wall, which may be part of the rake or the eave of the building.

FIELD — The job site or building site.

FIELD FABRICATION — Fabrication performed by the erection crew or others in the field.

FIELD WELDING — Welding performed at the job site, usually with gasoline-powered machines.

FILLER STRIP — Preformed neoprene material, resilient rubber or plastic used to close the ribs or corrugations of a panel.

FINAL ACCEPTANCE — The owner's acceptance of a completed project from a contractor.

FIXED JOINT — A connection between two members in such a manner as to cause them to act as a single continuous member, provides for transmission of forces from one member to the other without any movement in the connection itself.

FLANGE — That portion of a structural member normally projecting from the edges of the web of a member.

FLASHING — A sheet-metal closure that functions primarily to provide weather-tightness in a structure and secondarily to enhance appearance; the metalwork that prevents leakage over windows, door, etc., around chimneys, and at other roof details.

FOOTING — That bottom portion at the base of a wall or column used to distribute the load into the supporting soil.

FOUNDATION — The substructure that supports a building or other structure.

FRAMING — The structural steel members (columns, rafters, girts, purlins, brace rods, etc.) that go together to make up the skeleton of a structure ready for covering to be applied.

FURRING — Leveling up or building out of a part of wall or ceiling by wood, metal, or strips.

GLAZE (GLAZING) — The process of installing glass in window and door frames.

GRADE — The term used when referring to the ground elevation around a building or other structure.

GROUT — A mixture of cement, sand, and water used to solidly fill cracks and cavities; generally used under setting plates to obtain solid, uniform, full-bearing surface.

GUTTER — A channel member installed at the eave of the roof for the purpose of carrying water from the roof to the drains or downspouts.

HEAD — The top of a door, window, or frame.

IMPACT LOAD — The assumed load resulting from the motion of machinery, elevators, cranes, vehicles, and other similar moving equipment.

INSTRUCTIONS TO BIDDERS — A document stating the procedures to be followed by bidders.

INSULATION — Any material used in building construction for the protection from heat or cold.

INVITATION TO BID — An invitation to a selected list of contractors furnishing information on the submission of bids on a subject.

JAMB — The side of a door, window, or frame.

JOIST — Closely spaced beams supporting a floor or ceiling. They may be wood, steel, or concrete.

KIP — A unit of weight, force, or load equal to 1,000 pounds.

LAVATORY — A bathroom-type sink.

LIENS — Legal claims against an owner for amounts due those engaged in or supplying materials for the construction of the building.

LINTEL — The horizontal member placed over an opening to support the loads (weight) above it.

LIQUIDATED DAMAGES — An agreed-to sum chargeable against the contractor as reimbursement for damages suffered by the owner because of contractor's failure to fulfill contractual obligations.

LIVE LOAD — The load exerted on a member or structure due to all imposed loads except dead, wind, and seismic loads. Examples include snow, people, movable equipment, etc. This type of load is movable and does not necessarily exist on a given member of structure.

LOADS — Anything that causes an external force to be exerted on a structural member. Examples of different types are

a. *dead load:* in a building, the weight of all permanent constructions, such as floor, roof, framing, and covering members.

b. *impact load:* the assumed load resulting from the motion of machinery, elevators, craneways, vehicles, and other similar kinetic forces.

c. *roof live load:* all loads exerted on a roof (except dead, wind, and lateral loads) and applied to the horizontal projection of the building.

d. *seismic load:* the assumed lateral load due to the action of earthquakes and acting in any horizontal direction on the structural frame.

e. *wind load:* the load caused by wind blowing from any horizontal direction.

LOUVER — An opening provided with one or more slated, fixed, or movable fins to allow flow of air, but to exclude rain and sun or to provide privacy.

MULLION — The large vertical piece between windows. (It holds the window in place along the edge with which it makes contact.)

NONBEARING PARTITION — A partition which supports no weight except its own.

PARAPET — That portion of the vertical wall of a building that extends above the roof line at the intersection of the wall and roof.

PARTITION — A material or combination of materials used to divide a space into smaller spaces.

PERFORMANCE BOND — A bond that guarantees to the owner, within specified limits, that the contractor will perform the work in accordance with the contract documents.

PIER — A structure of masonry (concrete) used to support the bases of columns and bents. It carries the vertical load to a footing at the desired load-bearing soil.

PILASTER — A flat rectangular column attached to or built into a wall masonry or pier; structurally, a pier, but treated architecturally as a column with a capital, shaft, and base. It is used to provide strength for roof loads or support for the wall against lateral forces.

PRECAST CONCRETE — Concrete that is poured and cast in some position other than the one it will finally occupy; cast either on the job site and then put into place or away from the site to be transported to the site and erected.

PRESTRESSED CONCRETE — Concrete in which the reinforcing cables, wires, or rods are tensioned before there is load on the member.

PROGRESS PAYMENTS — Payments made during progress of the work, on account, for work completed and/or suitably stored.

PROGRESS SCHEDULE — A diagram showing proposed and actual times of starting and completion of the various operations in the project.

PUNCH LIST — A list prepared by the architect or engineer of the contractor's uncompleted work or work to be corrected.

PURLIN — Secondary horizontal structural members located on the roof extending between rafters, used as (light) beams for supporting the roof covering.

RAFTER — A primary roof support beam usually in an inclined position, running from the tops of the structural columns at the eave to the ridge or highest portion of the roof. It is used to support the purlins.

RECESS — A notch or cutout, usually referring to the blockout formed at the outside edge of a foundation, and providing support and serving as a closure at the bottom edge of wall panels.

REINFORCING STEEL — The steel placed in concrete to carry the tension, compression, and shear stresses.

RETAINAGE — A sum withheld from each payment to the contractor in accordance with the terms of the Owner-Contractor Agreement.

ROLLING DOORS — Doors that are supported on wheels that run on a track.

ROOF OVERHANG — A roof extension beyond the end or the side walls of a building.

ROOF PITCH — The angle or degree of slope of a roof from the eave to the ridge. The pitch can be found by dividing the height, or rise, by the span; for example, if the height is 8 feet and the span is 16 feet, the pitch is 8/16 or ½ and the angle of pitch is 45°. (See *Roof Slope.*)

ROOF SLOPE — The angle that a roof surface makes with the horizontal. Usually expressed as a certain rise in 12 inches of run.

SANDWICH PANEL — An integrated structural covering and insulating component consisting of a core material with inner and outer metal or wood skins.

SCHEDULE OF VALUES — A statement furnished to the architect by the contractor reflecting the amounts to be allotted for the principal divisions of the work. It is to serve as a guide for reviewing the contractor's periodic application for payment.

SEALANT — Any material that is used to close up cracks or joints.

SEPARATE CONTRACT — A contract between the owner and a contractor other than the general contractor for the construction of a portion of a project.

SHEATHING — Rough boarding (usually plywood) on outside of a wall or roof over which is placed siding or shingles.

SHIM — A piece of steel used to level or square beams or column baseplates.

SHIPPING LIST — A list that enumerates by part, number, or description each piece of material to be shipped.

SHOP DRAWINGS — Drawings that illustrate how specific portions of the work shall be fabricated and/or installed.

SILL — The lowest member beneath an opening such as a window or door; also, the horizontal framing members at floor level, such as sill girts or sill angles; the member at the bottom of a door or window opening.

SILL, LUG — A sill that projects into the masonry at each end of the sill. It must be installed as the building is being erected.

SILL, SLIP — A sill that is the same width as the opening—it will slip into place.

SKYLIGHT — An opening in a roof or ceiling for admitting daylight; also, the reinforced plastic panel or window fitted into such an opening.

SNOW LOAD — In locations subject to snow loads, as indicated by the average snow depth in the reports of the United States Weather Bureau, the design loads shall be modified accordingly.

SOFFIT — The underside of any subordinate member of a building, such as the undersurface of a roof overhang or canopy.

SOIL BORINGS — A boring made on the site in the general location of the proposed building to determine soil type, depth of the various types of soils, and water table level.

SOIL PRESSURE — The allowable soil pressure is the load per unit area a structure can safely exert on the substructure (soil) without exceeding reasonable values of footing settlements.

SPALL — A chip or fragment of concrete that has chipped, weathered, or otherwise broken from the main mass of concrete.

SPAN — The clear distance between supports of beams, girders, or trusses.

SPANDREL BEAM — A beam from column to column carrying an exterior wall and/or the outermost edge of an upper floor.

SPECIFICATIONS — A statement of particulars of a given job as to size of building, quality and performance of men and materials to be used, and the terms of the contract. A set of specifications generally indicates the design loads and design criteria.

SQUARE — One hundred square feet.

STOCK — A unit that is standard to its manufacturer. It is not custom-made.

STOOL — A shelf across the inside bottom of a window.

STUD — A vertical wall member to which exterior or interior covering or collateral material may be attached. Load-bearing studs are those which carry a portion of the loads from the floor, roof, or ceiling above as well as the collateral material on one or both sides. Nonload-bearing studs are used to support only the attached collateral materials and carry no load from the floor, roof, or ceiling above.

SUBCONTRACTOR — A separate contractor for a portion of the work (hired by the general contractor).

SUBSTANTIAL COMPLETION — For a project or specified area of a project, the date when the construction is sufficiently completed in accordance with the contract documents, as modified by any change orders agreed to by the parties, so that the owner can occupy the project or specified area of the project for the use for which it was intended.

SUPPLEMENTARY GENERAL CONDITIONS — One of the contract documents, prepared by the architect, that may modify provisions of the General Conditions of the contract.

TEMPERATURE REINFORCING — Lightweight deformed steel rods or wire mesh placed in concrete to resist possible cracks from expansion or contraction due to temperature changes.

TIME OF COMPLETION — The number of days (calendar or working) or the actual date by which completion of the work is required.

TRUSS — A structure made up of three or more members, with each member designed to carry basically a tension or a compression force. The entire structure in turn acts as a beam.

VENEER — A thin covering of valuable material over a less expensive body; for example, brick on a wood frame building.

WAINSCOT — Protective or decorative covering applied or built into the lower portion of a wall.

WALL BEARING — In cases where a floor, roof, or ceiling rests on a wall, the wall is designed to carry the load exerted. These types of walls are also referred to as load-bearing walls.

WALL COVERING — The exterior wall skin consisting of panels or sheets and including their attachment, trim, fascia, and weather sealants.

WALL NONBEARING — Wall not relied upon to support a structural system.

WATER CLOSET — More commonly known as a toilet.

WORKING DRAWING — The actual plans (drawings and illustrations) from which the building will be built. They show how the building is to be built and are included in the contract documents.

C

AIA GENERAL CONDITIONS AND STANDARD FORM OF AGREEMENT*

AIA Document A201

General Conditions of the Contract for Construction

*THIS DOCUMENT HAS IMPORTANT LEGAL CONSEQUENCES; CONSULTATION
WITH AN ATTORNEY IS ENCOURAGED WITH RESPECT TO ITS MODIFICATION*

1987 EDITION
TABLE OF ARTICLES

1. GENERAL PROVISIONS

2. OWNER

3. CONTRACTOR

4. ADMINISTRATION OF THE CONTRACT

5. SUBCONTRACTORS

6. CONSTRUCTION BY OWNER OR BY SEPARATE CONTRACTORS

7. CHANGES IN THE WORK

8. TIME

9. PAYMENTS AND COMPLETION

10. PROTECTION OF PERSONS AND PROPERTY

11. INSURANCE AND BONDS

12. UNCOVERING AND CORRECTION OF WORK

13. MISCELLANEOUS PROVISIONS

14. TERMINATION OR SUSPENSION OF THE CONTRACT

This document has been approved and endorsed by the Associated General Contractors of America.

Acceptance of Nonconforming Work 9.6.6, 9.9.3, **12.3**
Acceptance of Work 9.6.6, 9.8.2, 9.9.3, 9.10.1, 9.10.3
Access to Work **3.16**, 6.2.1, 12.1
Accident Prevention 4.2.3, 10
Acts and Omissions . . . 3.2.1, 3.2.2, 3.3.2, 3.12.8, 3.18, 4.2.3, 4.3.2,
    4.3.9, 8.3.1, 10.1.4, 10.2.5, 13.4.2, 13.7, 14.1
Addenda 1.1.1, 3.11
Additional Cost, Claims for 4.3.6, 4.3.7, 4.3.9, 6.1.1, 10.3
Additional Inspections and Testing 4.2.6, 9.8.2, 12.2.1, 13.5
Additional Time, Claims for 4.3.6, 4.3.8, 4.3.9, 8.3.2
ADMINISTRATION OF THE CONTRACT 3.3.3, **4**, 9.4, 9.5
Advertisement or Invitation to Bid 1.1.1
Aesthetic Effect 4.2.13, 4.5.1
Allowances **3.8**
All-risk Insurance 11.3.1.1
Applications for Payment . . 4.2.5, 7.3.7, 9.2, **9.3**, 9.4, 9.5.1, 9.6.3,
    9.8.3, 9.10.1, 9.10.3, 9.10.4, 11.1.3, 14.2.4
Approvals . . . . 2.4, 3.3.3, 3.5, 3.10.2, 3.12.4 through 3.12.8, 3.18.3,
    4.2.7, 9.3.2, 11.3.1.4, 13.4.2, 13.5
Arbitration 4.1.4, 4.3.2, 4.3.4, 4.4.4, **4.5**,
    8.3.1, 10.1.2, 11.3.9, 11.3.10
Architect **4.1**
Architect, Definition of 4.1.1
Architect, Extent of Authority 2.4, 3.12.6, 4.2, 4.3.2, 4.3.6,
    4.4, 5.2, 6.3, 7.1.2, 7.2.1, 7.3.6, 7.4, 9.2, 9.3.1,
    9.4, 9.5, 9.6.3, 9.8.2, 9.8.3, 9.10.1, 9.10.3, 12.1, 12.2.1,
    13.5.1, 13.5.2, 14.2.2, 14.2.4
Architect, Limitations of Authority and Responsibility . . 3.3.3, 3.12.8,
    3.12.11, 4.1.2, 4.2.1, 4.2.2, 4.2.3, 4.2.6, 4.2.7, 4.2.10, 4.2.12,
    4.2.13, 4.3.2, 5.2.1, 7.4, 9.4.2, 9.6.4, 9.6.6
Architect's Additional Services and Expenses 2.4, 9.8.2,
    11.3.1.1, 12.2.1, 12.2.4, 13.5.2, 13.5.3, 14.2.4
Architect's Administration of the Contract **4.2**, 4.3.6,
    4.3.7, 4.4, 9.4, 9.5
Architect's Approvals   2.4, 3.5.1, 3.10.2, 3.12.6, 3.12.8, 3.18.3, 4.2.7
Architect's Authority to Reject Work 3.5.1, 4.2.6, 12.1.2, 12.2.1
Architect's Copyright 1.3
Architect's Decisions 4.2.6, 4.2.7, 4.2.11, 4.2.12, 4.2.13,
    4.3.2, 4.3.6, 4.4.1, 4.4.4, 4.5, 6.3, 7.3.6, 7.3.8, 8.1.3, 8.3.1,
    9.2, 9.4, 9.5.1, 9.8.2, 9.9.1, 10.1.2, 13.5.2, 14.2.2, 14.2.4
Architect's Inspections 4.2.2, 4.2.9, 4.3.6, 9.4.2, 9.8.2,
    9.9.2, 9.10.1, 13.5
Architect's Instructions . . 4.2.6, 4.2.7, 4.2.8, 4.3.7, 7.4.1, 12.1, 13.5.2
Architect's Interpretations 4.2.11, 4.2.12, 4.3.7
Architect's On-Site Observations 4.2.2, 4.2.5, 4.3.6, 9.4.2,
    9.5.1, 9.10.1, 13.5
Architect's Project Representative 4.2.10
Architect's Relationship with Contractor 1.1.2, 3.2.1, 3.2.2,
    3.3.3, 3.5.1, 3.7.3, 3.11, 3.12.8, 3.12.11, 3.16, 3.18, 4.2.3, 4.2.4,
    4.2.6, 4.2.12, 5.2, 6.2.2, 7.3.4, 9.8.2, 11.3.7, 12.1, 13.5
Architect's Relationship with Subcontractors . . . 1.1.2, 4.2.3, 4.2.4,
    4.2.6, 9.6.3, 9.6.4, 11.3.7
Architect's Representations 9.4.2, 9.5.1, 9.10.1
Architect's Site Visits 4.2.2, 4.2.5, 4.2.9, 4.3.6, 9.4.2, 9.5.1,
    9.8.2, 9.9.2, 9.10.1, 13.5
Asbestos 10.1
Attorneys' Fees 3.18.1, 9.10.2, 10.1.4
Award of Separate Contracts 6.1.1
Award of Subcontracts and Other Contracts for
    Portions of the Work **5.2**
Basic Definitions **1.1**
Bidding Requirements 1.1.1, 1.1.7, 5.2.1, 11.4.1
Boiler and Machinery Insurance **11.3.2**
Bonds, Lien 9.10.2
Bonds, Performance and Payment 7.3.6.4, 9.10.3, 11.3.9, 11.4

Building Permit 3.7.1
Capitalization **1.4**
Certificate of Substantial Completion 9.8.2
Certificates for Payment 4.2.5, 4.2.9, 9.3.3, **9.4**, 9.5, 9.6.1,
    9.6.6, 9.7.1, 9.8.3, 9.10.1, 9.10.3, 13.7, 14.1.1.3, 14.2.4
Certificates of Inspection, Testing or Approval 3.12.11, 13.5.4
Certificates of Insurance 9.3.2, 9.10.2, 11.1.3
Change Orders 1.1.1, 2.4.1, 3.8.2.4, 3.11, 4.2.8, 4.3.3, 5.2.3,
    7.1, **7.2**, 7.3.2, 8.3.1, 9.3.1.1, 9.10.3, 11.3.1.2,
    11.3.4, 11.3.9, 12.1.2
Change Orders, Definition of 7.2.1
Changes **7.1**
CHANGES IN THE WORK 3.11, 4.2.8, **7**, 8.3.1, 9.3.1.1, 10.1.3
Claim, **Definition** of 4.3.1
Claims and Disputes **4.3**, 4.4, 4.5, 6.2.5, 8.3.2,
    9.3.1.2, 9.3.3, 9.10.4, 10.1.4
Claims and Timely Assertion of Claims 4.5.6
Claims for Additional Cost 4.3.6, **4.3.7**, 4.3.9, 6.1.1, 10.3
Claims for Additional Time 4.3.6, **4.3.8**, 4.3.9, 8.3.2
Claims for Concealed or Unknown Conditions 4.3.6
Claims for Damages . . . 3.18, 4.3.9, 6.1.1, 6.2.5, 8.3.2, 9.5.1.2, 10.1.4
Claims Subject to Arbitration 4.3.2, 4.4.4, 4.5.1
Cleaning Up **3.15**, 6.3
Commencement of Statutory Limitation Period **13.7**
Commencement of the Work, Conditions Relating to 2.1.2,
    2.2.1, 3.2.1, 3.2.2, 3.7.1, 3.10.1, 3.12.6, 4.3.7, 5.2.1,
    6.2.2, 8.1.2, 8.2.2, 9.2, 11.1.3, 11.3.6, 11.4.1
Commencement of the Work, Definition of 8.1.2
Communications Facilitating Contract
    Administration 3.9.1, 4.2.4, 5.2.1
Completion, Conditions Relating to 3.11, 3.15, 4.2.2, 4.2.9,
    4.3.2, 9.4.2, 9.8, 9.9.1, 9.10, 11.3.5, 12.2.2, 13.7.1
COMPLETION, PAYMENTS AND **9**
Completion, Substantial 4.2.9, 4.3.5.2, 8.1.1, 8.1.3, 8.2.3,
    9.8, 9.9.1, 12.2.2, 13.7
Compliance with Laws 1.3, 3.6, 3.7, 3.13, 4.1.1, 10.2.2, 11.1,
    11.3, 13.1, 13.5.1, 13.5.2, 13.6, 14.1.1, 14.2.1.3
Concealed or Unknown Conditions 4.3.6
Conditions of the Contract 1.1.1, 1.1.7, 6.1.1
Consent, Written 1.3.1, 3.12.8, 3.14.2, 4.1.2,
    4.3.4, 4.5.5, 9.3.2, 9.8.2, 9.9.1, 9.10.2, 9.10.3, 10.1.2, 10.1.3,
    11.3.1, 11.3.1.4, 11.3.11, 13.2, 13.4.2
CONSTRUCTION BY OWNER OR BY SEPARATE
    CONTRACTORS 1.1.4, **6**
Construction Change Directive, Definition of 7.3.1
Construction Change Directives 1.1.1, 4.2.8, 7.1, **7.3**, 9.3.1.1
Construction Schedules, Contractor's 3.10, 6.1.3
Contingent Assignment of Subcontracts **5.4**
Continuing Contract Performance **4.3.4**
Contract, Definition of 1.1.2
CONTRACT, TERMINATION OR
    SUSPENSION OF THE 4.3.7, 5.4.1.1, **14**
Contract Administration 3.3.3, 4, 9.4, 9.5
Contract Award and Execution, Conditions Relating to 3.7.1,
    3.10, 5.2, 9.2, 11.1.3, 11.3.6, 11.4.1
Contract Documents, The **1.1**, 1.2, 7
Contract Documents, Copies Furnished and Use of . . . 1.3, 2.2.5, 5.3
Contract Documents, Definition of 1.1.1
Contract Performance During Arbitration 4.3.4, 4.5.3
Contract Sum 3.8, 4.3.6, 4.3.7, 4.4.4, 5.2.3,
    6.1.3, 7.2, 7.3, **9.1**, 9.7, 11.3.1, 12.2.4, 12.3, 14.2.4
Contract Sum, Definition of **9.1**
Contract Time 4.3.6, 4.3.8, 4.4.4, 7.2.1.3, 7.3,
    8.2.1, 8.3.1, 9.7, 12.1.1
Contract Time, **Definition** of 8.1.1

AIA DOCUMENT A201 • GENERAL CONDITIONS OF THE CONTRACT FOR CONSTRUCTION • FOURTEENTH EDITION
AIA® • ©1987 THE AMERICAN INSTITUTE OF ARCHITECTS, 1735 NEW YORK AVENUE, N.W., WASHINGTON, D.C. 20006

CONTRACTOR **3**
Contractor, **Definition** of **3.1**, 6.1.2
Contractor's Bid 1.1.1
Contractor's Construction Schedules **3.10**, 6.1.3
Contractor's Employees 3.3.2, 3.4.2, 3.8.1, 3.9, 3.18, 4.2.3,
        4.2.6, 8.1.2, 10.2, 10.3, 11.1.1, 14.2.1.1
Contractor's Liability Insurance **11.1**
Contractor's Relationship with Separate Contractors
  and Owner's Forces 2.2.6, 3.12.5, 3.14.2, 4.2.4, 6, 12.2.5
Contractor's Relationship with Subcontractors 1.2.4, 3.3.2,
      3.18.1, 3.18.2, 5.2, 5.3, 5.4, 9.6.2, 11.3.7, 11.3.8, 14.2.1.2
Contractor's Relationship with the Architect 1.1.2, 3.2.1, 3.2.2,
    3.3.3, 3.5.1, 3.7.3, 3.11, 3.12.8 3.16, 3.18, 4.2.3, 4.2.4, 4.2.6,
        4.2.12, 5.2, 6.2.2, 7.3.4, 9.8.2, 11.3.7, 12.1, 13.5
Contractor's Representations . . 1.2.2, 3.5.1, 3.12.7, 6.2.2, 8.2.1, 9.3.3
Contractor's Responsibility for Those
  Performing the Work 3.3.2, 3.18, 4.2.3, 10
Contractor's Review of Contract Documents 1.2.2, 3.2, 3.7.3
Contractor's Right to Stop the Work 9.7
Contractor's Right to Terminate the Contract 14.1
Contractor's Submittals 3.10, 3.11, 3.12, 4.2.7, 5.2.1, 5.2.3,
        7.3.6, 9.2, 9.3.1, 9.8.2, 9.9.1, 9.10.2,
        9.10.3, 10.1.2, 11.4.2, 11.4.3
Contractor's Superintendent 3.9, 10.2.6
Contractor's Supervision and Construction Procedures 1.2.4,
        3.3, 3.4, 4.2.3, 8.2.2, 8.2.3, 10
Contractual Liability Insurance 11.1.1.7, 11.2.1
Coordination and Correlation 1.2.2, 1.2.4, 3.3.1,
        3.10, 3.12.7, 6.1.3, 6.2.1
Copies Furnished of Drawings and Specifications . . . 1.3, 2.2.5, 3.11
Correction of Work 2.3, 2.4, 4.2.1, 9.8.2,
        9.9.1, 12.1.2, 12.2, 13.7.1.3
Cost, Definition of 7.3.6, 14.3.5
Costs 2.4, 3.2.1, 3.7.4, 3.8.2, 3.15.2, 4.3.6, 4.3.7, 4.3.8.1, 5.2.3,
    6.1.1, 6.2.3, 6.3, 7.3.3.3, 7.3.6, 7.3.7, 9.7, 9.8.2, 9.10.2, 11.3.1.2,
    11.3.1.3, 11.3.4, 11.3.9, 12.1, 12.2.1, 12.2.4, 12.2.5, 13.5, 14
Cutting and Patching **3.14**, 6.2.6
Damage to Construction of Owner or Separate Contractors . 3.14.2,
        6.2.4, 9.5.1.5, 10.2.1.2, 10.2.5, 10.3, 11.1, 11.3, 12.2.5
Damage to the Work 3.14.2, 9.9.1, 10.2.1.2, 10.2.5, 10.3, 11.3
Damages, Claims for . . 3.18, 4.3.9, 6.1.1, 6.2.5, 8.3.2, 9.5.1.2, 10.1.4
Damages for Delay 6.1.1, 8.3.3, 9.5.1.6, 9.7
Date of Commencement of the Work, Definition of 8.1.2
Date of Substantial Completion, Definition of 8.1.3
Day, Definition of 8.1.4
Decisions of the Architect 4.2.6, 4.2.7, 4.2.11, 4.2.12, 4.2.13,
    4.3.2, 4.3.6, 4.4.1, 4.4.4, 4.5, 6.3, 7.3.6, 7.3.8, 8.1.3, 8.3.1, 9.2,
        9.4, 9.5.1, 9.8.2, 9.9.1, 10.1.2, 13.5.2, 14.2.2, 14.2.4
Decisions to Withhold Certification **9.5**, 9.7, 14.1.1.3
Defective or Nonconforming Work, Acceptance,
  Rejection and Correction of 2.3, 2.4, 3.5.1, 4.2.1,
        4.2.6, 4.3.5, 9.5.2, 9.8.2, 9.9.1, 10.2.5, 12, 13.7.1.3
Defective Work, Definition of 3.5.1
Definitions 1.1, 2.1.1, 3.1, 3.5.1, 3.12.1, 3.12.2, 3.12.3, 4.1.1,
        4.3.1, 5.1, 6.1.2, 7.2.1, 7.3.1, 7.3.6, 8.1, 9.1, 9.8.1
Delays and Extensions of Time 4.3.1, 4.3.8.1, 4.3.8.2,
        6.1.1, 6.2.3, 7.2.1, 7.3.1, 7.3.4, 7.3.5, 7.3.8,
        7.3.9, 8.1.1, **8.3**, 10.3.1, 14.1.1.4
Disputes 4.1.4, 4.3, 4.4, 4.5, 6.2.5, 6.3, 7.3.8, 9.3.1.2
Documents and Samples at the Site 3.11
Drawings, Definition of 1.1.5
Drawings and Specifications, Use and Ownership of 1.1.1, 1.3,
        2.2.5, 3.11, 5.3
Duty to Review Contract Documents and Field Conditions 3.2
Effective Date of Insurance 8.2.2, 11.1.2

Emergencies 4.3.7, **10.3**
Employees, Contractor's 3.3.2, 3.4.2, 3.8.1, 3.9, 3.18.1,
    3.18.2, 4.2.3, 4.2.6, 8.1.2, 10.2, 10.3, 11.1.1, 14.2.1.1
Equipment, Labor, Materials and 1.1.3, 1.1.6, 3.4, 3.5.1,
    3.8.2, 3.12.3, 3.12.7, 3.12.11, 3.13, 3.15.1, 4.2.7,
        6.2.1, 7.3.6, 9.3.2, 9.3.3, 11.3, 12.2.4, 14
Execution and Progress of the Work 1.1.3, 1.2.3, 3.2, 3.4.1,
    3.5.1, 4.2.2, 4.2.3, 4.3.4, 4.3.8, 6.2.2, 7.1.3,
        7.3.9, 8.2, 8.3, 9.5, 9.9.1, 10.2, 14.2, 14.3
Execution, Correlation and Intent of the
  Contract Documents **1.2**, 3.7.1
Extensions of Time 4.3.1, 4.3.8, 7.2.1.3, 8.3, 10.3.1
Failure of Payment by Contractor 9.5.1.3, 14.2.1.2
Failure of Payment by Owner 4.3.7, 9.7, 14.1.3
Faulty Work (See Defective or Nonconforming Work)
Final Completion and Final Payment 4.2.1, 4.2.9, 4.3.2,
    4.3.5, **9.10**, 11.1.2, 11.1.3, 11.3.5, 12.3.1, 13.7
Financial Arrangements, Owner's 2.2.1
Fire and Extended Coverage Insurance 11.3
GENERAL PROVISIONS **1**
Governing Law **13.1**
Guarantees (See Warranty and Warranties)
Hazardous Materials 10.1, 10.2.4
Identification of Contract Documents 1.2.1
Identification of Subcontractors and Suppliers 5.2.1
Indemnification 3.17, **3.18**, 9.10.2, 10.1.4, 11.3.1.2, 11.3.7
Information and Services Required of the Owner 2.1.2, **2.2**,
    4.3.4, 6.1.3, 6.1.4, 6.2.6, 9.3.2, 9.6.1, 9.6.4, 9.8.3, 9.9.2,
        9.10.3, 10.1.4, 11.2, 11.3, 13.5.1, 13.5.2
Injury or Damage to Person or Property **4.3.9**
Inspections 3.3.3, 3.3.4, 3.7.1, 4.2.2,
    4.2.6, 4.2.9, 4.3.6, 9.4.2, 9.8.2, 9.9.2, 9.10.1, 13.5
Instructions to Bidders 1.1.1
Instructions to the Contractor 3.8.1, 4.2.8, 5.2.1, 7, 12.1, 13.5.2
Insurance 4.3.9. 6.1.1, 7.3.6.4, 9.3.2, 9.8.2, 9.9.1, 9.10.2, 11
Insurance, Boiler and Machinery **11.3.2**
Insurance, Contractor's Liability **11.1**
Insurance, Effective Date of 8.2.2, 11.1.2
Insurance, Loss of Use **11.3.3**
Insurance, Owner's Liability **11.2**
Insurance, Property 10.2.5, **11.3**
Insurance, Stored Materials 9.3.2, 11.3.1.4
INSURANCE AND BONDS **11**
Insurance Companies, Consent to Partial Occupancy . . 9.9.1, 11.3.11
Insurance Companies, Settlement with 11.3.10
Intent of the Contract Documents 1.2.3, 3.12.4,
    4.2.6, 4.2.7, 4.2.12, 4.2.13, 7.4
Interest **13.6**
Interpretation 1.2.5, 1.4, **1.5**, 4.1.1, 4.3.1, 5.1, 6.1.2, 8.1.4
Interpretations, Written 4.2.11, 4.2.12, 4.3.7
Joinder and Consolidation of Claims Required 4.5.6
Judgment on Final Award 4.5.1, 4.5.4.1, **4.5.7**
Labor and Materials, Equipment 1.1.3, 1.1.6, **3.4**, 3.5.1, 3.8.2,
    3.12.2, 3.12.3, 3.12.7, 3.12.11, 3.13, 3.15.1,
        4.2.7, 6.2.1, 7.3.6, 9.3.2, 9.3.3, 12.2.4, 14
Labor Disputes 8.3.1
Laws and Regulations 1.3, 3.6, 3.7, 3.13, 4.1.1, 4.5.5, 4.5.7,
    9.9.1, 10.2.2, 11.1, 11.3, 13.1, 13.4, 13.5.1, 13.5.2, 13.6
Liens 2.1.2, 4.3.2, 4.3.5.1, 8.2.2, 9.3.3, 9.10.2
Limitation on Consolidation or Joinder **4.5.5**
Limitations, Statutes of 4.5.4.2, 12.2.6, 13.7
Limitations of Authority 3.3.1, 4.1.2, 4.2.1,
    4.2.3, 4.2.7, 4.2.10, 5.2.2, 5.2.4, 7.4, 11.3.10

Limitations of Liability 2.3, 3.2.1, 3.5.1, 3.7.3, 3.12.8, 3.12.11,
3.17, 3.18, 4.2.6, 4.2.7, 4.2.12, 6.2.2, 9.4.2, 9.6.4, 9.10.4,
10.1.4, 10.2.5, 11.1.2, 11.2.1, 11.3.7, 13.4.2, 13.5.2

Limitations of Time, General 2.2.1, 2.2.4, 3.2.1, 3.7.3,
3.8.2, 3.10, 3.12.5, 3.15.1, 4.2.1, 4.2.7, 4.2.11, 4.3.2,
4.3.3, 4.3.4, 4.3.6, 4.3.9, 4.5.4.2, 5.2.1, 5.2.3, 6.2.4, 7.3.4, 7.4,
8.2, 9.5, 9.6.2, 9.8, 9.9, 9.10, 11.1.3, 11.3.1, 11.3.2, 11.3.5,
11.3.6, 12.2.1, 12.2.2, 13.5, 13.7
Limitations of Time, Specific 2.1.2, 2.2.1, 2.4, 3.10, 3.11,
3.15.1, 4.2.1, 4.2.11, 4.3, 4.4, 4.5, 5.3, 5.4, 7.3.5, 7.3.9, 8.2,
9.2, 9.3.1, 9.3.3, 9.4.1, 9.6.1, 9.7, 9.8.2, 9.10.2, 11.1.3, 11.3.6,
11.3.10, 11.3.11, 12.2.2, 12.2.4, 12.2.6, 13.7, 14
Loss of Use Insurance **11.3.3**
Material Suppliers 1.3.1, 3.12.1, 4.2.4, 4.2.6, 5.2.1,
9.3.1, 9.3.1.2, 9.3.3, 9.4.2, 9.6.5, 9.10.4
Materials, Hazardous 10.1, 10.2.4
Materials, Labor, Equipment and 1.1.3, 1.1.6, 3.4, 3.5.1, 3.8.2,
3.12.2, 3.12.3, 3.12.7, 3.12.11, 3.13, 3.15.1, 4.2.7, 6.2.1,
7.3.6, 9.3.2, 9.3.3, 12.2.4, 14
Means, Methods, Techniques, Sequences and
Procedures of Construction 3.3.1, 4.2.3, 4.2.7, 9.4.2
Minor Changes in the Work 1.1.1, 4.2.8, 4.3.7, 7.1, **7.4**
MISCELLANEOUS PROVISIONS **13**
Modifications, Definition of 1.1.1
Modifications to the Contract 1.1.1, 1.1.2, 3.7.3, 3.11,
4.1.2, 4.2.1, 5.2.3, 7, 8.3.1, 9.7
Mutual Responsibility **6.2**
Nonconforming Work, Acceptance of **12.3**
Nonconforming Work, Rejection and Correction of 2.3.1,
4.3.5, 9.5.2, 9.8.2, 12, 13.7.1.3
Notice 2.3, 2.4, 3.2.1, 3.2.2, 3.7.3, 3.7.4, 3.9, 3.12.8,
3.12.9, 3.17, 4.3, 4.4.4, 4.5, 5.2.1, 5.3, 5.4.1.1, 8.2.2, 9.4.1,
9.5.1, 9.6.1, 9.7, 9.10, 10.1.2, 10.2.6, 11.1.3, 11.3, 12.2.2.1,
12.2.4, 13.3, 13.5.1, 13.5.2, 14
Notice, Written 2.3, 2.4, 3.9, 3.12.8, 3.12.9, 4.3,
4.4.4, 4.5, 5.2.1, 5.3, 5.4.1.1, 8.2.2, 9.4.1, 9.5.1, 9.7, 9.10,
10.1.2, 10.2.6, 11.1.3, 11.3, 12.2.2, 12.2.4, **13.3**, 13.5.2, 14
Notice of Testing and Inspections 13.5.1, 13.5.2
Notice to Proceed 8.2.2
Notices, Permits, Fees and 2.2.3, **3.7**, 3.13, 7.3.6.4, 10.2.2
Observations, Architect's On-Site 4.2.2, 4.2.5,
4.3.6, 9.4.2, 9.5.1, 9.10.1, 13.5
Observations, Contractor's 1.2.2, 3.2.2
Occupancy 9.6.6, 9.8.1, 9.9, 11.3.11
On-Site Inspections by the Architect 4.2.2, 4.2.9, 4.3.6,
9.4.2, 9.8.2, 9.9.2, 9.10.1
On-Site Observations by the Architect 4.2.2, 4.2.5, 4.3.6,
9.4.2, 9.5.1, 9.10.1, 13.5
Orders, Written 2.3, 3.9, 4.3.7, 7, 8.2.2, 11.3.9, 12.1,
12.2, 13.5.2, 14.3.1

OWNER **2**
Owner, **Definition** of **2.1**
Owner, Information and Services Required of the 2.1.2,
2.2, 4.3.4, 6, 9, 10.1.4, 11.2, 11.3, 13.5.1, 14.1.1.5, 14.1.3
Owner's Authority 3.8.1, 4.1.3, 4.2.9, 5.2.1, 5.2.4, 5.4.1,
7.3.1, 8.2.2, 9.3.1, 9.3.2, 11.4.1, 12.2.4, 13.5.2, 14.2, 14.3.1
Owner's Financial Capability 2.2.1, 14.1.1.5
Owner's Liability Insurance **11.2**
Owner's Loss of Use Insurance 11.3.3
Owner's Relationship with Subcontractors 1.1.2,
5.2.1, 5.4.1, 9.6.4
Owner's Right to Carry Out the Work 2.4, 12.2.4, 14.2.2.2
Owner's Right to Clean Up **6.3**

Owner's Right to Perform Construction and to
Award Separate Contracts **6.1**
Owner's Right to Stop the Work **2.3**, 4.3.7
Owner's Right to Suspend the Work 14.3
Owner's Right to Terminate the Contract 14.2
Ownership and Use of Architect's Drawings, Specifications
and Other Documents 1.1.1, **1.3**, 2.2.5, 5.3
Partial Occupancy or Use 9.6.6, **9.9**, 11.3.11
Patching, Cutting and **3.14**, 6.2.6
Patents, Royalties and **3.17**
Payment, Applications for 4.2.5, 9.2, **9.3**, 9.4,
9.5.1, 9.8.3, 9.10.1, 9.10.3, 9.10.4, 14.2.4
Payment, Certificates for 4.2.5, 4.2.9, 9.3.3, **9.4**, 9.5,
9.6.1, 9.6.6, 9.7.1, 9.8.3, 9.10.1, 9.10.3, 13.7, 14.1.1.3, 14.2.4
Payment, Failure of 4.3.7, 9.5.1.3,
9.7, 9.10.2, 14.1.1.3, 14.2.1.2
Payment, Final 4.2.1, 4.2.9, 4.3.2, 4.3.5, 9.10, 11.1.2,
11.1.3, 11.3.5, 12.3.1
Payment Bond, Performance Bond and 7.3.6.4,
9.10.3, 11.3.9, **11.4**
Payments, Progress 4.3.4, 9.3, 9.6,
9.8.3, 9.10.3, 13.6, 14.2.3
PAYMENTS AND COMPLETION **9,** 14
Payments to Subcontractors 5.4.2, 9.5.1.3,
9.6.2, 9.6.3, 9.6.4, 11.3.8, 14.2.1.2
PCB 10.1
Performance Bond and Payment Bond 7.3.6.4,
9.10.3, 11.3.9, 11.4
Permits, Fees and Notices 2.2.3, **3.7,** 3.13, 7.3.6.4, 10.2.2
PERSONS AND PROPERTY, PROTECTION OF **10**
Polychlorinated Biphenyl 10.1
Product Data, Definition of 3.12.2
Product Data and Samples, Shop Drawings . . . 3.11, **3.12,** 4.2.7
Progress and Completion 4.2.2, 4.3.4, **8.2**
Progress Payments 4.3.4, 9.3,
9.6, 9.8.3, 9.10.3, 13.6, 14.2.3
Project, Definition of the **1.1.4**
Project Manual, Definition of the **1.1.7**
Project Manuals 2.2.5
Project Representatives 4.2.10
Property Insurance 10.2.5, **11.3**
PROTECTION OF PERSONS AND PROPERTY **10**
Regulations and Laws 1.3, 3.6, 3.7, 3.13, 4.1.1, 4.5.5,
4.5.7, 10.2.2, 11.1, 11.3, 13.1, 13.4, 13.5.1, 13.5.2, 13.6, 14
Rejection of Work 3.5.1, 4.2.6, 12.2
Releases of Waivers and Liens 9.10.2
Representations 1.2.2, 3.5.1, 3.12.7,
6.2.2, 8.2.1, 9.3.3, 9.4.2, 9.5.1, 9.8.2, 9.10.1
Representatives 2.1.1, 3.1.1, 3.9,
4.1.1, 4.2.1, 4.2.10, 5.1.1, 5.1.2, 13.2.1
Resolution of Claims and Disputes **4.4,** 4.5
Responsibility for Those Performing the Work 3.3.2,
4.2.3, 6.1.3, 6.2, 10
Retainage 9.3.1, 9.6.2, 9.8.3, 9.9.1, 9.10.2, 9.10.3
Review of Contract Documents and Field
Conditions by Contractor 1.2.2, **3.2,** 3.7.3, 3.12.7
Review of Contractor's Submittals by
Owner and Architect 3.10.1, 3.10.2, 3.11, 3.12,
4.2.7, 4.2.9, 5.2.1, 5.2.3, 9.2, 9.8.2
Review of Shop Drawings, Product Data
and Samples by Contractor 3.12.5
Rights and Remedies 1.1.2, 2.3, 2.4, 3.5.1, 3.15.2,
4.2.6, 4.3.6, 4.5, 5.3, 6.1, 6.3, 7.3.1, 8.3.1, 9.5.1, 9.7, 10.2.5,
10.3, 12.2.2, 12.2.4, **13.4,** 14
Royalties and Patents 3.17

AIA DOCUMENT A201 • GENERAL CONDITIONS OF THE CONTRACT FOR CONSTRUCTION • FOURTEENTH EDITION
AIA® • ©1987 THE AMERICAN INSTITUTE OF ARCHITECTS, 1735 NEW YORK AVENUE, N.W., WASHINGTON, D.C. 20006

Rules and Notices for Arbitration **4.5.2**
Safety of Persons and Property **10.2**
Safety Precautions and Programs 4.2.3, 4.2.7, **10.1**
Samples, Definition of 3.12.3
Samples, Shop Drawings, Product Data and . . . 3.11, **3.12,** 4.2.7
Samples at the Site, Documents and **3.11**
Schedule of Values **9.2,** 9.3.1
Schedules, Construction 3.10
Separate Contracts and Contractors 1.1.4, 3.14.2, 4.2.4,
4.5.5, 6, 11.3.7, 12.1.2, 12.2.5
Shop Drawings, Definition of 3.12.1
Shop Drawings, Product Data and Samples 3.11, **3.12,** 4.2.7
Site, Use of **3.13,** 6.1.1, 6.2.1
Site Inspections 1.2.2, 3.3.4, 4.2.2, 4.2.9, 4.3.6, 9.8.2, 9.10.1, 13.5
Site Visits, Architect's 4.2.2, 4.2.5, 4.2.9, 4.3.6,
9.4.2, 9.5.1, 9.8.2, 9.9.2, 9.10.1, 13.5
Special Inspections and Testing 4.2.6, 12.2.1, 13.5
Specifications, Definition of the **1.1.6**
Specifications, The 1.1.1, **1.1.6,** 1.1.7, 1.2.4, 1.3, 3.11
Statutes of Limitations 4.5.4.2, 12.2.6, 13.7
Stopping the Work 2.3, 4.3.7, 9.7, 10.1.2, 10.3, 14.1
Stored Materials 6.2.1, 9.3.2, 10.2.1.2, 11.3.1.4, 12.2.4
Subcontractor, Definition of 5.1.1
SUBCONTRACTORS **5**
Subcontractors, Work by 1.2.4, 3.3.2, 3.12.1,
4.2.3, 5.3, 5.4
Subcontractual Relations **5.3,** 5.4, 9.3.1.2, 9.6.2,
9.6.3, 9.6.4, 10.2.1, 11.3.7, 11.3.8, 14.1.1, 14.2.1.2, 14.3.2
Submittals 1.3, 3.2.3, 3.10, 3.11, 3.12, 4.2.7, 5.2.1, 5.2.3,
7.3.6, 9.2, 9.3.1, 9.8.2, 9.9.1, 9.10.2, 9.10.3, 10.1.2, 11.1.3
Subrogation, Waivers of 6.1.1, 11.3.5, **11.3.7**
Substantial Completion 4.2.9, 4.3.5.2, 8.1.1, 8.1.3,
8.2.3, **9.8,** 9.9.1, 12.2.1, 12.2.2, 13.7
Substantial Completion, Definition of 9.8.1
Substitution of Subcontractors 5.2.3, 5.2.4
Substitution of the Architect 4.1.3
Substitutions of Materials 3.5.1
Sub-subcontractor, Definition of 5.1.2
Subsurface Conditions 4.3.6
Successors and Assigns **13.2**
Superintendent 3.9, 10.2.6
Supervision and Construction Procedures 1.2.4, **3.3,** 3.4,
4.2.3, 4.3.4, 6.1.3, 6.2.4, 7.1.3, 7.3.4, 8.2, 8.3.1, 10, 12, 14
Surety 4.4.1, 4.4.4, 5.4.1.2, 9.10.2, 9.10.3, 14.2.2
Surety, Consent of 9.9.1, 9.10.2, 9.10.3
Surveys 2.2.2, 3.18.3

Suspension by the Owner for Convenience **14.3**
Suspension of the Work 4.3.7, 5.4.2, 14.1.1.4, 14.3
Suspension or Termination of the Contract 4.3.7, 5.4.1.1, 14
Taxes **3.6,** 7.3.6.4
Termination by the Contractor **14.1**
Termination by the Owner for Cause 5.4.1.1, **14.2**
Termination of the Architect 4.1.3
Termination of the Contractor 14.2.2
TERMINATION OR SUSPENSION OF THE CONTRACT **14**
Tests and Inspections 3.3.3, 4.2.6, 4.2.9, 9.4.2, 12.2.1, **13.5**
TIME **8**
Time, Delays and Extensions of 4.3.8, 7.2.1, **8.3**
Time Limits, Specific 2.1.2, 2.2.1, 2.4, 3.10, 3.11, 3.15.1,
4.2.1, 4.2.11, 4.3, 4.4, 4.5, 5.3, 5.4, 7.3.5, 7.3.9, 8.2, 9.2, 9.3.1,
9.3.3, 9.4.1, 9.6.1, 9.7, 9.8.2, 9.10.2, 11.1.3, 11.3.6, 11.3.10,
11.3.11, 12.2.2, 12.2.4, 12.2.6, 13.7, 14
Time Limits on Claims 4.3.2, **4.3.3,** 4.3.6, 4.3.9, 4.4, 4.5
Title to Work 9.3.2, 9.3.3
UNCOVERING AND CORRECTION OF WORK **12**
Uncovering of Work **12.1**
Unforeseen Conditions 4.3.6, 8.3.1, 10.1
Unit Prices 7.1.4, 7.3.3.2
Use of Documents 1.1.1, 1.3, 2.2.5, 3.12.7, 5.3
Use of Site **3.13,** 6.1.1, 6.2.1
Values, Schedule of **9.2,** 9.3.1
Waiver of Claims: Final Payment **4.3.5,** 4.5.1, 9.10.3
Waiver of Claims by the Architect 13.4.2
Waiver of Claims by the Contractor 9.10.4, 11.3.7, 13.4.2
Waiver of Claims by the Owner 4.3.5, 4.5.1, 9.9.3,
9.10.3, 11.3.3, 11.3.7, 13.4.2
Waiver of Liens 9.10.2
Waivers of Subrogation 6.1.1, 11.3.5, 11.3.7
Warranty and Warranties **3.5,** 4.2.9,
4.3.5.3, 9.3.3, 9.8.2, 9.9.1, 12.2.2, 13.7.1.3
Weather Delays 4.3.8.2
When Arbitration May Be Demanded **4.5.4**
Work, Definition of 1.1.3
Written Consent 1.3.1, 3.12.8, 3.14.2, 4.1.2, 4.3.4,
4.5.5, 9.3.2, 9.8.2, 9.9.1, 9.10.2, 9.10.3, 10.1.2, 10.1.3,
11.3.1, 11.3.1.4, 11.3.11, 13.2, 13.4.2
Written Interpretations 4.2.11, 4.2.12, 4.3.7
Written Notice 2.3, 2.4, 3.9, 3.12.8, 3.12.9, 4.3, 4.4.4,
4.5, 5.2.1, 5.3, 5.4.1.1, 8.2.2, 9.4.1, 9.5.1, 9.7, 9.10, 10.1.2,
10.2.6, 11.1.3, 11.3, 12.2.2, 12.2.4, **13.3,** 13.5.2, 14
Written Orders 2.3, 3.9, 4.3.7,
7, 8.2.2, 11.3.9, 12.1, 12.2, 13.5.2, 14.3.1

ARTICLE 1

GENERAL PROVISIONS

1.1 BASIC DEFINITIONS

1.1.1 THE CONTRACT DOCUMENTS

The Contract Documents consist of the Agreement between Owner and Contractor (hereinafter the Agreement), Conditions of the Contract (General, Supplementary and other Conditions), Drawings, Specifications, addenda issued prior to execution of the Contract, other documents listed in the Agreement and Modifications issued after execution of the Contract. A Modification is (1) a written amendment to the Contract signed by both parties, (2) a Change Order, (3) a Construction Change Directive or (4) a written order for a minor change in the Work issued by the Architect. Unless specifically enumerated in the Agreement, the Contract Documents do not include other documents such as bidding requirements (advertisement or invitation to bid, Instructions to Bidders, sample forms, the Contractor's bid or portions of addenda relating to bidding requirements).

1.1.2 THE CONTRACT

The Contract Documents form the Contract for Construction. The Contract represents the entire and integrated agreement between the parties hereto and supersedes prior negotiations, representations or agreements, either written or oral. The Contract may be amended or modified only by a Modification. The Contract Documents shall not be construed to create a contractual relationship of any kind (1) between the Architect and Contractor, (2) between the Owner and a Subcontractor or Subsubcontractor or (3) between any persons or entities other than the Owner and Contractor. The Architect shall, however, be entitled to performance and enforcement of obligations under the Contract intended to facilitate performance of the Architect's duties.

1.1.3 THE WORK

The term "Work" means the construction and services required by the Contract Documents, whether completed or partially completed, and includes all other labor, materials, equipment and services provided or to be provided by the Contractor to fulfill the Contractor's obligations. The Work may constitute the whole or a part of the Project.

1.1.4 THE PROJECT

The Project is the total construction of which the Work performed under the Contract Documents may be the whole or a part and which may include construction by the Owner or by separate contractors.

1.1.5 THE DRAWINGS

The Drawings are the graphic and pictorial portions of the Contract Documents, wherever located and whenever issued, showing the design, location and dimensions of the Work, generally including plans, elevations, sections, details, schedules and diagrams.

1.1.6 THE SPECIFICATIONS

The Specifications are that portion of the Contract Documents consisting of the written requirements for materials, equip-

ment, construction systems, standards and workmanship for the Work, and performance of related services.

1.1.7 THE PROJECT MANUAL

The Project Manual is the volume usually assembled for the Work which may include the bidding requirements, sample forms, Conditions of the Contract and Specifications.

1.2 EXECUTION, CORRELATION AND INTENT

1.2.1 The Contract Documents shall be signed by the Owner and Contractor as provided in the Agreement. If either the Owner or Contractor or both do not sign all the Contract Documents, the Architect shall identify such unsigned Documents upon request.

1.2.2 Execution of the Contract by the Contractor is a representation that the Contractor has visited the site, become familiar with local conditions under which the Work is to be performed and correlated personal observations with requirements of the Contract Documents.

1.2.3 The intent of the Contract Documents is to include all items necessary for the proper execution and completion of the Work by the Contractor. The Contract Documents are complementary, and what is required by one shall be as binding as if required by all; performance by the Contractor shall be required only to the extent consistent with the Contract Documents and reasonably inferable from them as being necessary to produce the intended results.

1.2.4 Organization of the Specifications into divisions, sections and articles, and arrangement of Drawings shall not control the Contractor in dividing the Work among Subcontractors or in establishing the extent of Work to be performed by any trade.

1.2.5 Unless otherwise stated in the Contract Documents, words which have well-known technical or construction industry meanings are used in the Contract Documents in accordance with such recognized meanings.

1.3 OWNERSHIP AND USE OF ARCHITECT'S DRAWINGS, SPECIFICATIONS AND OTHER DOCUMENTS

1.3.1 The Drawings, Specifications and other documents prepared by the Architect are instruments of the Architect's service through which the Work to be executed by the Contractor is described. The Contractor may retain one contract record set. Neither the Contractor nor any Subcontractor, Subsubcontractor or material or equipment supplier shall own or claim a copyright in the Drawings, Specifications and other documents prepared by the Architect, and unless otherwise indicated the Architect shall be deemed the author of them and will retain all common law, statutory and other reserved rights, in addition to the copyright. All copies of them, except the Contractor's record set, shall be returned or suitably accounted for to the Architect, on request, upon completion of the Work. The Drawings, Specifications and other documents prepared by the Architect, and copies thereof furnished to the Contractor, are for use solely with respect to this Project. They are not to be used by the Contractor or any Subcontractor, Subsubcontractor or material or equipment supplier on other projects or for additions to this Project outside the scope of the

Work without the specific written consent of the Owner and Architect. The Contractor, Subcontractors, Sub-subcontractors and material or equipment suppliers are granted a limited license to use and reproduce applicable portions of the Drawings, Specifications and other documents prepared by the Architect appropriate to and for use in the execution of their Work under the Contract Documents. All copies made under this license shall bear the statutory copyright notice, if any, shown on the Drawings, Specifications and other documents prepared by the Architect. Submittal or distribution to meet official regulatory requirements or for other purposes in connection with this Project is not to be construed as publication in derogation of the Architect's copyright or other reserved rights.

1.4 CAPITALIZATION

1.4.1 Terms capitalized in these General Conditions include those which are (1) specifically defined, (2) the titles of numbered articles and identified references to Paragraphs, Subparagraphs and Clauses in the document or (3) the titles of other documents published by the American Institute of Architects.

1.5 INTERPRETATION

1.5.1 In the interest of brevity the Contract Documents frequently omit modifying words such as "all" and "any" and articles such as "the" and "an," but the fact that a modifier or an article is absent from one statement and appears in another is not intended to affect the interpretation of either statement.

ARTICLE 2

OWNER

2.1 DEFINITION

2.1.1 The Owner is the person or entity identified as such in the Agreement and is referred to throughout the Contract Documents as if singular in number. The term "Owner" means the Owner or the Owner's authorized representative.

2.1.2 The Owner upon reasonable written request shall furnish to the Contractor in writing information which is necessary and relevant for the Contractor to evaluate, give notice of or enforce mechanic's lien rights. Such information shall include a correct statement of the record legal title to the property on which the Project is located, usually referred to as the site, and the Owner's interest therein at the time of execution of the Agreement and, within five days after any change, information of such change in title, recorded or unrecorded.

2.2 INFORMATION AND SERVICES REQUIRED OF THE OWNER

2.2.1 The Owner shall, at the request of the Contractor, prior to execution of the Agreement and promptly from time to time thereafter, furnish to the Contractor reasonable evidence that financial arrangements have been made to fulfill the Owner's obligations under the Contract. *[Note: Unless such reasonable evidence were furnished on request prior to the execution of the Agreement, the prospective contractor would not be required to execute the Agreement or to commence the Work.]*

2.2.2 The Owner shall furnish surveys describing physical characteristics, legal limitations and utility locations for the site of the Project, and a legal description of the site.

2.2.3 Except for permits and fees which are the responsibility of the Contractor under the Contract Documents, the Owner shall secure and pay for necessary approvals, easements, assessments and charges required for construction, use or occupancy of permanent structures or for permanent changes in existing facilities.

2.2.4 Information or services under the Owner's control shall be furnished by the Owner with reasonable promptness to avoid delay in orderly progress of the Work.

2.2.5 Unless otherwise provided in the Contract Documents, the Contractor will be furnished, free of charge, such copies of Drawings and Project Manuals as are reasonably necessary for execution of the Work.

2.2.6 The foregoing are in addition to other duties and responsibilities of the Owner enumerated herein and especially those in respect to Article 6 (Construction by Owner or by Separate Contractors), Article 9 (Payments and Completion) and Article 11 (Insurance and Bonds).

2.3 OWNER'S RIGHT TO STOP THE WORK

2.3.1 If the Contractor fails to correct Work which is not in accordance with the requirements of the Contract Documents as required by Paragraph 12.2 or persistently fails to carry out Work in accordance with the Contract Documents, the Owner, by written order signed personally or by an agent specifically so empowered by the Owner in writing, may order the Contractor to stop the Work, or any portion thereof, until the cause for such order has been eliminated; however, the right of the Owner to stop the Work shall not give rise to a duty on the part of the Owner to exercise this right for the benefit of the Contractor or any other person or entity, except to the extent required by Subparagraph 6.1.3.

2.4 OWNER'S RIGHT TO CARRY OUT THE WORK

2.4.1 If the Contractor defaults or neglects to carry out the Work in accordance with the Contract Documents and fails within a seven-day period after receipt of written notice from the Owner to commence and continue correction of such default or neglect with diligence and promptness, the Owner may after such seven-day period give the Contractor a second written notice to correct such deficiencies within a second seven-day period. If the Contractor within such second seven-day period after receipt of such second notice fails to commence and continue to correct any deficiencies, the Owner may, without prejudice to other remedies the Owner may have, correct such deficiencies. In such case an appropriate Change Order shall be issued deducting from payments then or thereafter due the Contractor the cost of correcting such deficiencies, including compensation for the Architect's additional services and expenses made necessary by such default, neglect or failure. Such action by the Owner and amounts charged to the Contractor are both subject to prior approval of the Architect. If payments then or thereafter due the Contractor are not sufficient to cover such amounts, the Contractor shall pay the difference to the Owner.

ARTICLE 3

CONTRACTOR

3.1 DEFINITION

3.1.1 The Contractor is the person or entity identified as such in the Agreement and is referred to throughout the Contract Documents as if singular in number. The term "Contractor" means the Contractor or the Contractor's authorized representative.

3.2 REVIEW OF CONTRACT DOCUMENTS AND FIELD CONDITIONS BY CONTRACTOR

3.2.1 The Contractor shall carefully study and compare the Contract Documents with each other and with information furnished by the Owner pursuant to Subparagraph 2.2.2 and shall at once report to the Architect errors, inconsistencies or omissions discovered. The Contractor shall not be liable to the Owner or Architect for damage resulting from errors, inconsistencies or omissions in the Contract Documents unless the Contractor recognized such error, inconsistency or omission and knowingly failed to report it to the Architect. If the Contractor performs any construction activity knowing it involves a recognized error, inconsistency or omission in the Contract Documents without such notice to the Architect, the Contractor shall assume appropriate responsibility for such performance and shall bear an appropriate amount of the attributable costs for correction.

3.2.2 The Contractor shall take field measurements and verify field conditions and shall carefully compare such field measurements and conditions and other information known to the Contractor with the Contract Documents before commencing activities. Errors, inconsistencies or omissions discovered shall be reported to the Architect at once.

3.2.3 The Contractor shall perform the Work in accordance with the Contract Documents and submittals approved pursuant to Paragraph 3.12.

3.3 SUPERVISION AND CONSTRUCTION PROCEDURES

3.3.1 The Contractor shall supervise and direct the Work, using the Contractor's best skill and attention. The Contractor shall be solely responsible for and have control over construction means, methods, techniques, sequences and procedures and for coordinating all portions of the Work under the Contract, unless Contract Documents give other specific instructions concerning these matters.

3.3.2 The Contractor shall be responsible to the Owner for acts and omissions of the Contractor's employees, Subcontractors and their agents and employees, and other persons performing portions of the Work under a contract with the Contractor.

3.3.3 The Contractor shall not be relieved of obligations to perform the Work in accordance with the Contract Documents either by activities or duties of the Architect in the Architect's administration of the Contract, or by tests, inspections or approvals required or performed by persons other than the Contractor.

3.3.4 The Contractor shall be responsible for inspection of portions of Work already performed under this Contract to determine that such portions are in proper condition to receive subsequent Work.

3.4 LABOR AND MATERIALS

3.4.1 Unless otherwise provided in the Contract Documents, the Contractor shall provide and pay for labor, materials, equipment, tools, construction equipment and machinery, water, heat, utilities, transportation, and other facilities and services necessary for proper execution and completion of the Work, whether temporary or permanent and whether or not incorporated or to be incorporated in the Work.

3.4.2 The Contractor shall enforce strict discipline and good order among the Contractor's employees and other persons carrying out the Contract. The Contractor shall not permit employment of unfit persons or persons not skilled in tasks assigned to them.

3.5 WARRANTY

3.5.1 The Contractor warrants to the Owner and Architect that materials and equipment furnished under the Contract will be of good quality and new unless otherwise required or permitted by the Contract Documents, that the Work will be free from defects not inherent in the quality required or permitted, and that the Work will conform with the requirements of the Contract Documents. Work not conforming to these requirements, including substitutions not properly approved and authorized, may be considered defective. The Contractor's warranty excludes remedy for damage or defect caused by abuse, modifications not executed by the Contractor, improper or insufficient maintenance, improper operation, or normal wear and tear under normal usage. If required by the Architect, the Contractor shall furnish satisfactory evidence as to the kind and quality of materials and equipment.

3.6 TAXES

3.6.1 The Contractor shall pay sales, consumer, use and similar taxes for the Work or portions thereof provided by the Contractor which are legally enacted when bids are received or negotiations concluded, whether or not yet effective or merely scheduled to go into effect.

3.7 PERMITS, FEES AND NOTICES

3.7.1 Unless otherwise provided in the Contract Documents, the Contractor shall secure and pay for the building permit and other permits and governmental fees, licenses and inspections necessary for proper execution and completion of the Work which are customarily secured after execution of the Contract and which are legally required when bids are received or negotiations concluded.

3.7.2 The Contractor shall comply with and give notices required by laws, ordinances, rules, regulations and lawful orders of public authorities bearing on performance of the Work.

3.7.3 It is not the Contractor's responsibility to ascertain that the Contract Documents are in accordance with applicable laws, statutes, ordinances, building codes, and rules and regulations. However, if the Contractor observes that portions of the Contract Documents are at variance therewith, the Contractor shall promptly notify the Architect and Owner in writing, and necessary changes shall be accomplished by appropriate Modification.

3.7.4 If the Contractor performs Work knowing it to be contrary to laws, statutes, ordinances, building codes, and rules and regulations without such notice to the Architect and Owner, the Contractor shall assume full responsibility for such Work and shall bear the attributable costs.

3.8 ALLOWANCES

3.8.1 The Contractor shall include in the Contract Sum all allowances stated in the Contract Documents. Items covered by allowances shall be supplied for such amounts and by such persons or entities as the Owner may direct, but the Contractor shall not be required to employ persons or entities against which the Contractor makes reasonable objection.

3.8.2 Unless otherwise provided in the Contract Documents:

.1 materials and equipment under an allowance shall be selected promptly by the Owner to avoid delay in the Work;

.2 allowances shall cover the cost to the Contractor of materials and equipment delivered at the site and all required taxes, less applicable trade discounts;

AIA DOCUMENT A201 • GENERAL CONDITIONS OF THE CONTRACT FOR CONSTRUCTION • FOURTEENTH EDITION
AIA® • ©1987 THE AMERICAN INSTITUTE OF ARCHITECTS, 1735 NEW YORK AVENUE, N.W., WASHINGTON, D.C. 20006

.3 Contractor's costs for unloading and handling at the site, labor, installation costs, overhead, profit and other expenses contemplated for stated allowance amounts shall be included in the Contract Sum and not in the allowances;

.4 whenever costs are more than or less than allowances, the Contract Sum shall be adjusted accordingly by Change Order. The amount of the Change Order shall reflect (1) the difference between actual costs and the allowances under Clause 3.8.2.2 and (2) changes in Contractor's costs under Clause 3.8.2.3.

3.9 SUPERINTENDENT

3.9.1 The Contractor shall employ a competent superintendent and necessary assistants who shall be in attendance at the Project site during performance of the Work. The superintendent shall represent the Contractor, and communications given to the superintendent shall be as binding as if given to the Contractor. Important communications shall be confirmed in writing. Other communications shall be similarly confirmed on written request in each case.

3.10 CONTRACTOR'S CONSTRUCTION SCHEDULES

3.10.1 The Contractor, promptly after being awarded the Contract, shall prepare and submit for the Owner's and Architect's information a Contractor's construction schedule for the Work. The schedule shall not exceed time limits current under the Contract Documents, shall be revised at appropriate intervals as required by the conditions of the Work and Project, shall be related to the entire Project to the extent required by the Contract Documents, and shall provide for expeditious and practicable execution of the Work.

3.10.2 The Contractor shall prepare and keep current, for the Architect's approval, a schedule of submittals which is coordinated with the Contractor's construction schedule and allows the Architect reasonable time to review submittals.

3.10.3 The Contractor shall conform to the most recent schedules.

3.11 DOCUMENTS AND SAMPLES AT THE SITE

3.11.1 The Contractor shall maintain at the site for the Owner one record copy of the Drawings, Specifications, addenda, Change Orders and other Modifications, in good order and marked currently to record changes and selections made during construction, and in addition approved Shop Drawings, Product Data, Samples and similar required submittals. These shall be available to the Architect and shall be delivered to the Architect for submittal to the Owner upon completion of the Work.

3.12 SHOP DRAWINGS, PRODUCT DATA AND SAMPLES

3.12.1 Shop Drawings are drawings, diagrams, schedules and other data specially prepared for the Work by the Contractor or a Subcontractor, Sub-subcontractor, manufacturer, supplier or distributor to illustrate some portion of the Work.

3.12.2 Product Data are illustrations, standard schedules, performance charts, instructions, brochures, diagrams and other information furnished by the Contractor to illustrate materials or equipment for some portion of the Work.

3.12.3 Samples are physical examples which illustrate materials, equipment or workmanship and establish standards by which the Work will be judged.

3.12.4 Shop Drawings, Product Data, Samples and similar submittals are not Contract Documents. The purpose of their submittal is to demonstrate for those portions of the Work for which submittals are required the way the Contractor proposes to conform to the information given and the design concept expressed in the Contract Documents. Review by the Architect is subject to the limitations of Subparagraph 4.2.7.

3.12.5 The Contractor shall review, approve and submit to the Architect Shop Drawings, Product Data, Samples and similar submittals required by the Contract Documents with reasonable promptness and in such sequence as to cause no delay in the Work or in the activities of the Owner or of separate contractors. Submittals made by the Contractor which are not required by the Contract Documents may be returned without action.

3.12.6 The Contractor shall perform no portion of the Work requiring submittal and review of Shop Drawings, Product Data, Samples or similar submittals until the respective submittal has been approved by the Architect. Such Work shall be in accordance with approved submittals.

3.12.7 By approving and submitting Shop Drawings, Product Data, Samples and similar submittals, the Contractor represents that the Contractor has determined and verified materials, field measurements and field construction criteria related thereto, or will do so, and has checked and coordinated the information contained within such submittals with the requirements of the Work and of the Contract Documents.

3.12.8 The Contractor shall not be relieved of responsibility for deviations from requirements of the Contract Documents by the Architect's approval of Shop Drawings, Product Data, Samples or similar submittals unless the Contractor has specifically informed the Architect in writing of such deviation at the time of submittal and the Architect has given written approval to the specific deviation. The Contractor shall not be relieved of responsibility for errors or omissions in Shop Drawings, Product Data, Samples or similar submittals by the Architect's approval thereof.

3.12.9 The Contractor shall direct specific attention, in writing or on resubmitted Shop Drawings, Product Data, Samples or similar submittals, to revisions other than those requested by the Architect on previous submittals.

3.12.10 Informational submittals upon which the Architect is not expected to take responsive action may be so identified in the Contract Documents.

3.12.11 When professional certification of performance criteria of materials, systems or equipment is required by the Contract Documents, the Architect shall be entitled to rely upon the accuracy and completeness of such calculations and certifications.

3.13 USE OF SITE

3.13.1 The Contractor shall confine operations at the site to areas permitted by law, ordinances, permits and the Contract Documents and shall not unreasonably encumber the site with materials or equipment.

3.14 CUTTING AND PATCHING

3.14.1 The Contractor shall be responsible for cutting, fitting or patching required to complete the Work or to make its parts fit together properly.

3.14.2 The Contractor shall not damage or endanger a portion of the Work or fully or partially completed construction of the Owner or separate contractors by cutting, patching or otherwise altering such construction, or by excavation. The Contractor shall not cut or otherwise alter such construction by the

Owner or a separate contractor except with written consent of the Owner and of such separate contractor; such consent shall not be unreasonably withheld. The Contractor shall not unreasonably withhold from the Owner or a separate contractor the Contractor's consent to cutting or otherwise altering the Work.

3.15 CLEANING UP

3.15.1 The Contractor shall keep the premises and surrounding area free from accumulation of waste materials or rubbish caused by operations under the Contract. At completion of the Work the Contractor shall remove from and about the Project waste materials, rubbish, the Contractor's tools, construction equipment, machinery and surplus materials.

3.15.2 If the Contractor fails to clean up as provided in the Contract Documents, the Owner may do so and the cost thereof shall be charged to the Contractor.

3.16 ACCESS TO WORK

3.16.1 The Contractor shall provide the Owner and Architect access to the Work in preparation and progress wherever located.

3.17 ROYALTIES AND PATENTS

3.17.1 The Contractor shall pay all royalties and license fees. The Contractor shall defend suits or claims for infringement of patent rights and shall hold the Owner and Architect harmless from loss on account thereof, but shall not be responsible for such defense or loss when a particular design, process or product of a particular manufacturer or manufacturers is required by the Contract Documents. However, if the Contractor has reason to believe that the required design, process or product is an infringement of a patent, the Contractor shall be responsible for such loss unless such information is promptly furnished to the Architect.

3.18 INDEMNIFICATION

3.18.1 To the fullest extent permitted by law, the Contractor shall indemnify and hold harmless the Owner, Architect, Architect's consultants, and agents and employees of any of them from and against claims, damages, losses and expenses, including but not limited to attorneys' fees, arising out of or resulting from performance of the Work, provided that such claim, damage, loss or expense is attributable to bodily injury, sickness, disease or death, or to injury to or destruction of tangible property (other than the Work itself) including loss of use resulting therefrom, but only to the extent caused in whole or in part by negligent acts or omissions of the Contractor, a Subcontractor, anyone directly or indirectly employed by them or anyone for whose acts they may be liable, regardless of whether or not such claim, damage, loss or expense is caused in part by a party indemnified hereunder. Such obligation shall not be construed to negate, abridge, or reduce other rights or obligations of indemnity which would otherwise exist as to a party or person described in this Paragraph 3.18.

3.18.2 In claims against any person or entity indemnified under this Paragraph 3.18 by an employee of the Contractor, a Subcontractor, anyone directly or indirectly employed by them or anyone for whose acts they may be liable, the indemnification obligation under this Paragraph 3.18 shall not be limited by a limitation on amount or type of damages, compensation or benefits payable by or for the Contractor or a Subcontractor under workers' or workmen's compensation acts, disability benefit acts or other employee benefit acts.

3.18.3 The obligations of the Contractor under this Paragraph 3.18 shall not extend to the liability of the Architect, the Archi-

tect's consultants, and agents and employees of any of them arising out of (1) the preparation or approval of maps, drawings, opinions, reports, surveys, Change Orders, designs or specifications, or (2) the giving of or the failure to give directions or instructions by the Architect, the Architect's consultants, and agents and employees of any of them provided such giving or failure to give is the primary cause of the injury or damage.

ARTICLE 4

ADMINISTRATION OF THE CONTRACT

4.1 ARCHITECT

4.1.1 The Architect is the person lawfully licensed to practice architecture or an entity lawfully practicing architecture identified as such in the Agreement and is referred to throughout the Contract Documents as if singular in number. The term "Architect" means the Architect or the Architect's authorized representative.

4.1.2 Duties, responsibilities and limitations of authority of the Architect as set forth in the Contract Documents shall not be restricted, modified or extended without written consent of the Owner, Contractor and Architect. Consent shall not be unreasonably withheld.

4.1.3 In case of termination of employment of the Architect, the Owner shall appoint an architect against whom the Contractor makes no reasonable objection and whose status under the Contract Documents shall be that of the former architect.

4.1.4 Disputes arising under Subparagraphs 4.1.2 and 4.1.3 shall be subject to arbitration.

4.2 ARCHITECT'S ADMINISTRATION OF THE CONTRACT

4.2.1 The Architect will provide administration of the Contract as described in the Contract Documents, and will be the Owner's representative (1) during construction, (2) until final payment is due and (3) with the Owner's concurrence, from time to time during the correction period described in Paragraph 12.2. The Architect will advise and consult with the Owner. The Architect will have authority to act on behalf of the Owner only to the extent provided in the Contract Documents, unless otherwise modified by written instrument in accordance with other provisions of the Contract.

4.2.2 The Architect will visit the site at intervals appropriate to the stage of construction to become generally familiar with the progress and quality of the completed Work and to determine in general if the Work is being performed in a manner indicating that the Work, when completed, will be in accordance with the Contract Documents. However, the Architect will not be required to make exhaustive or continuous on-site inspections to check quality or quantity of the Work. On the basis of on-site observations as an architect, the Architect will keep the Owner informed of progress of the Work, and will endeavor to guard the Owner against defects and deficiencies in the Work.

4.2.3 The Architect will not have control over or charge of and will not be responsible for construction means, methods, techniques, sequences or procedures, or for safety precautions and programs in connection with the Work, since these are solely the Contractor's responsibility as provided in Paragraph 3.3. The Architect will not be responsible for the Contractor's failure to carry out the Work in accordance with the Contract Documents. The Architect will not have control over or charge of and will not be responsible for acts or omissions of the Contractor, Subcontractors, or their agents or employees, or of any other persons performing portions of the Work.

4.2.4 Communications Facilitating Contract Administration. Except as otherwise provided in the Contract Documents or when direct communications have been specially authorized, the Owner and Contractor shall endeavor to communicate through the Architect. Communications by and with the Architect's consultants shall be through the Architect. Communications by and with Subcontractors and material suppliers shall be through the Contractor. Communications by and with separate contractors shall be through the Owner.

4.2.5 Based on the Architect's observations and evaluations of the Contractor's Applications for Payment, the Architect will review and certify the amounts due the Contractor and will issue Certificates for Payment in such amounts.

4.2.6 The Architect will have authority to reject Work which does not conform to the Contract Documents. Whenever the Architect considers it necessary or advisable for implementation of the intent of the Contract Documents, the Architect will have authority to require additional inspection or testing of the Work in accordance with Subparagraphs 13.5.2 and 13.5.3, whether or not such Work is fabricated, installed or completed. However, neither this authority of the Architect nor a decision made in good faith either to exercise or not to exercise such authority shall give rise to a duty or responsibility of the Architect to the Contractor, Subcontractors, material and equipment suppliers, their agents or employees, or other persons performing portions of the Work.

4.2.7 The Architect will review and approve or take other appropriate action upon the Contractor's submittals such as Shop Drawings, Product Data and Samples, but only for the limited purpose of checking for conformance with information given and the design concept expressed in the Contract Documents. The Architect's action will be taken with such reasonable promptness as to cause no delay in the Work or in the activities of the Owner, Contractor or separate contractors, while allowing sufficient time in the Architect's professional judgment to permit adequate review. Review of such submittals is not conducted for the purpose of determining the accuracy and completeness of other details such as dimensions and quantities, or for substantiating instructions for installation or performance of equipment or systems, all of which remain the responsibility of the Contractor as required by the Contract Documents. The Architect's review of the Contractor's submittals shall not relieve the Contractor of the obligations under Paragraphs 3.3, 3.5 and 3.12. The Architect's review shall not constitute approval of safety precautions or, unless otherwise specifically stated by the Architect, of any construction means, methods, techniques, sequences or procedures. The Architect's approval of a specific item shall not indicate approval of an assembly of which the item is a component.

4.2.8 The Architect will prepare Change Orders and Construction Change Directives, and may authorize minor changes in the Work as provided in Paragraph 7.4.

4.2.9 The Architect will conduct inspections to determine the date or dates of Substantial Completion and the date of final completion, will receive and forward to the Owner for the Owner's review and records written warranties and related documents required by the Contract and assembled by the Contractor, and will issue a final Certificate for Payment upon compliance with the requirements of the Contract Documents.

4.2.10 If the Owner and Architect agree, the Architect will provide one or more project representatives to assist in carrying out the Architect's responsibilities at the site. The duties, responsibilities and limitations of authority of such project representatives shall be as set forth in an exhibit to be incorporated in the Contract Documents.

4.2.11 The Architect will interpret and decide matters concerning performance under and requirements of the Contract Documents on written request of either the Owner or Contractor. The Architect's response to such requests will be made with reasonable promptness and within any time limits agreed upon. If no agreement is made concerning the time within which interpretations required of the Architect shall be furnished in compliance with this Paragraph 4.2, then delay shall not be recognized on account of failure by the Architect to furnish such interpretations until 15 days after written request is made for them.

4.2.12 Interpretations and decisions of the Architect will be consistent with the intent of and reasonably inferable from the Contract Documents and will be in writing or in the form of drawings. When making such interpretations and decisions, the Architect will endeavor to secure faithful performance by both Owner and Contractor, will not show partiality to either and will not be liable for results of interpretations or decisions so rendered in good faith.

4.2.13 The Architect's decisions on matters relating to aesthetic effect will be final if consistent with the intent expressed in the Contract Documents.

4.3   CLAIMS AND DISPUTES

4.3.1 Definition. A Claim is a demand or assertion by one of the parties seeking, as a matter of right, adjustment or interpretation of Contract terms, payment of money, extension of time or other relief with respect to the terms of the Contract. The term "Claim" also includes other disputes and matters in question between the Owner and Contractor arising out of or relating to the Contract. Claims must be made by written notice. The responsibility to substantiate Claims shall rest with the party making the Claim.

4.3.2 Decision of Architect. Claims, including those alleging an error or omission by the Architect, shall be referred initially to the Architect for action as provided in Paragraph 4.4. A decision by the Architect, as provided in Subparagraph 4.4.4, shall be required as a condition precedent to arbitration or litigation of a Claim between the Contractor and Owner as to all such matters arising prior to the date final payment is due, regardless of (1) whether such matters relate to execution and progress of the Work or (2) the extent to which the Work has been completed. The decision by the Architect in response to a Claim shall not be a condition precedent to arbitration or litigation in the event (1) the position of Architect is vacant, (2) the Architect has not received evidence or has failed to render a decision within agreed time limits, (3) the Architect has failed to take action required under Subparagraph 4.4.4 within 30 days after the Claim is made, (4) 45 days have passed after the Claim has been referred to the Architect or (5) the Claim relates to a mechanic's lien.

4.3.3 Time Limits on Claims. Claims by either party must be made within 21 days after occurrence of the event giving rise to such Claim or within 21 days after the claimant first recognizes the condition giving rise to the Claim, whichever is later. Claims must be made by written notice. An additional Claim made after the initial Claim has been implemented by Change Order will not be considered unless submitted in a timely manner.

4.3.4 Continuing Contract Performance. Pending final resolution of a Claim including arbitration, unless otherwise agreed in writing the Contractor shall proceed diligently with performance of the Contract and the Owner shall continue to make payments in accordance with the Contract Documents.

4.3.5 Waiver of Claims: Final Payment. The making of final payment shall constitute a waiver of Claims by the Owner except those arising from:

   .1 liens, Claims, security interests or encumbrances arising out of the Contract and unsettled;

   .2 failure of the Work to comply with the requirements of the Contract Documents; or

   .3 terms of special warranties required by the Contract Documents.

4.3.6 Claims for Concealed or Unknown Conditions. If conditions are encountered at the site which are (1) subsurface or otherwise concealed physical conditions which differ materially from those indicated in the Contract Documents or (2) unknown physical conditions of an unusual nature, which differ materially from those ordinarily found to exist and generally recognized as inherent in construction activities of the character provided for in the Contract Documents, then notice by the observing party shall be given to the other party promptly before conditions are disturbed and in no event later than 21 days after first observance of the conditions. The Architect will promptly investigate such conditions and, if they differ materially and cause an increase or decrease in the Contractor's cost of, or time required for, performance of any part of the Work, will recommend an equitable adjustment in the Contract Sum or Contract Time, or both. If the Architect determines that the conditions at the site are not materially different from those indicated in the Contract Documents and that no change in the terms of the Contract is justified, the Architect shall so notify the Owner and Contractor in writing, stating the reasons. Claims by either party in opposition to such determination must be made within 21 days after the Architect has given notice of the decision. If the Owner and Contractor cannot agree on an adjustment in the Contract Sum or Contract Time, the adjustment shall be referred to the Architect for initial determination, subject to further proceedings pursuant to Paragraph 4.4.

4.3.7 Claims for Additional Cost. If the Contractor wishes to make Claim for an increase in the Contract Sum, written notice as provided herein shall be given before proceeding to execute the Work. Prior notice is not required for Claims relating to an emergency endangering life or property arising under Paragraph 10.3. If the Contractor believes additional cost is involved for reasons including but not limited to (1) a written interpretation from the Architect, (2) an order by the Owner to stop the Work where the Contractor was not at fault, (3) a written order for a minor change in the Work issued by the Architect, (4) failure of payment by the Owner, (5) termination of the Contract by the Owner, (6) Owner's suspension or (7) other reasonable grounds, Claim shall be filed in accordance with the procedure established herein.

4.3.8 Claims for Additional Time

4.3.8.1 If the Contractor wishes to make Claim for an increase in the Contract Time, written notice as provided herein shall be given. The Contractor's Claim shall include an estimate of cost and of probable effect of delay on progress of the Work. In the case of a continuing delay only one Claim is necessary.

4.3.8.2 If adverse weather conditions are the basis for a Claim for additional time, such Claim shall be documented by data substantiating that weather conditions were abnormal for the period of time and could not have been reasonably anticipated, and that weather conditions had an adverse effect on the scheduled construction.

4.3.9 Injury or Damage to Person or Property. If either party to the Contract suffers injury or damage to person or property because of an act or omission of the other party, of any of the other party's employees or agents, or of others for whose acts such party is legally liable, written notice of such injury or damage, whether or not insured, shall be given to the other party within a reasonable time not exceeding 21 days after first observance. The notice shall provide sufficient detail to enable the other party to investigate the matter. If a Claim for additional cost or time related to this Claim is to be asserted, it shall be filed as provided in Subparagraphs 4.3.7 or 4.3.8.

4.4   RESOLUTION OF CLAIMS AND DISPUTES

4.4.1 The Architect will review Claims and take one or more of the following preliminary actions within ten days of receipt of a Claim: (1) request additional supporting data from the claimant, (2) submit a schedule to the parties indicating when the Architect expects to take action, (3) reject the Claim in whole or in part, stating reasons for rejection, (4) recommend approval of the Claim by the other party or (5) suggest a compromise. The Architect may also, but is not obligated to, notify the surety, if any, of the nature and amount of the Claim.

4.4.2 If a Claim has been resolved, the Architect will prepare or obtain appropriate documentation.

4.4.3 If a Claim has not been resolved, the party making the Claim shall, within ten days after the Architect's preliminary response, take one or more of the following actions: (1) submit additional supporting data requested by the Architect, (2) modify the initial Claim or (3) notify the Architect that the initial Claim stands.

4.4.4 If a Claim has not been resolved after consideration of the foregoing and of further evidence presented by the parties or requested by the Architect, the Architect will notify the parties in writing that the Architect's decision will be made within seven days, which decision shall be final and binding on the parties but subject to arbitration. Upon expiration of such time period, the Architect will render to the parties the Architect's written decision relative to the Claim, including any change in the Contract Sum or Contract Time or both. If there is a surety and there appears to be a possibility of a Contractor's default, the Architect may, but is not obligated to, notify the surety and request the surety's assistance in resolving the controversy.

4.5   ARBITRATION

4.5.1 Controversies and Claims Subject to Arbitration. Any controversy or Claim arising out of or related to the Contract, or the breach thereof, shall be settled by arbitration in accordance with the Construction Industry Arbitration Rules of the American Arbitration Association, and judgment upon the award rendered by the arbitrator or arbitrators may be entered in any court having jurisdiction thereof, except controversies or Claims relating to aesthetic effect and except those waived as provided for in Subparagraph 4.3.5. Such controversies or Claims upon which the Architect has given notice and rendered a decision as provided in Subparagraph 4.4.4 shall be subject to arbitration upon written demand of either party. Arbitration may be commenced when 45 days have passed after a Claim has been referred to the Architect as provided in Paragraph 4.3 and no decision has been rendered.

AIA DOCUMENT A201 • GENERAL CONDITIONS OF THE CONTRACT FOR CONSTRUCTION • FOURTEENTH EDITION
AIA® • ©1987 THE AMERICAN INSTITUTE OF ARCHITECTS, 1735 NEW YORK AVENUE, N.W., WASHINGTON, D.C. 20006

4.5.2 Rules and Notices for Arbitration. Claims between the Owner and Contractor not resolved under Paragraph 4.4 shall, if subject to arbitration under Subparagraph 4.5.1, be decided by arbitration in accordance with the Construction Industry Arbitration Rules of the American Arbitration Association currently in effect, unless the parties mutually agree otherwise. Notice of demand for arbitration shall be filed in writing with the other party to the Agreement between the Owner and Contractor and with the American Arbitration Association, and a copy shall be filed with the Architect.

4.5.3 Contract Performance During Arbitration. During arbitration proceedings, the Owner and Contractor shall comply with Subparagraph 4.3.4.

4.5.4 When Arbitration May Be Demanded. Demand for arbitration of any Claim may not be made until the earlier of (1) the date on which the Architect has rendered a final written decision on the Claim, (2) the tenth day after the parties have presented evidence to the Architect or have been given reasonable opportunity to do so, if the Architect has not rendered a final written decision by that date, or (3) any of the five events described in Subparagraph 4.3.2.

4.5.4.1 When a written decision of the Architect states that (1) the decision is final but subject to arbitration and (2) a demand for arbitration of a Claim covered by such decision must be made within 30 days after the date on which the party making the demand receives the final written decision, then failure to demand arbitration within said 30 days' period shall result in the Architect's decision becoming final and binding upon the Owner and Contractor. If the Architect renders a decision after arbitration proceedings have been initiated, such decision may be entered as evidence, but shall not supersede arbitration proceedings unless the decision is acceptable to all parties concerned.

4.5.4.2 A demand for arbitration shall be made within the time limits specified in Subparagraphs 4.5.1 and 4.5.4 and Clause 4.5.4.1 as applicable, and in other cases within a reasonable time after the Claim has arisen, and in no event shall it be made after the date when institution of legal or equitable proceedings based on such Claim would be barred by the applicable statute of limitations as determined pursuant to Paragraph 13.7.

4.5.5 Limitation on Consolidation or Joinder. No arbitration arising out of or relating to the Contract Documents shall include, by consolidation or joinder or in any other manner, the Architect, the Architect's employees or consultants, except by written consent containing specific reference to the Agreement and signed by the Architect, Owner, Contractor and any other person or entity sought to be joined. No arbitration shall include, by consolidation or joinder or in any other manner, parties other than the Owner, Contractor, a separate contractor as described in Article 6 and other persons substantially involved in a common question of fact or law whose presence is required if complete relief is to be accorded in arbitration. No person or entity other than the Owner, Contractor or a separate contractor as described in Article 6 shall be included as an original third party or additional third party to an arbitration whose interest or responsibility is insubstantial. Consent to arbitration involving an additional person or entity shall not constitute consent to arbitration of a dispute not described therein or with a person or entity not named or described therein. The foregoing agreement to arbitrate and other agreements to arbitrate with an additional person or entity duly consented to by parties to the Agreement shall be specifically enforceable under applicable law in any court having jurisdiction thereof.

4.5.6 Claims and Timely Assertion of Claims. A party who files a notice of demand for arbitration must assert in the demand all Claims then known to that party on which arbitration is permitted to be demanded. When a party fails to include a Claim through oversight, inadvertence or excusable neglect, or when a Claim has matured or been acquired subsequently, the arbitrator or arbitrators may permit amendment.

4.5.7 Judgment on Final Award. The award rendered by the arbitrator or arbitrators shall be final, and judgment may be entered upon it in accordance with applicable law in any court having jurisdiction thereof.

ARTICLE 5

SUBCONTRACTORS

5.1   DEFINITIONS

5.1.1 A Subcontractor is a person or entity who has a direct contract with the Contractor to perform a portion of the Work at the site. The term "Subcontractor" is referred to throughout the Contract Documents as if singular in number and means a Subcontractor or an authorized representative of the Subcontractor. The term "Subcontractor" does not include a separate contractor or subcontractors of a separate contractor.

5.1.2 A Sub-subcontractor is a person or entity who has a direct or indirect contract with a Subcontractor to perform a portion of the Work at the site. The term "Sub-subcontractor" is referred to throughout the Contract Documents as if singular in number and means a Sub-subcontractor or an authorized representative of the Sub-subcontractor.

5.2   AWARD OF SUBCONTRACTS AND OTHER CONTRACTS FOR PORTIONS OF THE WORK

5.2.1 Unless otherwise stated in the Contract Documents or the bidding requirements, the Contractor, as soon as practicable after award of the Contract, shall furnish in writing to the Owner through the Architect the names of persons or entities (including those who are to furnish materials or equipment fabricated to a special design) proposed for each principal portion of the Work. The Architect will promptly reply to the Contractor in writing stating whether or not the Owner or the Architect, after due investigation, has reasonable objection to any such proposed person or entity. Failure of the Owner or Architect to reply promptly shall constitute notice of no reasonable objection.

5.2.2 The Contractor shall not contract with a proposed person or entity to whom the Owner or Architect has made reasonable and timely objection. The Contractor shall not be required to contract with anyone to whom the Contractor has made reasonable objection.

5.2.3 If the Owner or Architect has reasonable objection to a person or entity proposed by the Contractor, the Contractor shall propose another to whom the Owner or Architect has no reasonable objection. The Contract Sum shall be increased or decreased by the difference in cost occasioned by such change and an appropriate Change Order shall be issued. However, no increase in the Contract Sum shall be allowed for such change unless the Contractor has acted promptly and responsively in submitting names as required.

5.2.4 The Contractor shall not change a Subcontractor, person or entity previously selected if the Owner or Architect makes reasonable objection to such change.

5.3 SUBCONTRACTUAL RELATIONS

5.3.1 By appropriate agreement, written where legally required for validity, the Contractor shall require each Subcontractor, to the extent of the Work to be performed by the Subcontractor, to be bound to the Contractor by terms of the Contract Documents, and to assume toward the Contractor all the obligations and responsibilities which the Contractor, by these Documents, assumes toward the Owner and Architect. Each subcontract agreement shall preserve and protect the rights of the Owner and Architect under the Contract Documents with respect to the Work to be performed by the Subcontractor so that subcontracting thereof will not prejudice such rights, and shall allow to the Subcontractor, unless specifically provided otherwise in the subcontract agreement, the benefit of all rights, remedies and redress against the Contractor that the Contractor, by the Contract Documents, has against the Owner. Where appropriate, the Contractor shall require each Subcontractor to enter into similar agreements with Sub-sub-contractors. The Contractor shall make available to each proposed Subcontractor, prior to the execution of the subcontract agreement, copies of the Contract Documents to which the Subcontractor will be bound, and, upon written request of the Subcontractor, identify to the Subcontractor terms and conditions of the proposed subcontract agreement which may be at variance with the Contract Documents. Subcontractors shall similarly make copies of applicable portions of such documents available to their respective proposed Sub-subcontractors.

5.4 CONTINGENT ASSIGNMENT OF SUBCONTRACTS

5.4.1 Each subcontract agreement for a portion of the Work is assigned by the Contractor to the Owner provided that:

.1 assignment is effective only after termination of the Contract by the Owner for cause pursuant to Paragraph 14.2 and only for those subcontract agreements which the Owner accepts by notifying the Subcontractor in writing; and

.2 assignment is subject to the prior rights of the surety, if any, obligated under bond relating to the Contract.

5.4.2 If the Work has been suspended for more than 30 days, the Subcontractor's compensation shall be equitably adjusted.

ARTICLE 6

CONSTRUCTION BY OWNER OR BY SEPARATE CONTRACTORS

6.1 OWNER'S RIGHT TO PERFORM CONSTRUCTION AND TO AWARD SEPARATE CONTRACTS

6.1.1 The Owner reserves the right to perform construction or operations related to the Project with the Owner's own forces, and to award separate contracts in connection with other portions of the Project or other construction or operations on the site under Conditions of the Contract identical or substantially similar to these including those portions related to insurance and waiver of subrogation. If the Contractor claims that delay or additional cost is involved because of such action by the Owner, the Contractor shall make such Claim as provided elsewhere in the Contract Documents.

6.1.2 When separate contracts are awarded for different portions of the Project or other construction or operations on the site, the term "Contractor" in the Contract Documents in each case shall mean the Contractor who executes each separate Owner-Contractor Agreement.

6.1.3 The Owner shall provide for coordination of the activities of the Owner's own forces and of each separate contractor with the Work of the Contractor, who shall cooperate with them. The Contractor shall participate with other separate contractors and the Owner in reviewing their construction schedules when directed to do so. The Contractor shall make any revisions to the construction schedule and Contract Sum deemed necessary after a joint review and mutual agreement. The construction schedules shall then constitute the schedules to be used by the Contractor, separate contractors and the Owner until subsequently revised.

6.1.4 Unless otherwise provided in the Contract Documents, when the Owner performs construction or operations related to the Project with the Owner's own forces, the Owner shall be deemed to be subject to the same obligations and to have the same rights which apply to the Contractor under the Conditions of the Contract, including, without excluding others, those stated in Article 3, this Article 6 and Articles 10, 11 and 12.

6.2 MUTUAL RESPONSIBILITY

6.2.1 The Contractor shall afford the Owner and separate contractors reasonable opportunity for introduction and storage of their materials and equipment and performance of their activities and shall connect and coordinate the Contractor's construction and operations with theirs as required by the Contract Documents.

6.2.2 If part of the Contractor's Work depends for proper execution or results upon construction or operations by the Owner or a separate contractor, the Contractor shall, prior to proceeding with that portion of the Work, promptly report to the Architect apparent discrepancies or defects in such other construction that would render it unsuitable for such proper execution and results. Failure of the Contractor so to report shall constitute an acknowledgment that the Owner's or separate contractors' completed or partially completed construction is fit and proper to receive the Contractor's Work, except as to defects not then reasonably discoverable.

6.2.3 Costs caused by delays or by improperly timed activities or defective construction shall be borne by the party responsible therefor.

6.2.4 The Contractor shall promptly remedy damage wrongfully caused by the Contractor to completed or partially completed construction or to property of the Owner or separate contractors as provided in Subparagraph 10.2.5.

6.2.5 Claims and other disputes and matters in question between the Contractor and a separate contractor shall be subject to the provisions of Paragraph 4.3 provided the separate contractor has reciprocal obligations.

6.2.6 The Owner and each separate contractor shall have the same responsibilities for cutting and patching as are described for the Contractor in Paragraph 3.14.

6.3 OWNER'S RIGHT TO CLEAN UP

6.3.1 If a dispute arises among the Contractor, separate contractors and the Owner as to the responsibility under their respective contracts for maintaining the premises and surrounding area free from waste materials and rubbish as described in Paragraph 3.15, the Owner may clean up and allocate the cost among those responsible as the Architect determines to be just.

AIA DOCUMENT A201 • GENERAL CONDITIONS OF THE CONTRACT FOR CONSTRUCTION • FOURTEENTH EDITION
AIA® • ©1987 THE AMERICAN INSTITUTE OF ARCHITECTS, 1735 NEW YORK AVENUE, N.W., WASHINGTON, D.C. 20006

ARTICLE 7

CHANGES IN THE WORK

7.1 CHANGES

7.1.1 Changes in the Work may be accomplished after execution of the Contract, and without invalidating the Contract, by Change Order, Construction Change Directive or order for a minor change in the Work, subject to the limitations stated in this Article 7 and elsewhere in the Contract Documents.

7.1.2 A Change Order shall be based upon agreement among the Owner, Contractor and Architect; a Construction Change Directive requires agreement by the Owner and Architect and may or may not be agreed to by the Contractor; an order for a minor change in the Work may be issued by the Architect alone.

7.1.3 Changes in the Work shall be performed under applicable provisions of the Contract Documents, and the Contractor shall proceed promptly, unless otherwise provided in the Change Order, Construction Change Directive or order for a minor change in the Work.

7.1.4 If unit prices are stated in the Contract Documents or subsequently agreed upon, and if quantities originally contemplated are so changed in a proposed Change Order or Construction Change Directive that application of such unit prices to quantities of Work proposed will cause substantial inequity to the Owner or Contractor, the applicable unit prices shall be equitably adjusted.

7.2 CHANGE ORDERS

7.2.1 A Change Order is a written instrument prepared by the Architect and signed by the Owner, Contractor and Architect, stating their agreement upon all of the following:

    .1 a change in the Work;

    .2 the amount of the adjustment in the Contract Sum, if any; and

    .3 the extent of the adjustment in the Contract Time, if any.

7.2.2 Methods used in determining adjustments to the Contract Sum may include those listed in Subparagraph 7.3.3.

7.3 CONSTRUCTION CHANGE DIRECTIVES

7.3.1 A Construction Change Directive is a written order prepared by the Architect and signed by the Owner and Architect, directing a change in the Work and stating a proposed basis for adjustment, if any, in the Contract Sum or Contract Time, or both. The Owner may by Construction Change Directive, without invalidating the Contract, order changes in the Work within the general scope of the Contract consisting of additions, deletions or other revisions, the Contract Sum and Contract Time being adjusted accordingly.

7.3.2 A Construction Change Directive shall be used in the absence of total agreement on the terms of a Change Order.

7.3.3 If the Construction Change Directive provides for an adjustment to the Contract Sum, the adjustment shall be based on one of the following methods:

    .1 mutual acceptance of a lump sum properly itemized and supported by sufficient substantiating data to permit evaluation;

    .2 unit prices stated in the Contract Documents or subsequently agreed upon;

    .3 cost to be determined in a manner agreed upon by the parties and a mutually acceptable fixed or percentage fee; or

    .4 as provided in Subparagraph 7.3.6.

7.3.4 Upon receipt of a Construction Change Directive, the Contractor shall promptly proceed with the change in the Work involved and advise the Architect of the Contractor's agreement or disagreement with the method, if any, provided in the Construction Change Directive for determining the proposed adjustment in the Contract Sum or Contract Time.

7.3.5 A Construction Change Directive signed by the Contractor indicates the agreement of the Contractor therewith, including adjustment in Contract Sum and Contract Time or the method for determining them. Such agreement shall be effective immediately and shall be recorded as a Change Order.

7.3.6 If the Contractor does not respond promptly or disagrees with the method for adjustment in the Contract Sum, the method and the adjustment shall be determined by the Architect on the basis of reasonable expenditures and savings of those performing the Work attributable to the change, including, in case of an increase in the Contract Sum, a reasonable allowance for overhead and profit. In such case, and also under Clause 7.3.3.3, the Contractor shall keep and present, in such form as the Architect may prescribe, an itemized accounting together with appropriate supporting data. Unless otherwise provided in the Contract Documents, costs for the purposes of this Subparagraph 7.3.6 shall be limited to the following:

    .1 costs of labor, including social security, old age and unemployment insurance, fringe benefits required by agreement or custom, and workers' or workmen's compensation insurance;

    .2 costs of materials, supplies and equipment, including cost of transportation, whether incorporated or consumed;

    .3 rental costs of machinery and equipment, exclusive of hand tools, whether rented from the Contractor or others;

    .4 costs of premiums for all bonds and insurance, permit fees, and sales, use or similar taxes related to the Work; and

    .5 additional costs of supervision and field office personnel directly attributable to the change.

7.3.7 Pending final determination of cost to the Owner, amounts not in dispute may be included in Applications for Payment. The amount of credit to be allowed by the Contractor to the Owner for a deletion or change which results in a net decrease in the Contract Sum shall be actual net cost as confirmed by the Architect. When both additions and credits covering related Work or substitutions are involved in a change, the allowance for overhead and profit shall be figured on the basis of net increase, if any, with respect to that change.

7.3.8 If the Owner and Contractor do not agree with the adjustment in Contract Time or the method for determining it, the adjustment or the method shall be referred to the Architect for determination.

7.3.9 When the Owner and Contractor agree with the determination made by the Architect concerning the adjustments in the Contract Sum and Contract Time, or otherwise reach agreement upon the adjustments, such agreement shall be effective immediately and shall be recorded by preparation and execution of an appropriate Change Order.

7.4 MINOR CHANGES IN THE WORK

7.4.1 The Architect will have authority to order minor changes in the Work not involving adjustment in the Contract Sum or extension of the Contract Time and not inconsistent with the intent of the Contract Documents. Such changes shall be effected by written order and shall be binding on the Owner and Contractor. The Contractor shall carry out such written orders promptly.

ARTICLE 8

TIME

8.1 DEFINITIONS

8.1.1 Unless otherwise provided, Contract Time is the period of time, including authorized adjustments, allotted in the Contract Documents for Substantial Completion of the Work.

8.1.2 The date of commencement of the Work is the date established in the Agreement. The date shall not be postponed by the failure to act of the Contractor or of persons or entities for whom the Contractor is responsible.

8.1.3 The date of Substantial Completion is the date certified by the Architect in accordance with Paragraph 9.8.

8.1.4 The term "day" as used in the Contract Documents shall mean calendar day unless otherwise specifically defined.

8.2 PROGRESS AND COMPLETION

8.2.1 Time limits stated in the Contract Documents are of the essence of the Contract. By executing the Agreement the Contractor confirms that the Contract Time is a reasonable period for performing the Work.

8.2.2 The Contractor shall not knowingly, except by agreement or instruction of the Owner in writing, prematurely commence operations on the site or elsewhere prior to the effective date of insurance required by Article 11 to be furnished by the Contractor. The date of commencement of the Work shall not be changed by the effective date of such insurance. Unless the date of commencement is established by a notice to proceed given by the Owner, the Contractor shall notify the Owner in writing not less than five days or other agreed period before commencing the Work to permit the timely filing of mortgages, mechanic's liens and other security interests.

8.2.3 The Contractor shall proceed expeditiously with adequate forces and shall achieve Substantial Completion within the Contract Time.

8.3 DELAYS AND EXTENSIONS OF TIME

8.3.1 If the Contractor is delayed at any time in progress of the Work by an act or neglect of the Owner or Architect, or of an employee of either, or of a separate contractor employed by the Owner, or by changes ordered in the Work, or by labor disputes, fire, unusual delay in deliveries, unavoidable casualties or other causes beyond the Contractor's control, or by delay authorized by the Owner pending arbitration, or by other causes which the Architect determines may justify delay, then the Contract Time shall be extended by Change Order for such reasonable time as the Architect may determine.

8.3.2 Claims relating to time shall be made in accordance with applicable provisions of Paragraph 4.3.

8.3.3 This Paragraph 8.3 does not preclude recovery of damages for delay by either party under other provisions of the Contract Documents.

ARTICLE 9

PAYMENTS AND COMPLETION

9.1 CONTRACT SUM

9.1.1 The Contract Sum is stated in the Agreement and, including authorized adjustments, is the total amount payable by the Owner to the Contractor for performance of the Work under the Contract Documents.

9.2 SCHEDULE OF VALUES

9.2.1 Before the first Application for Payment, the Contractor shall submit to the Architect a schedule of values allocated to various portions of the Work, prepared in such form and supported by such data to substantiate its accuracy as the Architect may require. This schedule, unless objected to by the Architect, shall be used as a basis for reviewing the Contractor's Applications for Payment.

9.3 APPLICATIONS FOR PAYMENT

9.3.1 At least ten days before the date established for each progress payment, the Contractor shall submit to the Architect an itemized Application for Payment for operations completed in accordance with the schedule of values. Such application shall be notarized, if required, and supported by such data substantiating the Contractor's right to payment as the Owner or Architect may require, such as copies of requisitions from Subcontractors and material suppliers, and reflecting retainage if provided for elsewhere in the Contract Documents.

9.3.1.1 Such applications may include requests for payment on account of changes in the Work which have been properly authorized by Construction Change Directives but not yet included in Change Orders.

9.3.1.2 Such applications may not include requests for payment of amounts the Contractor does not intend to pay to a Subcontractor or material supplier because of a dispute or other reason.

9.3.2 Unless otherwise provided in the Contract Documents, payments shall be made on account of materials and equipment delivered and suitably stored at the site for subsequent incorporation in the Work. If approved in advance by the Owner, payment may similarly be made for materials and equipment suitably stored off the site at a location agreed upon in writing. Payment for materials and equipment stored on or off the site shall be conditioned upon compliance by the Contractor with procedures satisfactory to the Owner to establish the Owner's title to such materials and equipment or otherwise protect the Owner's interest, and shall include applicable insurance, storage and transportation to the site for such materials and equipment stored off the site.

9.3.3 The Contractor warrants that title to all Work covered by an Application for Payment will pass to the Owner no later than the time of payment. The Contractor further warrants that upon submittal of an Application for Payment all Work for which Certificates for Payment have been previously issued and payments received from the Owner shall, to the best of the Contractor's knowledge, information and belief, be free and clear of liens, claims, security interests or encumbrances in favor of the Contractor, Subcontractors, material suppliers, or other persons or entities making a claim by reason of having provided labor, materials and equipment relating to the Work.

9.4 CERTIFICATES FOR PAYMENT

9.4.1 The Architect will, within seven days after receipt of the Contractor's Application for Payment, either issue to the

AIA DOCUMENT A201 • GENERAL CONDITIONS OF THE CONTRACT FOR CONSTRUCTION • FOURTEENTH EDITION
AIA® • ©1987 THE AMERICAN INSTITUTE OF ARCHITECTS, 1735 NEW YORK AVENUE, N.W., WASHINGTON, D.C. 20006

Owner a Certificate for Payment, with a copy to the Contractor, for such amount as the Architect determines is properly due, or notify the Contractor and Owner in writing of the Architect's reasons for withholding certification in whole or in part as provided in Subparagraph 9.5.1.

9.4.2 The issuance of a Certificate for Payment will constitute a representation by the Architect to the Owner, based on the Architect's observations at the site and the data comprising the Application for Payment, that the Work has progressed to the point indicated and that, to the best of the Architect's knowledge, information and belief, quality of the Work is in accordance with the Contract Documents. The foregoing representations are subject to an evaluation of the Work for conformance with the Contract Documents upon Substantial Completion, to results of subsequent tests and inspections, to minor deviations from the Contract Documents correctable prior to completion and to specific qualifications expressed by the Architect. The issuance of a Certificate for Payment will further constitute a representation that the Contractor is entitled to payment in the amount certified. However, the issuance of a Certificate for Payment will not be a representation that the Architect has (1) made exhaustive or continuous on-site inspections to check the quality or quantity of the Work, (2) reviewed construction means, methods, techniques, sequences or procedures, (3) reviewed copies of requisitions received from Subcontractors and material suppliers and other data requested by the Owner to substantiate the Contractor's right to payment or (4) made examination to ascertain how or for what purpose the Contractor has used money previously paid on account of the Contract Sum.

9.5 DECISIONS TO WITHHOLD CERTIFICATION

9.5.1 The Architect may decide not to certify payment and may withhold a Certificate for Payment in whole or in part, to the extent reasonably necessary to protect the Owner, if in the Architect's opinion the representations to the Owner required by Subparagraph 9.4.2 cannot be made. If the Architect is unable to certify payment in the amount of the Application, the Architect will notify the Contractor and Owner as provided in Subparagraph 9.4.1. If the Contractor and Architect cannot agree on a revised amount, the Architect will promptly issue a Certificate for Payment for the amount for which the Architect is able to make such representations to the Owner. The Architect may also decide not to certify payment or, because of subsequently discovered evidence or subsequent observations, may nullify the whole or a part of a Certificate for Payment previously issued, to such extent as may be necessary in the Architect's opinion to protect the Owner from loss because of:

    .1 defective Work not remedied;

    .2 third party claims filed or reasonable evidence indicating probable filing of such claims;

    .3 failure of the Contractor to make payments properly to Subcontractors or for labor, materials or equipment;

    .4 reasonable evidence that the Work cannot be completed for the unpaid balance of the Contract Sum;

    .5 damage to the Owner or another contractor;

    .6 reasonable evidence that the Work will not be completed within the Contract Time, and that the unpaid balance would not be adequate to cover actual or liquidated damages for the anticipated delay; or

    .7 persistent failure to carry out the Work in accordance with the Contract Documents.

9.5.2 When the above reasons for withholding certification are removed, certification will be made for amounts previously withheld.

9.6 PROGRESS PAYMENTS

9.6.1 After the Architect has issued a Certificate for Payment, the Owner shall make payment in the manner and within the time provided in the Contract Documents, and shall so notify the Architect.

9.6.2 The Contractor shall promptly pay each Subcontractor, upon receipt of payment from the Owner, out of the amount paid to the Contractor on account of such Subcontractor's portion of the Work, the amount to which said Subcontractor is entitled, reflecting percentages actually retained from payments to the Contractor on account of such Subcontractor's portion of the Work. The Contractor shall, by appropriate agreement with each Subcontractor, require each Subcontractor to make payments to Sub-subcontractors in similar manner.

9.6.3 The Architect will, on request, furnish to a Subcontractor, if practicable, information regarding percentages of completion or amounts applied for by the Contractor and action taken thereon by the Architect and Owner on account of portions of the Work done by such Subcontractor.

9.6.4 Neither the Owner nor Architect shall have an obligation to pay or to see to the payment of money to a Subcontractor except as may otherwise be required by law.

9.6.5 Payment to material suppliers shall be treated in a manner similar to that provided in Subparagraphs 9.6.2, 9.6.3 and 9.6.4.

9.6.6 A Certificate for Payment, a progress payment, or partial or entire use or occupancy of the Project by the Owner shall not constitute acceptance of Work not in accordance with the Contract Documents.

9.7 FAILURE OF PAYMENT

9.7.1 If the Architect does not issue a Certificate for Payment, through no fault of the Contractor, within seven days after receipt of the Contractor's Application for Payment, or if the Owner does not pay the Contractor within seven days after the date established in the Contract Documents the amount certified by the Architect or awarded by arbitration, then the Contractor may, upon seven additional days' written notice to the Owner and Architect, stop the Work until payment of the amount owing has been received. The Contract Time shall be extended appropriately and the Contract Sum shall be increased by the amount of the Contractor's reasonable costs of shut-down, delay and start-up, which shall be accomplished as provided in Article 7.

9.8 SUBSTANTIAL COMPLETION

9.8.1 Substantial Completion is the stage in the progress of the Work when the Work or designated portion thereof is sufficiently complete in accordance with the Contract Documents so the Owner can occupy or utilize the Work for its intended use.

9.8.2 When the Contractor considers that the Work, or a portion thereof which the Owner agrees to accept separately, is substantially complete, the Contractor shall prepare and submit to the Architect a comprehensive list of items to be completed or corrected. The Contractor shall proceed promptly to complete and correct items on the list. Failure to include an item on such list does not alter the responsibility of the Contractor to complete all Work in accordance with the Contract Documents. Upon receipt of the Contractor's list, the Architect will make an inspection to determine whether the Work or designated portion thereof is substantially complete. If the Architect's inspection discloses any item, whether or not included on the Contractor's list, which is not in accordance with the requirements of the Contract Documents, the Contractor shall, before issuance of the Certificate of Substantial Completion, complete or correct such item upon notification by the Architect. The Contractor shall then submit a request for another inspection by the Architect to determine Substantial Completion. When the Work or designated portion thereof is substantially complete, the Architect will prepare a Certificate of Substantial Completion which shall establish the date of Substantial Completion, shall establish responsibilities of the Owner and Contractor for security, maintenance, heat, utilities, damage to the Work and insurance, and shall fix the time within which the Contractor shall finish all items on the list accompanying the Certificate. Warranties required by the Contract Documents shall commence on the date of Substantial Completion of the Work or designated portion thereof unless otherwise provided in the Certificate of Substantial Completion. The Certificate of Substantial Completion shall be submitted to the Owner and Contractor for their written acceptance of responsibilities assigned to them in such Certificate.

9.8.3 Upon Substantial Completion of the Work or designated portion thereof and upon application by the Contractor and certification by the Architect, the Owner shall make payment, reflecting adjustment in retainage, if any, for such Work or portion thereof as provided in the Contract Documents.

9.9 PARTIAL OCCUPANCY OR USE

9.9.1 The Owner may occupy or use any completed or partially completed portion of the Work at any stage when such portion is designated by separate agreement with the Contractor, provided such occupancy or use is consented to by the insurer as required under Subparagraph 11.3.11 and authorized by public authorities having jurisdiction over the Work. Such partial occupancy or use may commence whether or not the portion is substantially complete, provided the Owner and Contractor have accepted in writing the responsibilities assigned to each of them for payments, retainage if any, security, maintenance, heat, utilities, damage to the Work and insurance, and have agreed in writing concerning the period for correction of the Work and commencement of warranties required by the Contract Documents. When the Contractor considers a portion substantially complete, the Contractor shall prepare and submit a list to the Architect as provided under Subparagraph 9.8.2. Consent of the Contractor to partial occupancy or use shall not be unreasonably withheld. The stage of the progress of the Work shall be determined by written agreement between the Owner and Contractor or, if no agreement is reached, by decision of the Architect.

9.9.2 Immediately prior to such partial occupancy or use, the Owner, Contractor and Architect shall jointly inspect the area to be occupied or portion of the Work to be used in order to determine and record the condition of the Work.

9.9.3 Unless otherwise agreed upon, partial occupancy or use of a portion or portions of the Work shall not constitute acceptance of Work not complying with the requirements of the Contract Documents.

9.10 FINAL COMPLETION AND FINAL PAYMENT

9.10.1 Upon receipt of written notice that the Work is ready for final inspection and acceptance and upon receipt of a final Application for Payment, the Architect will promptly make such inspection and, when the Architect finds the Work acceptable under the Contract Documents and the Contract fully performed, the Architect will promptly issue a final Certificate for Payment stating that to the best of the Architect's knowledge, information and belief, and on the basis of the Architect's observations and inspections, the Work has been completed in accordance with terms and conditions of the Contract Documents and that the entire balance found to be due the Contractor and noted in said final Certificate is due and payable. The Architect's final Certificate for Payment will constitute a further representation that conditions listed in Subparagraph 9.10.2 as precedent to the Contractor's being entitled to final payment have been fulfilled.

9.10.2 Neither final payment nor any remaining retained percentage shall become due until the Contractor submits to the Architect (1) an affidavit that payrolls, bills for materials and equipment, and other indebtedness connected with the Work for which the Owner or the Owner's property might be responsible or encumbered (less amounts withheld by Owner) have been paid or otherwise satisfied, (2) a certificate evidencing that insurance required by the Contract Documents to remain in force after final payment is currently in effect and will not be cancelled or allowed to expire until at least 30 days' prior written notice has been given to the Owner, (3) a written statement that the Contractor knows of no substantial reason that the insurance will not be renewable to cover the period required by the Contract Documents, (4) consent of surety, if any, to final payment and (5), if required by the Owner, other data establishing payment or satisfaction of obligations, such as receipts, releases and waivers of liens, claims, security interests or encumbrances arising out of the Contract, to the extent and in such form as may be designated by the Owner. If a Subcontractor refuses to furnish a release or waiver required by the Owner, the Contractor may furnish a bond satisfactory to the Owner to indemnify the Owner against such lien. If such lien remains unsatisfied after payments are made, the Contractor shall refund to the Owner all money that the Owner may be compelled to pay in discharging such lien, including all costs and reasonable attorneys' fees.

9.10.3 If, after Substantial Completion of the Work, final completion thereof is materially delayed through no fault of the Contractor or by issuance of Change Orders affecting final completion, and the Architect so confirms, the Owner shall, upon application by the Contractor and certification by the Architect, and without terminating the Contract, make payment of the balance due for that portion of the Work fully completed and accepted. If the remaining balance for Work not fully completed or corrected is less than retainage stipulated in the Contract Documents, and if bonds have been furnished, the written consent of surety to payment of the balance due for that portion of the Work fully completed and accepted shall be submitted by the Contractor to the Architect prior to certification of such payment. Such payment shall be made under terms and conditions governing final payment, except that it shall not constitute a waiver of claims. The making of final payment shall constitute a waiver of claims by the Owner as provided in Subparagraph 4.3.5.

9.10.4 Acceptance of final payment by the Contractor, a Subcontractor or material supplier shall constitute a waiver of claims by that payee except those previously made in writing and identified by that payee as unsettled at the time of final Application for Payment. Such waivers shall be in addition to the waiver described in Subparagraph 4.3.5.

AIA DOCUMENT A201 • GENERAL CONDITIONS OF THE CONTRACT FOR CONSTRUCTION • FOURTEENTH EDITION
AIA® • ©1987 THE AMERICAN INSTITUTE OF ARCHITECTS, 1735 NEW YORK AVENUE, N.W., WASHINGTON, D.C. 20006

ARTICLE 10

PROTECTION OF PERSONS AND PROPERTY

10.1 SAFETY PRECAUTIONS AND PROGRAMS

10.1.1 The Contractor shall be responsible for initiating, maintaining and supervising all safety precautions and programs in connection with the performance of the Contract.

10.1.2 In the event the Contractor encounters on the site material reasonably believed to be asbestos or polychlorinated biphenyl (PCB) which has not been rendered harmless, the Contractor shall immediately stop Work in the area affected and report the condition to the Owner and Architect in writing. The Work in the affected area shall not thereafter be resumed except by written agreement of the Owner and Contractor if in fact the material is asbestos or polychlorinated biphenyl (PCB) and has not been rendered harmless. The Work in the affected area shall be resumed in the absence of asbestos or polychlorinated biphenyl (PCB), or when it has been rendered harmless, by written agreement of the Owner and Contractor, or in accordance with final determination by the Architect on which arbitration has not been demanded, or by arbitration under Article 4.

10.1.3 The Contractor shall not be required pursuant to Article 7 to perform without consent any Work relating to asbestos or polychlorinated biphenyl (PCB).

10.1.4 To the fullest extent permitted by law, the Owner shall indemnify and hold harmless the Contractor, Architect, Architect's consultants and agents and employees of any of them from and against claims, damages, losses and expenses, including but not limited to attorneys' fees, arising out of or resulting from performance of the Work in the affected area if in fact the material is asbestos or polychlorinated biphenyl (PCB) and has not been rendered harmless, provided that such claim, damage, loss or expense is attributable to bodily injury, sickness, disease or death, or to injury to or destruction of tangible property (other than the Work itself) including loss of use resulting therefrom, but only to the extent caused in whole or in part by negligent acts or omissions of the Owner, anyone directly or indirectly employed by the Owner or anyone for whose acts the Owner may be liable, regardless of whether or not such claim, damage, loss or expense is caused in part by a party indemnified hereunder. Such obligation shall not be construed to negate, abridge, or reduce other rights or obligations of indemnity which would otherwise exist as to a party or person described in this Subparagraph 10.1.4.

10.2 SAFETY OF PERSONS AND PROPERTY

10.2.1 The Contractor shall take reasonable precautions for safety of, and shall provide reasonable protection to prevent damage, injury or loss to:

    .1 employees on the Work and other persons who may be affected thereby;

    .2 the Work and materials and equipment to be incorporated therein, whether in storage on or off the site, under care, custody or control of the Contractor or the Contractor's Subcontractors or Sub-subcontractors; and

    .3 other property at the site or adjacent thereto, such as trees, shrubs, lawns, walks, pavements, roadways, structures and utilities not designated for removal, relocation or replacement in the course of construction.

10.2.2 The Contractor shall give notices and comply with applicable laws, ordinances, rules, regulations and lawful orders of public authorities bearing on safety of persons or property or their protection from damage, injury or loss.

10.2.3 The Contractor shall erect and maintain, as required by existing conditions and performance of the Contract, reasonable safeguards for safety and protection, including posting danger signs and other warnings against hazards, promulgating safety regulations and notifying owners and users of adjacent sites and utilities.

10.2.4 When use or storage of explosives or other hazardous materials or equipment or unusual methods are necessary for execution of the Work, the Contractor shall exercise utmost care and carry on such activities under supervision of properly qualified personnel.

10.2.5 The Contractor shall promptly remedy damage and loss (other than damage or loss insured under property insurance required by the Contract Documents) to property referred to in Clauses 10.2.1.2 and 10.2.1.3 caused in whole or in part by the Contractor, a Subcontractor, a Sub-subcontractor, or anyone directly or indirectly employed by any of them, or by anyone for whose acts they may be liable and for which the Contractor is responsible under Clauses 10.2.1.2 and 10.2.1.3, except damage or loss attributable to acts or omissions of the Owner or Architect or anyone directly or indirectly employed by either of them, or by anyone for whose acts either of them may be liable, and not attributable to the fault or negligence of the Contractor. The foregoing obligations of the Contractor are in addition to the Contractor's obligations under Paragraph 3.18.

10.2.6 The Contractor shall designate a responsible member of the Contractor's organization at the site whose duty shall be the prevention of accidents. This person shall be the Contractor's superintendent unless otherwise designated by the Contractor in writing to the Owner and Architect.

10.2.7 The Contractor shall not load or permit any part of the construction or site to be loaded so as to endanger its safety.

10.3 EMERGENCIES

10.3.1 In an emergency affecting safety of persons or property, the Contractor shall act, at the Contractor's discretion, to prevent threatened damage, injury or loss. Additional compensation or extension of time claimed by the Contractor on account of an emergency shall be determined as provided in Paragraph 4.3 and Article 7.

ARTICLE 11

INSURANCE AND BONDS

11.1 CONTRACTOR'S LIABILITY INSURANCE

11.1.1 The Contractor shall purchase from and maintain in a company or companies lawfully authorized to do business in the jurisdiction in which the Project is located such insurance as will protect the Contractor from claims set forth below which may arise out of or result from the Contractor's operations under the Contract and for which the Contractor may be legally liable, whether such operations be by the Contractor or by a Subcontractor or by anyone directly or indirectly employed by any of them, or by anyone for whose acts any of them may be liable:

    .1 claims under workers' or workmen's compensation, disability benefit and other similar employee benefit acts which are applicable to the Work to be performed;

.2 claims for damages because of bodily injury, occupational sickness or disease, or death of the Contractor's employees;

.3 claims for damages because of bodily injury, sickness or disease, or death of any person other than the Contractor's employees;

.4 claims for damages insured by usual personal injury liability coverage which are sustained (1) by a person as a result of an offense directly or indirectly related to employment of such person by the Contractor, or (2) by another person;

.5 claims for damages, other than to the Work itself, because of injury to or destruction of tangible property, including loss of use resulting therefrom;

.6 claims for damages because of bodily injury, death of a person or property damage arising out of ownership, maintenance or use of a motor vehicle; and

.7 claims involving contractual liability insurance applicable to the Contractor's obligations under Paragraph 3.18.

11.1.2 The insurance required by Subparagraph 11.1.1 shall be written for not less than limits of liability specified in the Contract Documents or required by law, whichever coverage is greater. Coverages, whether written on an occurrence or claims-made basis, shall be maintained without interruption from date of commencement of the Work until date of final payment and termination of any coverage required to be maintained after final payment.

11.1.3 Certificates of Insurance acceptable to the Owner shall be filed with the Owner prior to commencement of the Work. These Certificates and the insurance policies required by this Paragraph 11.1 shall contain a provision that coverages afforded under the policies will not be cancelled or allowed to expire until at least 30 days' prior written notice has been given to the Owner. If any of the foregoing insurance coverages are required to remain in force after final payment and are reasonably available, an additional certificate evidencing continuation of such coverage shall be submitted with the final Application for Payment as required by Subparagraph 9.10.2. Information concerning reduction of coverage shall be furnished by the Contractor with reasonable promptness in accordance with the Contractor's information and belief.

11.2 OWNER'S LIABILITY INSURANCE

11.2.1 The Owner shall be responsible for purchasing and maintaining the Owner's usual liability insurance. Optionally, the Owner may purchase and maintain other insurance for self-protection against claims which may arise from operations under the Contract. The Contractor shall not be responsible for purchasing and maintaining this optional Owner's liability insurance unless specifically required by the Contract Documents.

11.3 PROPERTY INSURANCE

11.3.1 Unless otherwise provided, the Owner shall purchase and maintain, in a company or companies lawfully authorized to do business in the jurisdiction in which the Project is located, property insurance in the amount of the initial Contract Sum as well as subsequent modifications thereto for the entire Work at the site on a replacement cost basis without voluntary deductibles. Such property insurance shall be maintained, unless otherwise provided in the Contract Documents or otherwise agreed in writing by all persons and entities who are beneficiaries of such insurance, until final payment has been made as provided in Paragraph 9.10 or until no person or entity

other than the Owner has an insurable interest in the property required by this Paragraph 11.3 to be covered, whichever is earlier. This insurance shall include interests of the Owner, the Contractor, Subcontractors and Sub-subcontractors in the Work.

11.3.1.1 Property insurance shall be on an all-risk policy form and shall insure against the perils of fire and extended coverage and physical loss or damage including, without duplication of coverage, theft, vandalism, malicious mischief, collapse, false-work, temporary buildings and debris removal including demolition occasioned by enforcement of any applicable legal requirements, and shall cover reasonable compensation for Architect's services and expenses required as a result of such insured loss. Coverage for other perils shall not be required unless otherwise provided in the Contract Documents.

11.3.1.2 If the Owner does not intend to purchase such property insurance required by the Contract and with all of the coverages in the amount described above, the Owner shall so inform the Contractor in writing prior to commencement of the Work. The Contractor may then effect insurance which will protect the interests of the Contractor, Subcontractors and Sub-subcontractors in the Work, and by appropriate Change Order the cost thereof shall be charged to the Owner. If the Contractor is damaged by the failure or neglect of the Owner to purchase or maintain insurance as described above, without so notifying the Contractor, then the Owner shall bear all reasonable costs properly attributable thereto.

11.3.1.3 If the property insurance requires minimum deductibles and such deductibles are identified in the Contract Documents, the Contractor shall pay costs not covered because of such deductibles. If the Owner or insurer increases the required minimum deductibles above the amounts so identified or if the Owner elects to purchase this insurance with voluntary deductible amounts, the Owner shall be responsible for payment of the additional costs not covered because of such increased or voluntary deductibles. If deductibles are not identified in the Contract Documents, the Owner shall pay costs not covered because of deductibles.

11.3.1.4 Unless otherwise provided in the Contract Documents, this property insurance shall cover portions of the Work stored off the site after written approval of the Owner at the value established in the approval, and also portions of the Work in transit.

11.3.2 Boiler and Machinery Insurance. The Owner shall purchase and maintain boiler and machinery insurance required by the Contract Documents or by law, which shall specifically cover such insured objects during installation and until final acceptance by the Owner; this insurance shall include interests of the Owner, Contractor, Subcontractors and Sub-subcontractors in the Work, and the Owner and Contractor shall be named insureds.

11.3.3 Loss of Use Insurance. The Owner, at the Owner's option, may purchase and maintain such insurance as will insure the Owner against loss of use of the Owner's property due to fire or other hazards, however caused. The Owner waives all rights of action against the Contractor for loss of use of the Owner's property, including consequential losses due to fire or other hazards however caused.

11.3.4 If the Contractor requests in writing that insurance for risks other than those described herein or for other special hazards be included in the property insurance policy, the Owner shall, if possible, include such insurance, and the cost thereof shall be charged to the Contractor by appropriate Change Order.

AIA DOCUMENT A201 • GENERAL CONDITIONS OF THE CONTRACT FOR CONSTRUCTION • FOURTEENTH EDITION
AIA® • ©1987 THE AMERICAN INSTITUTE OF ARCHITECTS, 1735 NEW YORK AVENUE, N.W., WASHINGTON, D.C. 20006

11.3.5 If during the Project construction period the Owner insures properties, real or personal or both, adjoining or adjacent to the site by property insurance under policies separate from those insuring the Project, or if after final payment property insurance is to be provided on the completed Project through a policy or policies other than those insuring the Project during the construction period, the Owner shall waive all rights in accordance with the terms of Subparagraph 11.3.7 for damages caused by fire or other perils covered by this separate property insurance. All separate policies shall provide this waiver of subrogation by endorsement or otherwise.

11.3.6 Before an exposure to loss may occur, the Owner shall file with the Contractor a copy of each policy that includes insurance coverages required by this Paragraph 11.3. Each policy shall contain all generally applicable conditions, definitions, exclusions and endorsements related to this Project. Each policy shall contain a provision that the policy will not be cancelled or allowed to expire until at least 30 days' prior written notice has been given to the Contractor.

11.3.7 Waivers of Subrogation. The Owner and Contractor waive all rights against (1) each other and any of their subcontractors, sub-subcontractors, agents and employees, each of the other, and (2) the Architect, Architect's consultants, separate contractors described in Article 6, if any, and any of their subcontractors, sub-subcontractors, agents and employees, for damages caused by fire or other perils to the extent covered by property insurance obtained pursuant to this Paragraph 11.3 or other property insurance applicable to the Work, except such rights as they have to proceeds of such insurance held by the Owner as fiduciary. The Owner or Contractor, as appropriate, shall require of the Architect, Architect's consultants, separate contractors described in Article 6, if any, and the subcontractors, sub-subcontractors, agents and employees of any of them, by appropriate agreements, written where legally required for validity, similar waivers each in favor of other parties enumerated herein. The policies shall provide such waivers of subrogation by endorsement or otherwise. A waiver of subrogation shall be effective as to a person or entity even though that person or entity would otherwise have a duty of indemnification, contractual or otherwise, did not pay the insurance premium directly or indirectly, and whether or not the person or entity had an insurable interest in the property damaged.

11.3.8 A loss insured under Owner's property insurance shall be adjusted by the Owner as fiduciary and made payable to the Owner as fiduciary for the insureds, as their interests may appear, subject to requirements of any applicable mortgagee clause and of Subparagraph 11.3.10. The Contractor shall pay Subcontractors their just shares of insurance proceeds received by the Contractor, and by appropriate agreements, written where legally required for validity, shall require Subcontractors to make payments to their Sub-subcontractors in similar manner.

11.3.9 If required in writing by a party in interest, the Owner as fiduciary shall, upon occurrence of an insured loss, give bond for proper performance of the Owner's duties. The cost of required bonds shall be charged against proceeds received as fiduciary. The Owner shall deposit in a separate account proceeds so received, which the Owner shall distribute in accordance with such agreement as the parties in interest may reach, or in accordance with an arbitration award in which case the procedure shall be as provided in Paragraph 4.5. If after such loss no other special agreement is made, replacement of damaged property shall be covered by appropriate Change Order.

11.3.10 The Owner as fiduciary shall have power to adjust and settle a loss with insurers unless one of the parties in interest shall object in writing within five days after occurrence of loss to the Owner's exercise of this power; if such objection be made, arbitrators shall be chosen as provided in Paragraph 4.5. The Owner as fiduciary shall, in that case, make settlement with insurers in accordance with directions of such arbitrators. If distribution of insurance proceeds by arbitration is required, the arbitrators will direct such distribution.

11.3.11 Partial occupancy or use in accordance with Paragraph 9.9 shall not commence until the insurance company or companies providing property insurance have consented to such partial occupancy or use by endorsement or otherwise. The Owner and the Contractor shall take reasonable steps to obtain consent of the insurance company or companies and shall, without mutual written consent, take no action with respect to partial occupancy or use that would cause cancellation, lapse or reduction of insurance.

11.4 PERFORMANCE BOND AND PAYMENT BOND

11.4.1 The Owner shall have the right to require the Contractor to furnish bonds covering faithful performance of the Contract and payment of obligations arising thereunder as stipulated in bidding requirements or specifically required in the Contract Documents on the date of execution of the Contract.

11.4.2 Upon the request of any person or entity appearing to be a potential beneficiary of bonds covering payment of obligations arising under the Contract, the Contractor shall promptly furnish a copy of the bonds or shall permit a copy to be made.

ARTICLE 12

UNCOVERING AND CORRECTION OF WORK

12.1 UNCOVERING OF WORK

12.1.1 If a portion of the Work is covered contrary to the Architect's request or to requirements specifically expressed in the Contract Documents, it must, if required in writing by the Architect, be uncovered for the Architect's observation and be replaced at the Contractor's expense without change in the Contract Time.

12.1.2 If a portion of the Work has been covered which the Architect has not specifically requested to observe prior to its being covered, the Architect may request to see such Work and it shall be uncovered by the Contractor. If such Work is in accordance with the Contract Documents, costs of uncovering and replacement shall, by appropriate Change Order, be charged to the Owner. If such Work is not in accordance with the Contract Documents, the Contractor shall pay such costs unless the condition was caused by the Owner or a separate contractor in which event the Owner shall be responsible for payment of such costs.

12.2 CORRECTION OF WORK

12.2.1 The Contractor shall promptly correct Work rejected by the Architect or failing to conform to the requirements of the Contract Documents, whether observed before or after Substantial Completion and whether or not fabricated, installed or completed. The Contractor shall bear costs of correcting such rejected Work, including additional testing and inspections and compensation for the Architect's services and expenses made necessary thereby.

12.2.2 If, within one year after the date of Substantial Completion of the Work or designated portion thereof, or after the date for commencement of warranties established under Subparagraph 9.9.1, or by terms of an applicable special warranty required by the Contract Documents, any of the Work is found to be not in accordance with the requirements of the Contract Documents, the Contractor shall correct it promptly after receipt of written notice from the Owner to do so unless the Owner has previously given the Contractor a written acceptance of such condition. This period of one year shall be extended with respect to portions of Work first performed after Substantial Completion by the period of time between Substantial Completion and the actual performance of the Work. This obligation under this Subparagraph 12.2.2 shall survive acceptance of the Work under the Contract and termination of the Contract. The Owner shall give such notice promptly after discovery of the condition.

12.2.3 The Contractor shall remove from the site portions of the Work which are not in accordance with the requirements of the Contract Documents and are neither corrected by the Contractor nor accepted by the Owner.

12.2.4 If the Contractor fails to correct nonconforming Work within a reasonable time, the Owner may correct it in accordance with Paragraph 2.4. If the Contractor does not proceed with correction of such nonconforming Work within a reasonable time fixed by written notice from the Architect, the Owner may remove it and store the salvable materials or equipment at the Contractor's expense. If the Contractor does not pay costs of such removal and storage within ten days after written notice, the Owner may upon ten additional days' written notice sell such materials and equipment at auction or at private sale and shall account for the proceeds thereof, after deducting costs and damages that should have been borne by the Contractor, including compensation for the Architect's services and expenses made necessary thereby. If such proceeds of sale do not cover costs which the Contractor should have borne, the Contract Sum shall be reduced by the deficiency. If payments then or thereafter due the Contractor are not sufficient to cover such amount, the Contractor shall pay the difference to the Owner.

12.2.5 The Contractor shall bear the cost of correcting destroyed or damaged construction, whether completed or partially completed, of the Owner or separate contractors caused by the Contractor's correction or removal of Work which is not in accordance with the requirements of the Contract Documents.

12.2.6 Nothing contained in this Paragraph 12.2 shall be construed to establish a period of limitation with respect to other obligations which the Contractor might have under the Contract Documents. Establishment of the time period of one year as described in Subparagraph 12.2.2 relates only to the specific obligation of the Contractor to correct the Work, and has no relationship to the time within which the obligation to comply with the Contract Documents may be sought to be enforced, nor to the time within which proceedings may be commenced to establish the Contractor's liability with respect to the Contractor's obligations other than specifically to correct the Work.

12.3 ACCEPTANCE OF NONCONFORMING WORK

12.3.1 If the Owner prefers to accept Work which is not in accordance with the requirements of the Contract Documents, the Owner may do so instead of requiring its removal and correction, in which case the Contract Sum will be reduced as appropriate and equitable. Such adjustment shall be effected whether or not final payment has been made.

ARTICLE 13

MISCELLANEOUS PROVISIONS

13.1 GOVERNING LAW

13.1.1 The Contract shall be governed by the law of the place where the Project is located.

13.2 SUCCESSORS AND ASSIGNS

13.2.1 The Owner and Contractor respectively bind themselves, their partners, successors, assigns and legal representatives to the other party hereto and to partners, successors, assigns and legal representatives of such other party in respect to covenants, agreements and obligations contained in the Contract Documents. Neither party to the Contract shall assign the Contract as a whole without written consent of the other. If either party attempts to make such an assignment without such consent, that party shall nevertheless remain legally responsible for all obligations under the Contract.

13.3 WRITTEN NOTICE

13.3.1 Written notice shall be deemed to have been duly served if delivered in person to the individual or a member of the firm or entity or to an officer of the corporation for which it was intended, or if delivered at or sent by registered or certified mail to the last business address known to the party giving notice.

13.4 RIGHTS AND REMEDIES

13.4.1 Duties and obligations imposed by the Contract Documents and rights and remedies available thereunder shall be in addition to and not a limitation of duties, obligations, rights and remedies otherwise imposed or available by law.

13.4.2 No action or failure to act by the Owner, Architect or Contractor shall constitute a waiver of a right or duty afforded them under the Contract, nor shall such action or failure to act constitute approval of or acquiescence in a breach thereunder, except as may be specifically agreed in writing.

13.5 TESTS AND INSPECTIONS

13.5.1 Tests, inspections and approvals of portions of the Work required by the Contract Documents or by laws, ordinances, rules, regulations or orders of public authorities having jurisdiction shall be made at an appropriate time. Unless otherwise provided, the Contractor shall make arrangements for such tests, inspections and approvals with an independent testing laboratory or entity acceptable to the Owner, or with the appropriate public authority, and shall bear all related costs of tests, inspections and approvals. The Contractor shall give the Architect timely notice of when and where tests and inspections are to be made so the Architect may observe such procedures. The Owner shall bear costs of tests, inspections or approvals which do not become requirements until after bids are received or negotiations concluded.

13.5.2 If the Architect, Owner or public authorities having jurisdiction determine that portions of the Work require additional testing, inspection or approval not included under Subparagraph 13.5.1, the Architect will, upon written authorization from the Owner, instruct the Contractor to make arrangements for such additional testing, inspection or approval by an entity acceptable to the Owner, and the Contractor shall give timely notice to the Architect of when and where tests and inspections are to be made so the Architect may observe such procedures.

The Owner shall bear such costs except as provided in Subparagraph 13.5.3.

13.5.3 If such procedures for testing, inspection or approval under Subparagraphs 13.5.1 and 13.5.2 reveal failure of the portions of the Work to comply with requirements established by the Contract Documents, the Contractor shall bear all costs made necessary by such failure including those of repeated procedures and compensation for the Architect's services and expenses.

13.5.4 Required certificates of testing, inspection or approval shall, unless otherwise required by the Contract Documents, be secured by the Contractor and promptly delivered to the Architect.

13.5.5 If the Architect is to observe tests, inspections or approvals required by the Contract Documents, the Architect will do so promptly and, where practicable, at the normal place of testing.

13.5.6 Tests or inspections conducted pursuant to the Contract Documents shall be made promptly to avoid unreasonable delay in the Work.

13.6   INTEREST

13.6.1 Payments due and unpaid under the Contract Documents shall bear interest from the date payment is due at such rate as the parties may agree upon in writing or, in the absence thereof, at the legal rate prevailing from time to time at the place where the Project is located.

13.7   COMMENCEMENT OF STATUTORY LIMITATION PERIOD

13.7.1 As between the Owner and Contractor:

    .1 Before Substantial Completion. As to acts or failures to act occurring prior to the relevant date of Substantial Completion, any applicable statute of limitations shall commence to run and any alleged cause of action shall be deemed to have accrued in any and all events not later than such date of Substantial Completion;

    .2 Between Substantial Completion and Final Certificate for Payment. As to acts or failures to act occurring subsequent to the relevant date of Substantial Completion and prior to issuance of the final Certificate for Payment, any applicable statute of limitations shall commence to run and any alleged cause of action shall be deemed to have accrued in any and all events not later than the date of issuance of the final Certificate for Payment; and

    .3 After Final Certificate for Payment. As to acts or failures to act occurring after the relevant date of issuance of the final Certificate for Payment, any applicable statute of limitations shall commence to run and any alleged cause of action shall be deemed to have accrued in any and all events not later than the date of any act or failure to act by the Contractor pursuant to any warranty provided under Paragraph 3.5, the date of any correction of the Work or failure to correct the Work by the Contractor under Paragraph 12.2, or the date of actual commission of any other act or failure to perform any duty or obligation by the Contractor or Owner, whichever occurs last.

ARTICLE 14

TERMINATION OR SUSPENSION OF THE CONTRACT

14.1   TERMINATION BY THE CONTRACTOR

14.1.1 The Contractor may terminate the Contract if the Work is stopped for a period of 30 days through no act or fault of the Contractor or a Subcontractor, Sub-subcontractor or their agents or employees or any other persons performing portions of the Work under contract with the Contractor, for any of the following reasons:

    .1 issuance of an order of a court or other public authority having jurisdiction;

    .2 an act of government, such as a declaration of national emergency, making material unavailable;

    .3 because the Architect has not issued a Certificate for Payment and has not notified the Contractor of the reason for withholding certification as provided in Subparagraph 9.4.1, or because the Owner has not made payment on a Certificate for Payment within the time stated in the Contract Documents;

    .4 if repeated suspensions, delays or interruptions by the Owner as described in Paragraph 14.3 constitute in the aggregate more than 100 percent of the total number of days scheduled for completion, or 120 days in any 365-day period, whichever is less; or

    .5 the Owner has failed to furnish to the Contractor promptly, upon the Contractor's request, reasonable evidence as required by Subparagraph 2.2.1.

14.1.2 If one of the above reasons exists, the Contractor may, upon seven additional days' written notice to the Owner and Architect, terminate the Contract and recover from the Owner payment for Work executed and for proven loss with respect to materials, equipment, tools, and construction equipment and machinery, including reasonable overhead, profit and damages.

14.1.3 If the Work is stopped for a period of 60 days through no act or fault of the Contractor or a Subcontractor or their agents or employees or any other persons performing portions of the Work under contract with the Contractor because the Owner has persistently failed to fulfill the Owner's obligations under the Contract Documents with respect to matters important to the progress of the Work, the Contractor may, upon seven additional days' written notice to the Owner and the Architect, terminate the Contract and recover from the Owner as provided in Subparagraph 14.1.2.

14.2   TERMINATION BY THE OWNER FOR CAUSE

14.2.1 The Owner may terminate the Contract if the Contractor:

    .1 persistently or repeatedly refuses or fails to supply enough properly skilled workers or proper materials;

    .2 fails to make payment to Subcontractors for materials or labor in accordance with the respective agreements between the Contractor and the Subcontractors;

    .3 persistently disregards laws, ordinances, or rules, regulations or orders of a public authority having jurisdiction; or

    .4 otherwise is guilty of substantial breach of a provision of the Contract Documents.

14.2.2 When any of the above reasons exist, the Owner, upon certification by the Architect that sufficient cause exists to justify such action, may without prejudice to any other rights or remedies of the Owner and after giving the Contractor and the Contractor's surety, if any, seven days' written notice, terminate employment of the Contractor and may, subject to any prior rights of the surety:

    .1 take possession of the site and of all materials, equipment, tools, and construction equipment and machinery thereon owned by the Contractor;

    .2 accept assignment of subcontracts pursuant to Paragraph 5.4; and

    .3 finish the Work by whatever reasonable method the Owner may deem expedient.

14.2.3 When the Owner terminates the Contract for one of the reasons stated in Subparagraph 14.2.1, the Contractor shall not be entitled to receive further payment until the Work is finished.

14.2.4 If the unpaid balance of the Contract Sum exceeds costs of finishing the Work, including compensation for the Architect's services and expenses made necessary thereby, such excess shall be paid to the Contractor. If such costs exceed the unpaid balance, the Contractor shall pay the difference to the Owner. The amount to be paid to the Contractor or Owner, as the case may be, shall be certified by the Architect, upon application, and this obligation for payment shall survive termination of the Contract.

14.3 SUSPENSION BY THE OWNER FOR CONVENIENCE

14.3.1 The Owner may, without cause, order the Contractor in writing to suspend, delay or interrupt the Work in whole or in part for such period of time as the Owner may determine.

14.3.2 An adjustment shall be made for increases in the cost of performance of the Contract, including profit on the increased cost of performance, caused by suspension, delay or interruption. No adjustment shall be made to the extent:

    .1 that performance is, was or would have been so suspended, delayed or interrupted by another cause for which the Contractor is responsible; or

    .2 that an equitable adjustment is made or denied under another provision of this Contract.

14.3.3 Adjustments made in the cost of performance may have a mutually agreed fixed or percentage fee.

AIA DOCUMENT A201 • GENERAL CONDITIONS OF THE CONTRACT FOR CONSTRUCTION • FOURTEENTH EDITION
AIA® • ©1987 THE AMERICAN INSTITUTE OF ARCHITECTS, 1735 NEW YORK AVENUE, N.W., WASHINGTON, D.C. 20006

AIA Document A101

Standard Form of Agreement Between Owner and Contractor

where the basis of payment is a

STIPULATED SUM

1987 EDITION

THIS DOCUMENT HAS IMPORTANT LEGAL CONSEQUENCES; CONSULTATION WITH AN ATTORNEY IS ENCOURAGED WITH RESPECT TO ITS COMPLETION OR MODIFICATION.

The 1987 Edition of AIA Document A201, General Conditions of the Contract for Construction, is adopted in this document by reference. Do not use with other general conditions unless this document is modified.

This document has been approved and endorsed by The Associated General Contractors of America.

AGREEMENT

made as of the   day of   in the year of
Nineteen Hundred and

BETWEEN the Owner:
(Name and address)

and the Contractor:
(Name and address)

The Project is:
(Name and location)

The Architect is:
(Name and address)

The Owner and Contractor agree as set forth below.

ARTICLE 1
THE CONTRACT DOCUMENTS

The Contract Documents consist of this Agreement, Conditions of the Contract (General, Supplementary and other Conditions), Drawings, Specifications, Addenda issued prior to execution of this Agreement, other documents listed in this Agreement and Modifications issued after execution of this Agreement; these form the Contract, and are as fully a part of the Contract as if attached to this Agreement or repeated herein. The Contract represents the entire and integrated agreement between the parties hereto and supersedes prior negotiations, representations or agreements, either written or oral. An enumeration of the Contract Documents, other than Modifications, appears in Article 9.

ARTICLE 2
THE WORK OF THIS CONTRACT

The Contractor shall execute the entire Work described in the Contract Documents, except to the extent specifically indicated in the Contract Documents to be the responsibility of others, or as follows:

ARTICLE 3
DATE OF COMMENCEMENT AND SUBSTANTIAL COMPLETION

3.1 The date of commencement is the date from which the Contract Time of Paragraph 3.2 is measured, and shall be the date of this Agreement, as first written above, unless a different date is stated below or provision is made for the date to be fixed in a notice to proceed issued by the Owner.

(Insert the date of commencement, if it differs from the date of this Agreement or, if applicable, state that the date will be fixed in a notice to proceed.)

Unless the date of commencement is established by a notice to proceed issued by the Owner, the Contractor shall notify the Owner in writing not less than five days before commencing the Work to permit the timely filing of mortgages, mechanic's liens and other security interests.

3.2 The Contractor shall achieve Substantial Completion of the entire Work not later than

(Insert the calendar date or number of calendar days after the date of commencement. Also insert any requirements for earlier Substantial Completion of certain portions of the Work, if not stated elsewhere in the Contract Documents.)

, subject to adjustments of this Contract Time as provided in the Contract Documents.

(Insert provisions, if any, for liquidated damages relating to failure to complete on time.)

ARTICLE 4
CONTRACT SUM

4.1 The Owner shall pay the Contractor in current funds for the Contractor's performance of the Contract the Contract Sum of
                                                                                                          Dollars
($), subject to additions and deductions as provided in the Contract Documents.

4.2 The Contract Sum is based upon the following alternates, if any, which are described in the Contract Documents and are hereby accepted by the Owner:

(State the numbers or other identification of accepted alternates. If decisions on other alternates are to be made by the Owner subsequent to the execution of this Agreement, attach a schedule of such other alternates showing the amount for each and the date until which that amount is valid.)

4.3 Unit prices, if any, are as follows:

ARTICLE 5
PROGRESS PAYMENTS

5.1 Based upon Applications for Payment submitted to the Architect by the Contractor and Certificates for Payment issued by the Architect, the Owner shall make progress payments on account of the Contract Sum to the Contractor as provided below and elsewhere in the Contract Documents.

5.2 The period covered by each Application for Payment shall be one calendar month ending on the last day of the month, or as follows:

5.3 Provided an Application for Payment is received by the Architect not later than the
day of a month, the Owner shall make payment to the Contractor not later than the
the day of the month. If an Application for Payment is received by the Architect after the application date fixed above, payment shall be made by the Owner not later than
days after the Architect receives the Application for Payment.

5.4 Each Application for Payment shall be based upon the Schedule of Values submitted by the Contractor in accordance with the Contract Documents. The Schedule of Values shall allocate the entire Contract Sum among the various portions of the Work and be prepared in such form and supported by such data to substantiate its accuracy as the Architect may require. This Schedule, unless objected to by the Architect, shall be used as a basis for reviewing the Contractor's Applications for Payment.

5.5 Applications for Payment shall indicate the percentage of completion of each portion of the Work as of the end of the period covered by the Application for Payment.

5.6 Subject to the provisions of the Contract Documents, the amount of each progress payment shall be computed as follows:

5.6.1 Take that portion of the Contract Sum properly allocable to completed Work as determined by multiplying the percentage completion of each portion of the Work by the share of the total Contract Sum allocated to that portion of the Work in the Schedule of Values, less retainage of percent
(%). Pending final determination of cost to the Owner of changes in the Work, amounts not in dispute may be included as provided in Subparagraph 7.3.7 of the General Conditions even though the Contract Sum has not yet been adjusted by Change Order;

5.6.2 Add that portion of the Contract Sum properly allocable to materials and equipment delivered and suitably stored at the site for subsequent incorporation in the completed construction (or, if approved in advance by the Owner, suitably stored off the site at a location agreed upon in writing), less retainage of
percent (%);

5.6.3 Subtract the aggregate of previous payments made by the Owner; and

5.6.4 Subtract amounts, if any, for which the Architect has withheld or nullified a Certificate for Payment as provided in Paragraph 9.5 of the General Conditions.

5.7 The progress payment amount determined in accordance with Paragraph 5.6 shall be further modified under the following circumstances:

5.7.1 Add, upon Substantial Completion of the Work, a sum sufficient to increase the total payments to
percent (%) of the Contract
Sum, less such amounts as the Architect shall determine for incomplete Work and unsettled claims; and

5.7.2 Add, if final completion of the Work is thereafter materially delayed through no fault of the Contractor, any additional amounts payable in accordance with Subparagraph 9.10.3 of the General Conditions.

5.8 Reduction or limitation of retainage, if any, shall be as follows:

(If it is intended, prior to Substantial Completion of the entire Work, to reduce or limit the retainage resulting from the percentages inserted in Subparagraphs 5.6.1 and 5.6.2 above, and this is not explained elsewhere in the Contract Documents, insert here provisions for such reduction or limitation.)

ARTICLE 6
FINAL PAYMENT

Final payment, constituting the entire unpaid balance of the Contract Sum, shall be made by the Owner to the Contractor when (1) the Contract has been fully performed by the Contractor except for the Contractor's responsibility to correct nonconforming Work as provided in Subparagraph 12.2.2 of the General Conditions and to satisfy other requirements, if any, which necessarily survive final payment; and (2) a final Certificate for Payment has been issued by the Architect; such final payment shall be made by the Owner not more than 30 days after the issuance of the Architect's final Certificate for Payment, or as follows:

ARTICLE 7
MISCELLANEOUS PROVISIONS

7.1 Where reference is made in this Agreement to a provision of the General Conditions or another Contract Document, the reference refers to that provision as amended or supplemented by other provisions of the Contract Documents.

7.2 Payments due and unpaid under the Contract shall bear interest from the date payment is due at the rate stated below, or in the absence thereof, at the legal rate prevailing from time to time at the place where the Project is located.

(Insert rate of interest agreed upon, if any.)

(Usury laws and requirements under the Federal Truth in Lending Act, similar state and local consumer credit laws and other regulations at the Owner's and Contractor's principal places of business, the location of the Project and elsewhere may affect the validity of this provision. Legal advice should be obtained with respect to deletions or modifications, and also regarding requirements such as written disclosures or waivers.)

7.3 Other provisions:

ARTICLE 8
TERMINATION OR SUSPENSION

8.1 The Contract may be terminated by the Owner or the Contractor as provided in Article 14 of the General Conditions.

8.2 The Work may be suspended by the Owner as provided in Article 14 of the General Conditions.

ARTICLE 9
ENUMERATION OF CONTRACT DOCUMENTS

9.1 The Contract Documents, except for Modifications issued after execution of this Agreement, are enumerated as follows:

9.1.1 The Agreement is this executed Standard Form of Agreement Between Owner and Contractor, AIA Document A101, 1987 Edition.

9.1.2 The General Conditions are the General Conditions of the Contract for Construction, AIA Document A201, 1987 Edition.

9.1.3 The Supplementary and other Conditions of the Contract are those contained in the Project Manual dated
, and are as follows:

Document	Title	Pages

9.1.4 The Specifications are those contained in the Project Manual dated as in Subparagraph 9.1.3, and are as follows:
(Either list the Specifications here or refer to an exhibit attached to this Agreement.)

Section	Title	Pages

9.1.5 The Drawings are as follows, and are dated   unless a different date is shown below:
(Either list the Drawings here or refer to an exhibit attached to this Agreement.)

Number   **Title**   **Date**

9.1.6 The Addenda, if any, are as follows:

Number   **Date**   **Pages**

Portions of Addenda relating to bidding requirements are not part of the Contract Documents unless the bidding requirements are also enumerated in this Article 9.

9.1.7 Other documents, if any, forming part of the Contract Documents are as follows:

(List here any additional documents which are intended to form part of the Contract Documents. The General Conditions provide that bidding requirements such as advertisement or invitation to bid, Instructions to Bidders, sample forms and the Contractor's bid are not part of the Contract Documents unless enumerated in this Agreement. They should be listed here only if intended to be part of the Contract Documents.)

This Agreement is entered into as of the day and year first written above and is executed in at least three original copies of which one is to be delivered to the Contractor, one to the Architect for use in the administration of the Contract, and the remainder to the Owner.

OWNER   CONTRACTOR

_____   _____
(Signature)   *(Signature)*

_____   _____
(Printed name and title)   *(Printed name and title)*

D
CONVERSIONS

USEFUL CONVERSION FACTORS

Cement

$$1 \text{ sack} = 1 \text{ c.f.} = 94 \text{ pounds}$$
$$1 \text{ barrel} = 4 \text{ sacks} = 376 \text{ pounds}$$

Water

$$1 \text{ U.S. gallon} = 231 \text{ c.i.} = 0.1337 \text{ c.f.} = 8.35 \text{ pounds}$$
$$1 \text{ c.f.} = 7.5 \text{ U.S. gallons} = 62.4 \text{ pounds}$$

Aggregate

$$1 \text{ ton} = 2,000 \text{ pounds} = 19 \text{ c.f.} = 0.70 \text{ c.y.}$$
$$1 \text{ c.y.} = 27 \text{ c.f.} = 2,800 \text{ pounds} = 1.40 \text{ tons (approximate)}$$

Ready-Mix Concrete

$$1 \text{ c.y.} = 27 \text{ c.f.} = 2 \text{ tons} = 4,000 \text{ pounds (approximate)}$$
$$1 \text{ ton} = 2,000 \text{ pounds} = 0.50 \text{ c.y.} = 13.5 \text{ c.f.}$$

Concrete: weights (approximate)

$$1 \text{ c.f.} = 145\# \text{ (heavyweight)}$$
$$1 \text{ c.f.} = 115\# \text{ (lightweight)}$$

CUBE

V (Volume) $= l \times w \times h$

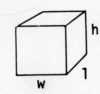

CYLINDER

$V = $ Area of $a \times b$

CONE

$$V = \frac{\text{Area of } a \times b}{3}$$

RECTANGLE

A (area) $= b \times h$

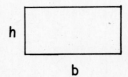

CIRCLE

$A = \pi r^2$ or $0.785\ d^2$
$\pi = 3.14159$

circumference $= \pi d$

use $\pi = 3.14$

RIGHT TRIANGLE

$$A = \frac{bh}{2}$$

hypotenuse $= c = \sqrt{b^2 + h^2}$

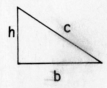

SQUARE

$A = s^2$ or $s \times s$ (side squared)

PARALLELOGRAM

$A = bh$

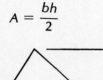

ANY TRIANGLE

$$A = \frac{bh}{2}$$

TRAPEZOID

$$A = \left(\frac{a + b}{2}\right) h$$

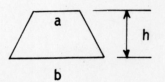

LENGTHS

$$12 \text{ inches} = 1 \text{ foot}$$
$$3 \text{ feet} = 1 \text{ yard}$$

AREAS

$$144 \text{ s.i.} = 1 \text{ s.f.}$$
$$9 \text{ s.f.} = 1 \text{ s.y.}$$
$$100 \text{ s.f.} = 1 \text{ square}$$

VOLUME

$$1,728 \text{ c.i.} = 1 \text{ c.f.}$$
$$27 \text{ c.f.} = 1 \text{ c.y.}$$

$$1 \text{ c.f.} = 7.4850 \text{ gallons}$$
$$1 \text{ gallon} = 231 \text{ c.i.}$$
$$1 \text{ gallon} = 8.33 \text{ pounds}$$
$$1 \text{ c.f.} = 62.3 \text{ pounds}$$

1 millimeter (mm) = 0.0394 inch
1 centimeter (cm) = 0.3937 inch
1 decimeter (dm) = 3.937 inches
1 meter (m)   = 39.37 inches
1 meter (m)   = 1.1 yards
1 kilometer (km) = 0.621 mile

1 square cm = 0.155 s.f.
1 square m = 10.764 s.f. = 1.196 s.y.

1 cubic cm = 0.06 c.i.
1 cubic dm = 61.02 c.i.
1 cubic m = 1.308 c.y.

INCHES REDUCED TO DECIMALS

INCHES	DECIMAL
½	0.041
1	0.083
1½	0.125
2	0.167
2½	0.209
3	0.250
3½	0.292
4	0.333
4½	0.375
5	0.417
5½	0.458
6	0.500
6½	0.542
7	0.583
7½	0.625
8	0.667
8½	0.708
9	0.750
9½	0.792
10	0.833
10½	0.875
11	0.917
11½	0.958
12	1.00

COMMON FRACTIONS STATED AS DECIMALS

FRACTION	DECIMAL
1/16	0.0625
1/8	0.125
3/16	0.1875
1/4	0.250
5/16	0.3125
3/8	0.375
7/16	0.4375
1/2	0.500
9/16	0.5625
5/8	0.625
11/16	0.6875
3/4	0.750
13/16	0.8125
7/8	0.875
15/16	0.9375
16/16	1.00

FRACTIONS OF *INCHES* REDUCED TO DECIMALS OF A *FOOT*

FRACTION	DECIMAL
1/16	0.00520
1/8	0.0104
3/16	0.0156
1/4	0.0208
5/16	0.0260
3/8	0.0312
7/16	0.0364
1/2	0.0416
9/16	0.0468
5/8	0.0520
11/16	0.0572
3/4	0.0624
13/16	0.0676
7/8	0.0728
15/16	0.0780
16/16	0.0833

E

DRAWINGS AND OUTLINE SPECIFICATIONS OF A SMALL COMMERCIAL BUILDING

OUTLINE SPECIFICATIONS—JOY BUILDING

Section 1—Excavation

Topsoil excavation shall be carried 5 feet beyond all walls.
Excavations shall be carried to the depths shown on the drawings.
All excess earth, not needed for fill, shall be removed from the premises.

Section 2—Concrete

Concrete for footings shall develop a compressive strength of 3,000 psi at 28 days.
All other concrete shall develop a compressive strength of 2,500 psi at 28 days unless otherwise specified.
Reinforcing bars shall be A-15 Intermediate grade.
Reinforcing mesh shall be 4 × 4 8/8 welded wire mesh.
Vapor Barrier shall be 6 mil polyethylene.
Expansion joint filler shall be ½" × 4" asphalt impregnated fiberboard.
Concrete lintels shall be precast, 5,000 psi at 28 days.

Section 3—Masonry

Concrete masonry units shall be 16" in length.
Masonry wall reinforcement shall be ladder design and installed every 3rd course.
Stone shall be bluestone with stone thickness ¾" to 2". All mortar joints ⅜" thick.
Exterior window sills shall be bluestone, 1¼" thick, slip in type.

Section 4—Structural

Complete structural system as indicated on the drawings.
All required holes in the roof deck to be made by roof deck installer.

Furnish all required accessories as indicated on the drawings.
Furnish and install in compliance with all manufacturer's specifications.

Section 5—Door Frames, Doors and Windows

Door frames 16-gauge steel.
Doors shall be flush, solid core, five-ply construction with a rotary birch veneer. Prefinish interior doors with semi-gloss lacquer and exterior doors with polyurethane varnish.
Allow $200 for the purchase of hardware.
Install hardware in accordance with the manufacturer's specifications.
Windows shall be custom made, ⅛-inch thick aluminum, 2-inch × 4-inch frames, glazing material as indicated on the drawings.

Section 6—Roofing

Roofing shall be five-ply, 20-year bond.
Cant strips shall be asphalt impregnated fiberboard.
Insulation shall be rigid, perlite.
Slag surface (400 lbs. per sq.)

Section 7—Finishes

All exposed exterior masonry finishes shall receive two coats of silicone spray.
All exposed interior finishes of drywall shall receive two coats of an approved oil base paint.
All vinyl tile shall be 1/8″ thick, sheet vinyl, to be selected from the Foamcraft collection, Aztec, as manufactured by GAF Corp. or approved equal.
Bases shall be 1/8″ thick, 2 1/2″ high, color to be selected.
All concrete slabs shall be machine trowel finished.

Section 8—Drywall and Wetwall

Furnish and install complete systems with any required accessories as indicated on the drawings.

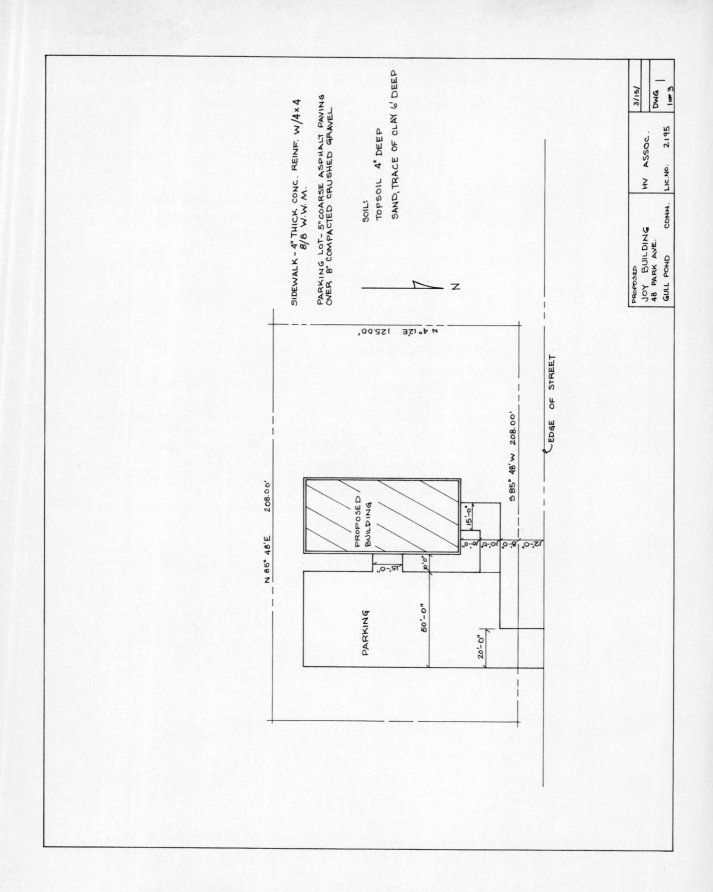

SIDEWALK - 4" THICK CONC. REINF. W/4 x 4
8/8 W.W.M.

PARKING LOT- 5"COARSE ASPHALT PAVING
OVER 8" COMPACTED CRUSHED GRAVEL

SOIL:
TOPSOIL 4" DEEP
SAND, TRACE OF CLAY 6' DEEP

N

N 4°12'E 125.00'

EDGE OF STREET

S 85° 48'W 208.00'

N 85° 48'E 208.00'

PROPOSED BUILDING

PARKING

15'-0"

15'-0"

9'-0"

9'-0"

10'-0"

9'-0"

12'-0"

9'-0"

50'-0"

20'-0"

PROPOSED
JOY BUILDING
48 PARK AVE.
GULL POND   CONN.

HV ASSOC.

LIC.NO. 2195

3/15/

DWG
1 OF 3

appendix e

392

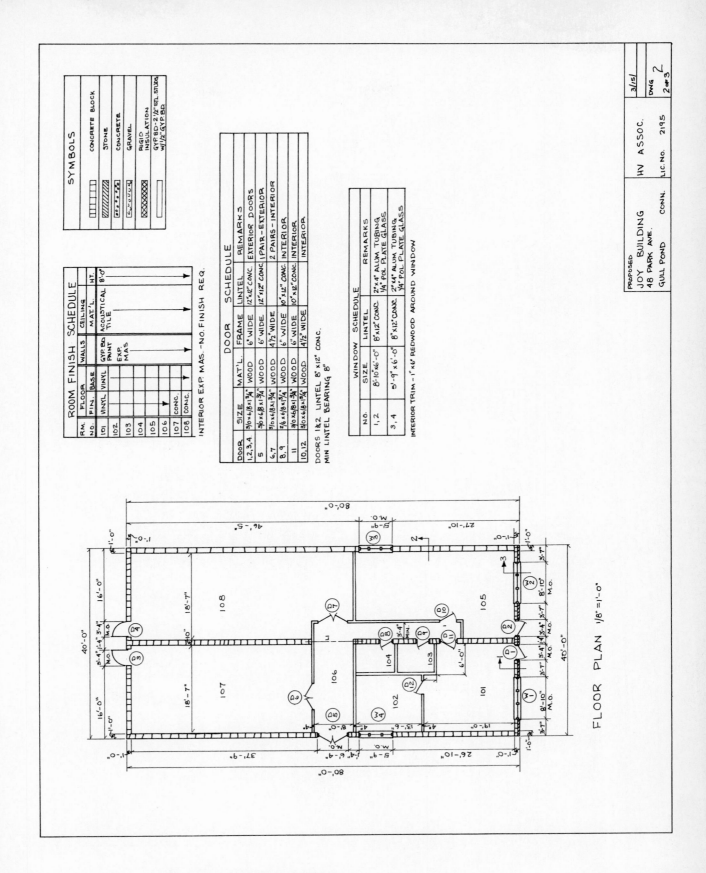

PROPOSED
JOY BUILDING
48 PARK AVE.
GULL POND   CONN.
HV ASSOC.
LIC. NO.   2195

SYMBOLS

	CONCRETE BLOCK
	STONE
	CONCRETE
	GRAVEL
	RIGID INSULATION
	GYP BD-2½" ST. STUDS w/½" GYP BD

ROOM FINISH SCHEDULE

RM. NO.	FLOOR		BASE		WALLS		CEILING	
	FIN.	VINYL		VINYL	GYP BD PAINT	EXP. MAS.	MAT'L.	HT.
101							ACOUSTICAL TILE	8'-0"
102								
103								
104								
105								
106								
107	CONC.							
108	CONC.							

INTERIOR EXP. MAS.–NO. FINISH REQ.

DOOR SCHEDULE

DOOR	SIZE	MAT'L.	FRAME	LINTEL	REMARKS
1,2,3,4	3/0×6/8×1¾	WOOD	6" WIDE	12"×12" CONC.	EXTERIOR DOORS
5	3/0×6/8×1¾	WOOD	6" WIDE	12"×12" CONC.	1 PAIR-EXTERIOR
6,7	3/0×6/8×1¾	WOOD	4½" WIDE		2 PAIRS-INTERIOR
8,9	2/6×6/8×1¾	WOOD	6" WIDE	10"×12" CONC.	INTERIOR
11	3/0×6/8×1¾	WOOD	6" WIDE	10"×12" CONC.	INTERIOR
10,12	3/0×6/8×1¾	WOOD	4½" WIDE		INTERIOR

DOORS 1&2 LINTEL 8"×12" CONC.
MIN LINTEL BEARING 8"

WINDOW SCHEDULE

NO.	SIZE	LINTEL	REMARKS
1,2	8'-10"×6'-0"	8"×12" CONC.	2"×4" ALUM TUBING ¼" POL PLATE GLASS
3,4	5'-9"×6'-0"	8"×12" CONC.	2"×4" ALUM TUBING ¼" POL PLATE GLASS

INTERIOR TRIM–1"×6" REDWOOD AROUND WINDOW

FLOOR PLAN 1/8"=1'-0"

small commercial building

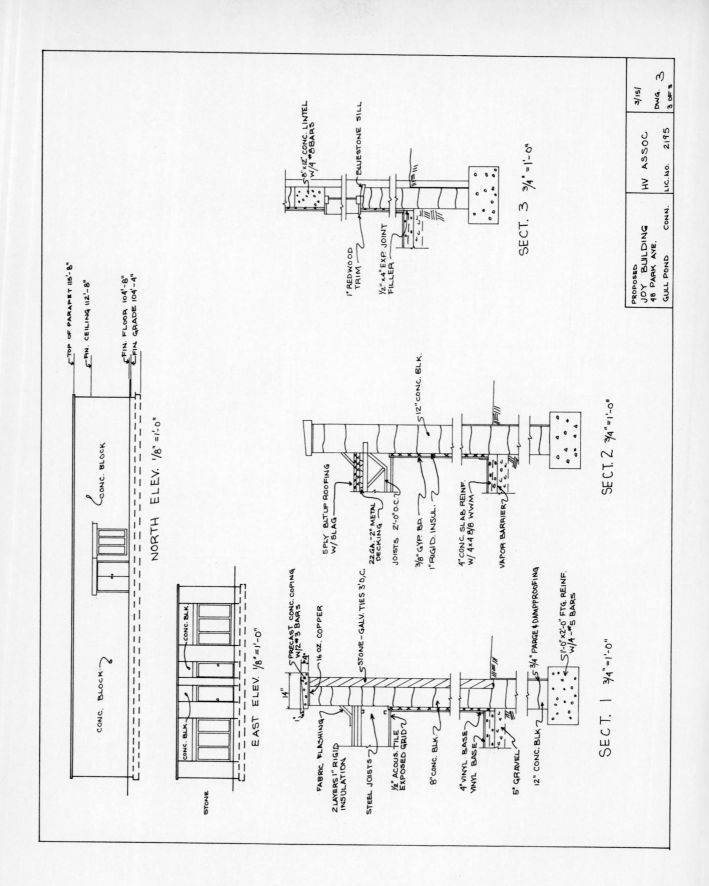

NORTH ELEV. 1/8"=1'-0"

TOP OF PARAPET 115'-8"
FIN. CEILING 112'-8"
FIN. FLOOR 104'-8"
FIN. GRADE 104'-4"

CONC. BLOCK

CONC. BLOCK

EAST ELEV. 1/8"=1'-0"

CONC. BLK.

CONC. BLK.

STONE

SECT. 1 3/4"=1'-0"

PRECAST CONC. COPING W/2 #3 BARS
16 OZ. COPPER
STONE-GALV. TIES 3'O.C.
FABRIC FLASHING
2 LAYERS 1" RIGID INSULATION
STEEL JOISTS
1/2" ACOUS. TILE EXPOSED GRID
8" CONC. BLK.
4" VINYL BASE
VINYL BASE
6" GRAVEL
12" CONC. BLK.
3/4" PARGE & DAMPPROOFING
1'-0"x2'-0" FTG. REINF. W/4-#5 BARS

SECT. 2 3/4"=1'-0"

5 PLY BLT-UP ROOFING W/ SLAG
22 GA.-2" METAL DECKING
JOISTS 2'-0"O.C.
3/8" GYP. BD.
1" RIGID. INSUL.
12" CONC. BLK.
4" CONC. SLAB REINF. W/ 4x4 8/8 WWM
VAPOR BARRIER

SECT. 3 3/4"=1'-0"

8"x12" CONC. LINTEL W/4 #8 BARS
BLUESTONE SILL
1" REDWOOD TRIM
1/2"x4" EXP. JOINT FILLER

PROPOSED
JOY BUILDING
48 PARK AVE.
GULL POND   CONN.

HV ASSOC
LIC. NO. 2,195
CONN.

3/15/
DWG. 3
3 OF 3

appendix e

394

F
SMALL COMMERCIAL
BUILDING

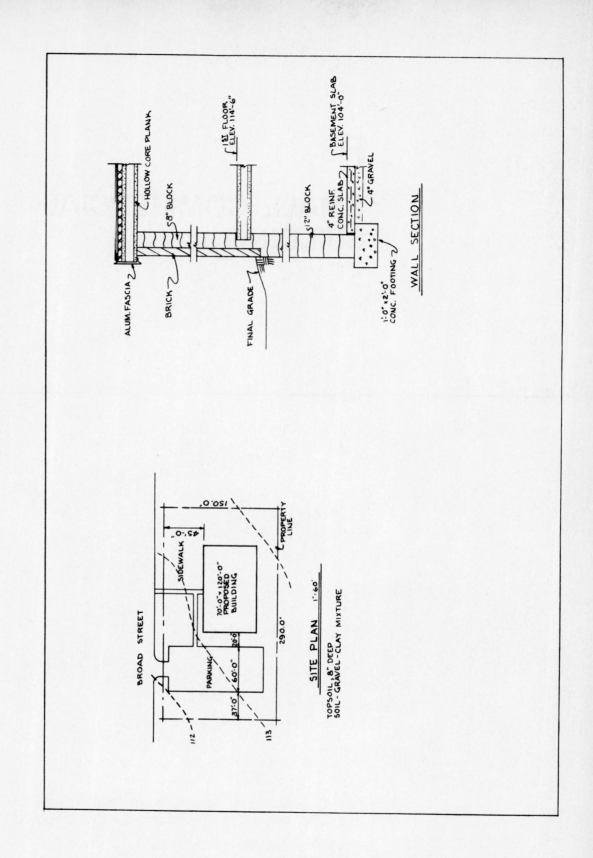

WALL SECTION

ALUM. FASCIA

HOLLOW CORE PLANK

8" BLOCK

1st FLOOR ELEV. 114'-6"

BRICK

FINAL GRADE

12" BLOCK

BASEMENT SLAB ELEV. 104'-0"

4" REINF. CONC. SLAB

4" GRAVEL

1'-0" x 2'-0" CONC. FOOTING

SITE PLAN 1":60'

BROAD STREET

SIDEWALK

150.0'

45'-0"

70'-0" x 120'-0" PROPOSED BUILDING

PROPERTY LINE

PARKING

37'-0"   60'-0"   20'-0"

290.0'

112

113

TOPSOIL 8" DEEP
SOIL - GRAVEL - CLAY MIXTURE

appendix f

396

G

DRAWINGS AND OUTLINE SPECIFICATIONS OF A RESIDENCE

OUTLINE SPECIFICATIONS

Section 1—Excavation

Topsoil excavation shall be carried 5 feet beyond all walls.

Excavation shall be carried to the depths shown on the drawings. All excess earth, not needed for fill, shall be removed from the premises.

Section 2—Concrete

Concrete for footings shall develop a compressive strength of 2,500 psi at 28 days. Reinforcing bars shall be A-15 Intermediate Grade.

Section 3—Masonry

Concrete masonry units shall be 16 inches in length.

Mortar joints shall be gray mortar, 3/8-inch thick.

Section 4—Framing

All wood framing shall be sized in accordance with the drawings and building code requirements.

All wood framing shall be kiln dried.

Section 5—Door frames, door, windows

Door frames shall be ponderosa pine.

Interior doors shall be 1 3/8 inches thick, hollow core, birch.

Exterior door—allow $150 for exterior doors.

Doors shall be of the sizes indicated on the drawings.

Windows shall be wood, of the size and type indicated on the drawings.

Hardware—allow $150.

Section 6—Roofing

Roofing felt shall be 15 lb., 2 layers required.

Roofing shingles shall be asphalt, 235 lbs. per square, class C wind resistant, as manufactured by Johns-Manville or approved equal.

Section 7—Finishes

All exposed exterior wood finish shall receive one coat of approved stain.

All interior drywall finishes shall receive two coats of approved alkyd paint.

Floors shall receive 1/8-inch thick, 12 inches × 12 inches from Royal Stoneglow collection manufactured by GAF Corp. or approved equal.

Base shall be 1/8-inch vinyl, 2 1/2 inches high.

Section 8—Drywall

Furnish and install in accordance with the drawings.

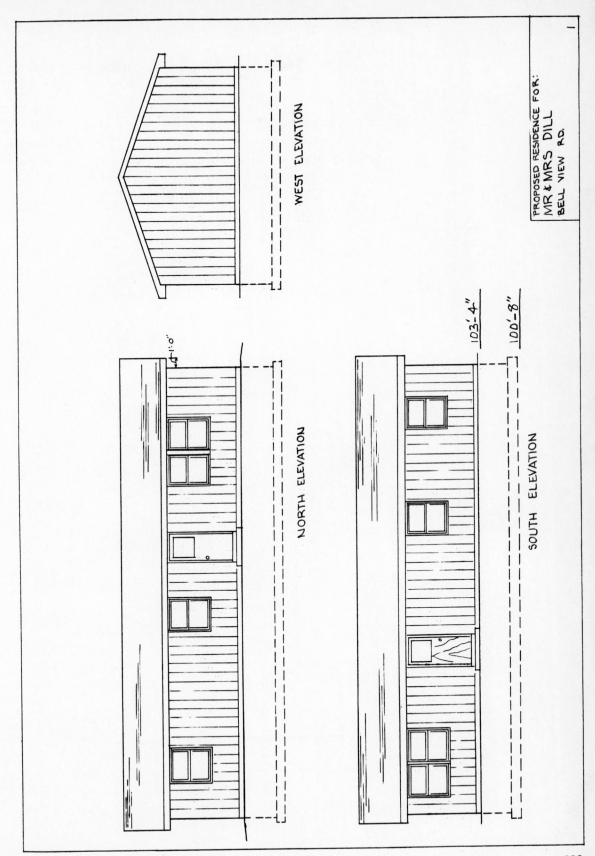

WEST ELEVATION

PROPOSED RESIDENCE FOR:
MR & MRS DILL
BELL VIEW RD.

4'-0"

NORTH ELEVATION

103'-4"

100'-8"

SOUTH ELEVATION

residence

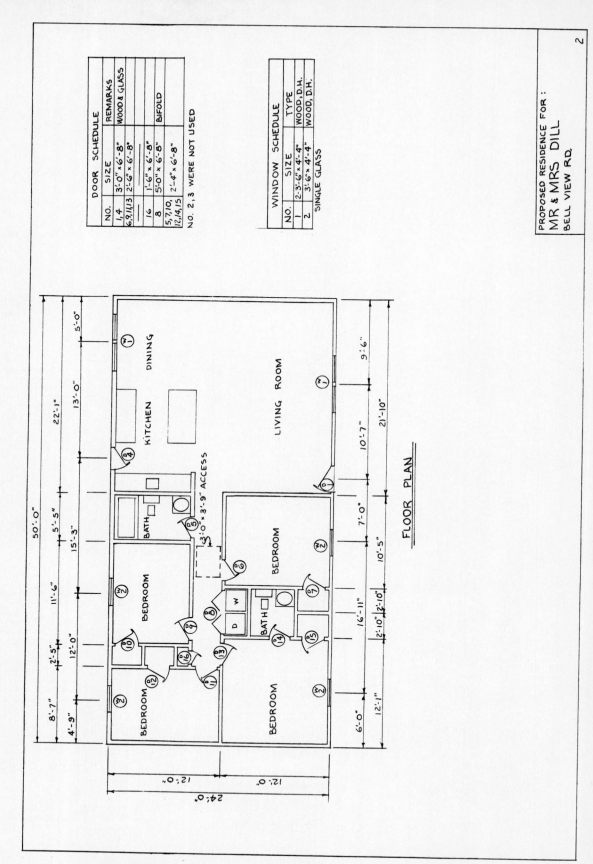

FLOOR PLAN

DOOR SCHEDULE

NO.	SIZE	REMARKS
1,4	3'-0" x 6'-8"	WOOD & GLASS
6,9,11,13	2'-6" x 6'-8"	
16	1'-6" x 6'-8"	
8	5'-0" x 6'-8"	BIFOLD
5,7,10, 12,14,15	2'-4" x 6'-8"	

NO. 2, 3 WERE NOT USED

WINDOW SCHEDULE

NO.	SIZE	TYPE
1	2'-3'-6" x 4'-4"	WOOD, D.H.
2	3'-6" 4'-4"	WOOD, D.H.
	SINGLE GLASS	

PROPOSED RESIDENCE FOR :
MR & MRS DILL
BELL VIEW RD.

2

appendix g

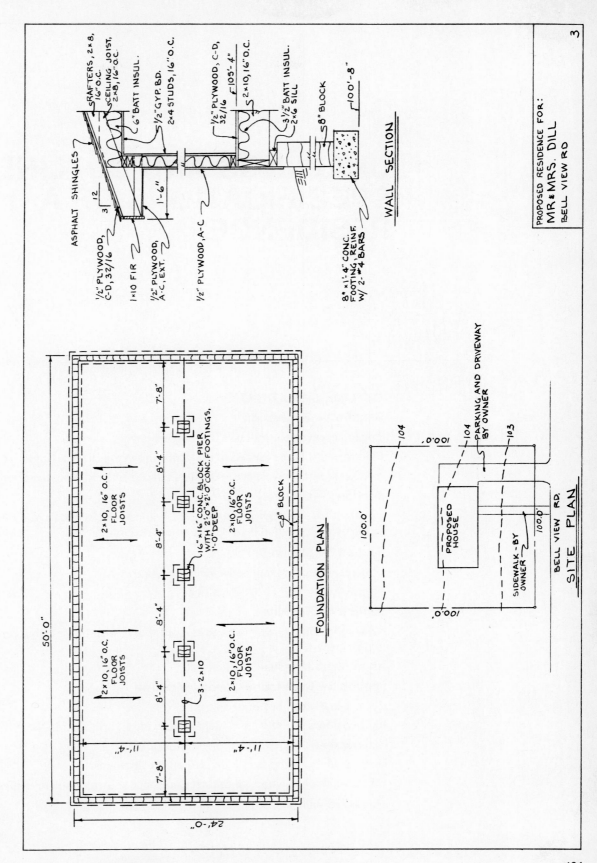

ASPHALT SHINGLES

RAFTERS, 2×8, 16" O.C.
CEILING JOIST, 2×8, 16" O.C.
6" BATT INSUL.
5½" GYP. BD.
2×4 STUDS, 16" O.C.

½" PLYWOOD, C-D, 32/16
2×10, 16" O.C.

3½" BATT INSUL.
2×6 SILL
8" BLOCK
100'-8"

½" PLYWOOD, C-D, 32/16

1×10 FIR

½" PLYWOOD, A-C, EXT.

½" PLYWOOD, A-C

12
3

1'-6"

105'-4"

WALL SECTION

8"×1'-4" CONC. FOOTING, REINF. W/ 2-#4 BARS

PROPOSED RESIDENCE FOR:
MR. & MRS. DILL
BELL VIEW RD

3

FOUNDATION PLAN

50'-0"

7'-8"
8'-4"
8'-4"
8'-4"
8'-4"
7'-8"

2×10, 16" O.C. FLOOR JOISTS
2×10, 16" O.C. FLOOR JOISTS
2×10, 16" O.C. FLOOR JOISTS
2×10, 16" O.C. FLOOR JOISTS

16"×16" CONC. BLOCK PIER WITH 2'-0"×2'-0" CONC. FOOTINGS, 1'-0" DEEP

8" BLOCK

3-2×10

11'-4"
11'-4"
24'-0"

SITE PLAN

PROPOSED HOUSE

PARKING AND DRIVEWAY BY OWNER

SIDEWALK - BY OWNER

BELL VIEW RD.

100.0'
100.0'
100.0'
100.0'

104
104
103

residence

H

DRAWINGS AND OUTLINE SPECIFICATIONS OF A RESIDENCE

OUTLINE SPECIFICATIONS

Section 1—Excavation

Topsoil excavation shall be carried 4 feet beyond all walls.

Excavation shall be carried to the depth shown on the drawings.

All excess earth, not needed for fill, shall be removed from the premises.

Section 2—Concrete

Concrete for footings shall develop a compressive strength of 3,000 psi at 28 days. Reinforcing bars shall be A-15 Intermediate Grade.

Section 3—Masonry

Concrete masonry shall be 16 inches in length.

Mortar joints shall be 3/8-inch thick, gray.

Section 4—Framing

All wood framing shall be sized in accordance with the drawings and building code requirements.

All wood framing shall be kiln dried.

Section 5—Door frames, doors, windows

Door frames shall be ponderosa pine.

Interior doors shall be 1 3/8-inches thick, birch, hollow core.

Exterior doors—allow $200 for exterior doors.

Doors shall be of the sizes indicated on the drawings.

Windows shall be wood, of the size and shape shown on the drawings.

Hardware—allow $250.

Sliding glass doors shall have insulating glass, anodized aluminum frames and a screen.

Section 6—Roofing

Roofing felt shall be 15 lbs.

Roofing shingles shall be 300 lbs. per square, asphalt, Johns-Manville or equal.

Section 7—Finishes

All exposed exterior wood finish shall receive one coat of approved stain.

All interior drywall finishes shall receive two coats of approved alkyd paint.

Resilient tile shall be 3/32 inch, vinyl, 12 inches × 12 inches from royal Stoneglow collection by GAF Corp.

Base shall be 1/8-inch vinyl, 4 inches high.

Carpeting shall be selected from the Opening Night Collection by Lees Carpets.

Section 8—Drywall

Furnish and install in accordance with the drawings.

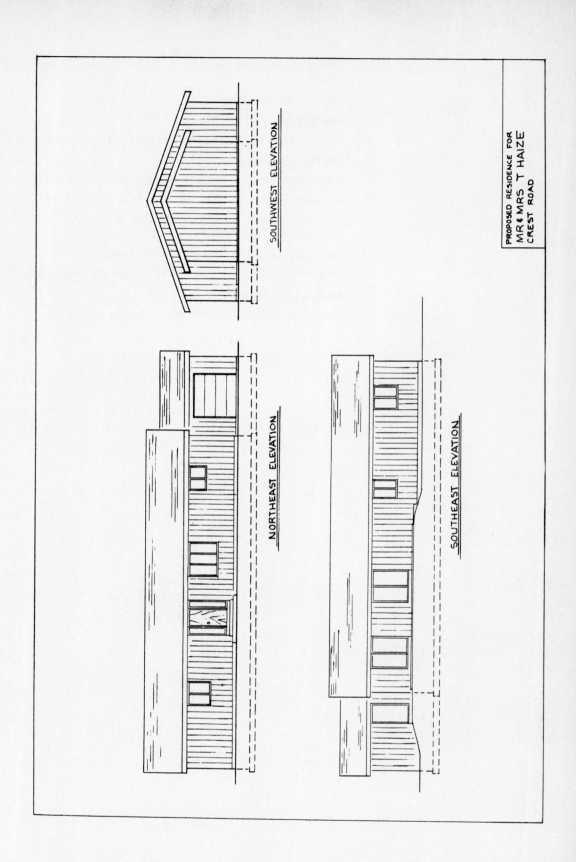

SOUTHWEST ELEVATION

NORTHEAST ELEVATION

SOUTHEAST ELEVATION

PROPOSED RESIDENCE FOR
MR & MRS T HAIZE
CREST ROAD

appendix h

404

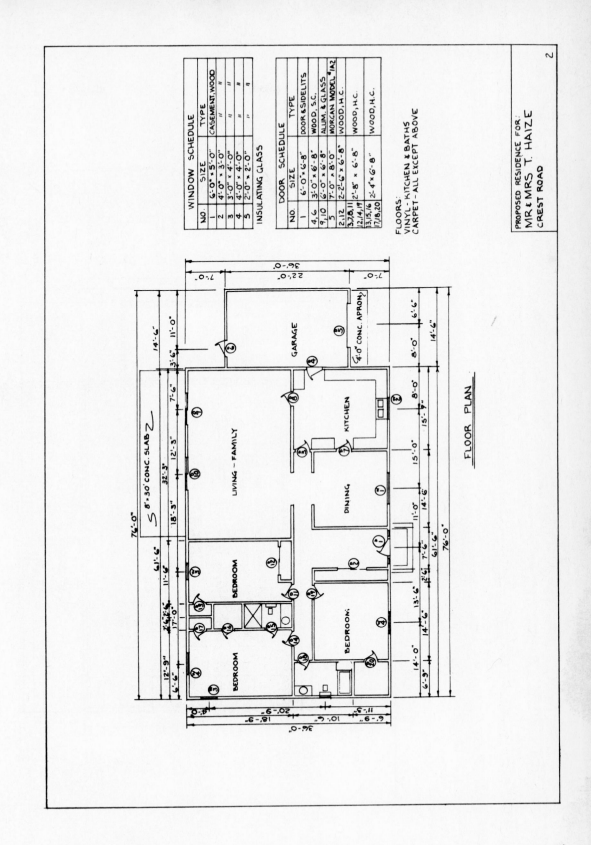

FLOOR PLAN

WINDOW SCHEDULE

NO.	SIZE	TYPE
1	6'-0" × 5'-0"	CASEMENT, WOOD
2	4'-0" × 3'-0"	"
3	3'-0" × 4'-0"	"
4	4'-0" × 4'-0"	"
5	2'-0" × 2'-0"	"

INSULATING GLASS

DOOR SCHEDULE

NO.	SIZE	TYPE
1	6'-0" × 6'-8"	DOOR & SIDELITS
4,6	3'-0" × 6'-8"	WOOD, S.C.
9,10	6'-0" × 6'-8"	ALUM. & GLASS
5	7'-0" × 8'-0"	MORGAN MODEL #1A2
2,12	2'-2'-6" × 6'-8"	
3,7,8,11	2'-8" × 6'-8"	WOOD, H.C.
12,14,19		WOOD, H.C.
13,15,16	2'-4" × 6'-8"	WOOD, H.C.
17,18,20		

FLOORS:
VINYL - KITCHEN & BATHS
CARPET - ALL EXCEPT ABOVE

PROPOSED RESIDENCE FOR:
MR & MRS T. HAIZE
CREST ROAD

2

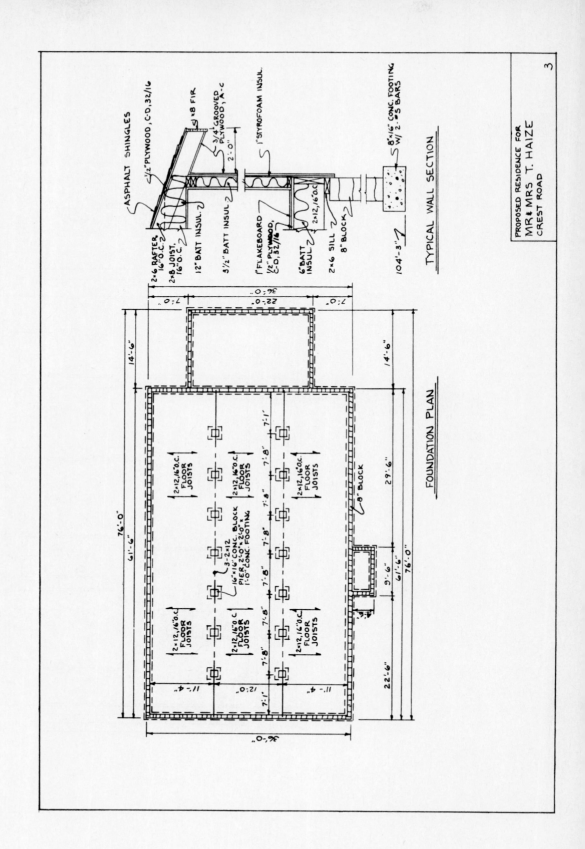

ASPHALT SHINGLES

½" PLYWOOD, C-D, 32/16

1" × 8 FIR

¾" GROOVED PLYWOOD, A-C

1" STYROFOAM INSUL.

2×6 RAFTER, 16"0.C.

2×8 JOIST, 16"0.C.

12" BATT INSUL.

3½" BATT INSUL.

1" FLAKEBOARD

½" PLYWOOD, C-D, 32/16

6" BATT INSUL.

2×6 SILL

8" BLOCK

2×12, 16"0.C.

8"×16" CONC. FOOTING W/ 2-#5 BARS

104'-3"

TYPICAL WALL SECTION

36'-0"

7'-0"   22'-0"   7'-0"

14'-6"

4'-6"

76'-0"

61'-6"

2×12, 16"0.C. FLOOR JOISTS

2×12, 16"0.C. FLOOR JOISTS

16"×16" CONC. BLOCK PIER, 2'-0"×2'-0"× 1'-0" CONC. FOOTING

3-2×12

2×12, 16"0.C. FLOOR JOISTS

7'-1"

7'-8"   7'-8"   7'-8"   7'-8"   7'-8"   7'-8"

8" BLOCK

29'-6"

6'-6"

5'-6"

76'-0"

22'-6"

2×12, 16"0.C. FLOOR JOISTS

2×12, 16"0.C. FLOOR JOISTS

2×12, 16"0.C. FLOOR JOISTS

11'-4"   12'-0"   11'-4"

7'-1"

36'-0"

FOUNDATION PLAN

PROPOSED RESIDENCE FOR
MR & MRS T. HAIZE
CREST ROAD

3

appendix h

406

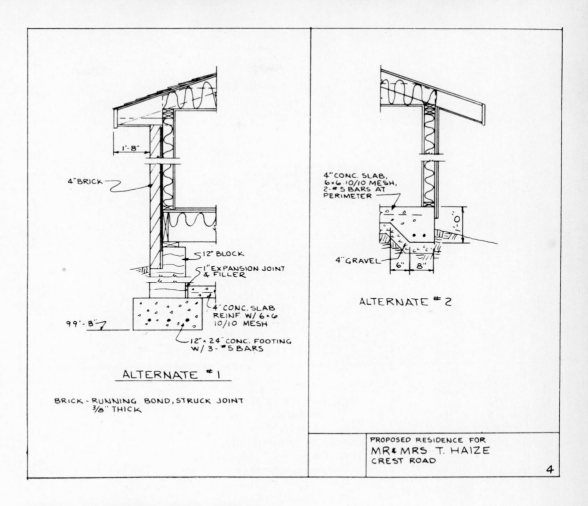

ALTERNATE #1

4" BRICK

1'-8"

12" BLOCK

1" EXPANSION JOINT & FILLER

4" CONC. SLAB REINF W/ 6×6 10/10 MESH

99'-8"

12"×24" CONC. FOOTING W/ 3-#5 BARS

BRICK - RUNNING BOND, STRUCK JOINT 3/8" THICK

ALTERNATE #2

4" CONC. SLAB, 6×6 10/10 MESH, 2-#5 BARS AT PERIMETER

4" GRAVEL

6" 8"

PROPOSED RESIDENCE FOR
MR & MRS T. HAIZE
CREST ROAD

4

NOTE: Finished floor elevation 109'-9"
Garage floor elevation 107'-7"
Exterior grade elevation 106'-9"
Ceiling height 8'-0"
Roof slope 4 in 12
Assume topsoil is 6" deep
Garage floor: 4" concrete slab, rein-
force with 6×6, 10/10 welded wire
mesh, over 4 mil polyethylene vapor
barrier over 6" gravel.

INDEX

Advertisement for Bidders, 10, 11, 27 (*see also* Contract documents)
Agreement, 15-19
Agreement, provisions, 18, 19
  acceptance, 19
  contract sum, 18, 19
  costs, 18, 19
  final payment, 19
  materials, 19
  payments, 19
  retainage, 19
  schedule of values, 19
  scope of work, 18
  work in place, 19
Agreements, 15-19
  AIA, 15
  bonus, 18
  completion time, 18
  cost-plus-fee, 15, 17, 18
  fixed fee, 18
  lump sum, 15-17
  owner–contractor, 15, 34
  penalty, 18
  percentage fee, 17
  sliding scale fee, 18
  type of, 15
  unit price, 15, 16, 17
Air conditioning, 331-335 (*see also* HVAC)
American Concrete Institute, 125

American Institute of Architects, General Conditions, 352-71
American Plywood Association, 210
American Society for Testing and Materials (ASTM), 151, 155
Asbestos shingles, 312, 313 (*see also* Roofing)
Asphalt paving, 120-122 (*see also* Paving, asphalt)
Asphalt shingles, 311, 312 (*see also* Roofing)

Backfill, 108-114 (*see also* Excavation, backfill)
Bearing piles, 118-120 (*see also* Piles)
Bid bond, 20, 23
Bidding information, 12, 13
   advertisement for bids, 11
   AGC, 11
   competitive, 1
   Dodge Reports, 11
   prime contractors, 11
   reporting services, 11
Bonds, 19-24
   obtaining, 21, 23, 24
   statutory, 20
   surety, 19
   types of, 20-24
      bid bond, 20, 23
      labor, 21
      license, 21
      lien, 21
      material, 21
      performance, 20, 21
      permit, 21
      roofing, 205
      subcontract, 21
Borrow, 114, 118 (*see also* Excavation)
Brick, 160-165 (*see also* Masonry)
Building areas, 91, 92
   perimeters, 91, 92
Bush hammering, 139

Carpeting, 303, 304 (*see also* Flooring)
Cements (*see* Concrete)
Ceramic tile, 305, 306 (*see also* Flooring)
Clay masonry, 160-165
Clay tile, 160, 165
Compaction (*see* Excavation)
Competitive bidding, 1
Computers, 13, 57-60
Concrete, 125-150
   accessories, 137, 138
   admixtures, 125
   block, 154-160 (*see also* Masonry)
      accessories, 160
   curing, 141-143
   estimating, 126-130
   field batching, 125

Concrete (*cont.*):
   finishing, 138-141
      equipment, 139
      estimating, 139, 140
      labor, 140, 141
      types of, 138, 139
   forms, 144-149
      checklist, 148, 149
      design, 144, 145
      estimating, 145-147
      labor, 147
      liners, 147, 148
      metal, 145
      miscellaneous, 145
      textures, 147
      wood, 144, 145
   job-mixed, 125
   labor, 129
   masonry, 151-180 (*see also* Masonry)
   precast, 188-198
      aggregates, 189
      beams, 194, 195
      checklist, 197, 198
      columns, 194, 195
      costs, 197
      estimating, 190-196
      finishes, 194
      lintels, 194-196
      miscellaneous, 196
      reinforcing, 188
      slabs, 193, 194
      specifications, 189
      tees, 190-193
   ready-mixed, 125
   reinforcing, 130-135
      bars, 130-133
      bends, 130, 131
      estimating, 133, 135
      labor, 135
      lapping, 131-133
      rib lath, 132, 135
      shop drawings, 131
      splicing, 131
      steel, decking, 133
      supports, 133
      waste, 133
      wire mesh, 131-133
   sealing compounds, 143
   transporting, 143
   vapor barrier, 136
Construction documents, 9-12
Contingencies, 60, 61
Contract documents, 9-12
   addenda, 9, 36, 37
   Advertisement for Bidders, 10, 11, 27

Contract documents (*cont.*):
  agreements, 9, 15-19
  availability, 11, 12
  bidding period, 11
  deposit, 11
  General Conditions, 9, 352-371
  material suppliers, 11
  Notice to Bidders, 11
  SCAN, 11
  specifications, 9
  subcontractors, 11
  working drawings, 9
Contractor, prime, 14
Contracts, 14-15
  separate, 14, 15
  single, 14, 15
  systems, 14, 15
Conversions, 377-381
Coping, 173
Cost analysis, 8, 9
  methods, 8, 9
    area, 8, 9
    detailed, 8
    preliminary, 8, 9
    volume, 8, 9
Critical path method (CPM), 57-60
Curing, 141-143
  aggregates, 142
  cold weather, 142
  covers, 141, 142
  enclosures, 142
  estimating, 142
  hot weather, 141, 142
  hydration, 141
  moisture, 141-143
  period, 141-143
  specifications, 143
  unions, 142
  water, 142
Curtain wall, 282-287
  accessories, 285-287
  estimating, 283-285
  flashing, 287
  frames, 282-287
    aluminum, 284, 285
    wood, 283
  glass, 296
  glazing, 285, 286, 296
  lintel, 287
  sill, 286

Dampproofing, 184-187
  checklist, 186, 187
  painting, 185, 186
    estimating, 185, 186

Dampproofing (*cont.*):
  parging, 186
    equipment, 186
    estimating, 186
Decking, wood, 209 (*see also* Wood decking)
Detailed estimate, 8
De-watering, 116
Doors, 287-296
  accessories, 287, 295
  acoustical, 290
  checklist, 295
  estimating, 292
  exterior, 287
  face veneer, 287
  fire-rated, 289
  folding, 291
  frames, 293, 294
    finishes, 294
    heads, 294
    jambs, 294
    materials, 294
    types of, 294
  hardware, 291, 294, 295
    cash allowance, 294, 295
    finished, 294
    rough, 294
    specifications, 294
    templates, 294
  hollow core, 287, 288
  interior, 287
  materials, 287
  metal, 288
  operation, 287
  overhead, 291
  prefinished, 292
  prehung, 291, 292
  premachining, 291, 292
  schedules, 291
  shop drawings, 291
  solid core, 287
  specifications, 287
  swing, 288
Drywall, 260-274
  accessories, 260
  adhesives, 267
  assemblies, 265
  blankets, 271
  checklist, 281
  estimating, 270-273
  fasteners, 260, 269
  fireproofing, 268
  labor, 272, 273
  metal studs, 261, 262
  specifications, 260
  supporting construction, 260-264

Drywall (*cont.*):
  suspended ceilings, 264, 265
  tape and compound, 270
  wallboard, 266, 267
    estimating, 266, 267
    labor, 268
    specifications, 266, 268
    types of, 266-268

Electrical, 321-325
  checklist, 325
  codes, 321
  contracts, 321-325
    separate, 321, 323
    single, 321-324
  coordination, 323-325
  drawings, 322
  finish, 322
  fixtures, 322
  installation, 322
  rough, 322
  specifications, 321
Equipment, 75-81
  backfill, 108
  cost accounting, 80, 81
  depreciation, 77
  excavation, 94
  insurance, 79
  interest, 78, 79
  mobilization, 76, 81
  operating costs, 76, 77
    fuel, 76, 77
    tires, 77
  ownership, 75-79
    costs, 75-79
  rentals, 75, 79, 80
    costs, 79, 80
  repair, 79
  salvage value, 79
  selection, 75, 76
  taxes, 79
Estimating (*see also each material*):
  computers, 13
  cost manuals, 13
  detailed, 8
  information sources, 12, 13
Excavation, 83-124
  areas, 91, 92
  backfill, 108-114
    equipment, 108
    estimating, 108-114
    labor, 113, 114
    variables, 108
  borrow, 114, 118
  checklist, 117, 118

Excavation, (*cont.*)
  compaction, 114
  cutting, 85-91
  equipment, 84, 85
  excess, 114, 115
    specifications, 114
  fill, 85-91
  general, 94-105
    estimating, 94-105
    labor, 105
    slope of bank, 94
  grading, 85-91
  landscaping (*see* Landscaping)
  mass, 94-105
  perimeter, 91, 92
  rock, 116, 117
    blasting, 116, 117
    classifications, 116
    drills, 116
  shrinkage, 84
  soil borings, 84
  special, 105-108
    estimating, 105-108
    labor, 107, 108
  specifications, 83, 84
  subcontractors, 117
  swell, 84, 114, 117
  topsoil, 92-94
    equipment, 93
    estimating, 92-94
    labor, 94
    removal, 92-94
    specifications, 115
    spreading, 115
  variables, 83
Excess, 114, 115
Expansion joints, 310
  covers, 136
  fillers, 136, 137

Field personnel, 71, 72
  reports, 71, 72
  uses, 71, 72
Fireproofing (*see* Drywall, fireproofing)
Flashing, 171, 310, 318
Flooring, 298-306
  carpeting, 303, 304
  checklist, 308, 309
  resilient, 299-303
    estimating, 300-303
    materials, 299
  tile, 305, 306
  wood, 298, 299
Forms, concrete 144-149 (*see also* Concrete)
Frame, construction, 213-258 (*see also* Wood frame)

General Conditions, 11, 352-371
General excavation, 94-105 (*see also* Excavation)
Glass, 285, 296

Hardware, 286, 291, 294, 295
  door, 291, 294, 295
Hardwood floors, 298-299
Heating, ventilating, air conditioning (HVAC), 331-335
  checklist, 334, 335
  codes, 331
  contracts, 331, 332
    separate, 331, 332
    single, 331, 332
    subcontractors, 332
  coordination, 332, 333

Insulation, 255, 310, 318, 319
  roofing, 310, 318, 319
Interest, 78, 79

Labor, 70-73 (*see also each material*)
  bond, 21
  diagrams, 72, 73
  fringe benefits, 71
  pricing, 70-72
    man-hours, 72
    productivity, 70
    work units, 72
  unions, 71
  unit of time, 70
  variables, 70, 71
  wages, 71
Laminated wood, 151, 152
Landscaping, 115, 116
Lath, plaster, 277-279 (*see also* Wetwall)
License bond, 21
Lien bond, 21
Lintels, 173, 224, 225
Liquid roofing, 318 (*see also* Roofing)
Lump sum, 15, 16, 17

Masonry, 151-180
  accessories, 168
  bonds, 152, 153
  bricks, (*see* Clay)
  checklist, 179, 180
  clay, 160-165
    cleaning, 178
    estimating, 162-165
    finishes, 160, 162
    labor, 164, 165
    specifications, 162
    types of, 162
  cleaning, 177, 178

Masonry (*cont.*)
  cold weather, 178, 179
  concrete block, 154-160
    cleaning, 178
    estimating, 157-160
    labor, 159, 160
    specifications, 151, 155, 157
    types, 151
    waste, 157, 158
  control joints, 171
  coping, 173
  equipment, 178
  flashing, 171
  labor, 152
  lintels, 173
  mortar, 167, 168
    aggregate, 167
    estimating, 167, 168
  reinforcing, 168-171, 176, 177
  sills, 173
  specifications, 151, 152
  stone, 166, 167
    accessories, 167
    cleaning, 178
    dressing, 167
    materials, 151
    mortar, 167, 168
    specifications, 167
    waste, 168
  subcontractors, 179
  wall ties, 171, 173
  weep holes, 171
Mass excavation, 94-105 (*see also* Excavation)
Material bond, 21
Mobilization, 81
Mortar, 167, 168 (*see also* Masonry)
Mortar joints, 152
  shapes, 155

Overhead, 52-67, 117
  direct, 54
  general, 52-54
    depreciation, 53
    miscellaneous, 53
    office, 52
    salaries, 53
  job, 52, 54
    checklist, 67-69
    cleanup, 56
    enclosures, 55
    heat, 55
    miscellaneous, 56
    office, 54
    photographs, 56
    salaries, 54
    sanitary, 55

Overhead (*cont.*):
  job (*cont.*):
    surveys, 56
    utilities, 55
    water, 55, 56

Painting, 306-309
  checklist, 309
  estimating, 307, 308
  specifications, 306, 307
Paving, asphalt, 120-122
  classification, 120
  cold weather, 121
  equipment, 121
  specifications, 120
Payment, progress, 19
  application for, 19
Performance bonds, 20, 21
Permit bond, 21
Piles, 118-120
  checklist, 120
  estimating, 118-120
  soil borings, 119
  specifications, 119, 120
  types, 118-120
Plaster, 273-281 (*see also* Wetwall)
Plumbing, 326-331
  checklist, 329
  codes, 326
  contracts, 326, 328
    separate, 326, 328
    single, 326, 328
  coordination, 328, 329
  drawings, 326
  rough, 326
  specifications, 326
Plywood, 209, 210, 231, 232, 244, 245, 249
  sheathing, 231, 232
  systems, 209, 210
    checklist, 210
Portland Cement Association, 125
Precast concrete, 188-198 (*see also* Concrete)
Profit, 117, 336, 337
Pumping, 116

Quantity take-off, 2, 6, 7
Quarry tile, 305, 306

Rates, interest, 78, 79
Reinforcing, 130-135, 176, 179 (*see also* Concrete *and* Masonry)
Resilient flooring, 299-303 (*see also* Flooring)
Rock, 116, 117 (*see also* Excavation)
Roofing, 310-320
  areas, 310, 311

Roofing (*cont.*)
  built-up, 313-315
    aggregates, 314, 315
    bitumens, 314
    caulking, 315
    estimating, 314, 315
    felts, 313-315
    insulation, 314
    materials, 314
    specifications, 313
  cant strip, 310
  checklist, 320
  corrugated, 316
    estimating, 316
    materials, 316
  curbs, 310
  decking, 310
  equipment, 319, 320
  expansion joints, 310
  flashing, 310, 318
  insulation, 310, 318, 319
    estimating, 318, 319
    specifications, 318
  labor, 319
  liquid, 318
  metal, 316
  shingles, 311-313
    asbestos, 312, 313
    asphalt, 311, 312
    estimating, 312, 313
    wood, 313
  slate, 316, 317
  tile, 317
  trim, 319

Sandblasting, 138, 139
Scheduling, 56-60
Screen, window, 286
Separate contracts, 14, 15
Sheathing, 231, 232
Shingles, 311-313 (*see also* Roofing)
Sills, 216-218
Site investigation, 45, 46
Special excavation, 105-108 (*see also* Excavation)
Specifications, 1, 26-38 (*see also material being estimated*)
Steel, structural, 198-206
  checklist, 205, 206
  decking, 203-205
  estimating, 198, 199, 203, 205
  framing, 198, 199
  joists, 199-203
  miscellaneous, 205
  specifications, 201, 203
  subcontractors, 205
Stone masonry, 166, 167 (*see also* Masonry)
Structural systems, 188-212

Subcontractor bonds, 21
Subcontractors, 117, 179, 332
Supplementary General Conditions, 9
Suspended ceilings, 264, 265
Swell (*see* Excavation)

Tile:
    ceramic, 305, 306
    clay, 160, 165
    quarry, 305, 306
Topsoil, 92-94, 115 (*see also* Excavation)
Trusses, wood, 206, 207 (*see also* Wood)

Unions, 70

Vapor barrier, 135, 136
    estimating, 135, 136
    labor, 136
Ventilating, 234-238

Wall ties, 171, 173
Waterproofing, 181-184
    checklist, 186, 187
    hydrostatic, 181
    integral, 184
        estimating, 184
    membrane, 181-183
        equipment, 183
        estimating, 182-183
        labor, 183
    metallic, 184
    methods, 181
Wetwall, 260-267, 273-281
    accessories, 260, 280
        estimate, 280
    assembly, 265
    checklist, 280, 281
    estimating, 275-277
    fasteners, 260
    furring, 263, 264
    labor, 276, 277
    lath, 277-279
        estimating, 278, 279
        types, 271-279
    metal studs, 261, 262
    plaster, 273-281
        acoustical, 275
    coats, 274, 275
        equipment, 276, 277
        estimating 275
        Keene cement, 275
        ornamental, 275
        specifications, 275

Wetwall (*cont.*):
    stucco, 275
    supporting construction, 260-264, 273
    suspended ceilings, 264, 265
Windows, 282-287, 296
    accessories, 285-287
    estimating, 283-285
    flashing, 287
    frames, 282-287
        aluminum, 284, 285
        wood, 283
    glass, 296
    glazing, 285, 286, 296
    hardware, 286
    lintel, 287
    mullion, 286
    screens, 286
    sill, 286
Wood:
    board measure, 214, 215
    decking, 209
    frame, estimating, 213-258
    framing, ceiling, 247-249
        headers, 248, 249
        joists, 247
        trimmers, 248, 249
    framing, floor, 215-232
        bridging, 216, 230, 231
        girders, 215, 216, 223
        headers, 216, 224-230
        joist, 216, 218-223
        sheathing, 216, 231, 232
        sill, 216-218
        trimmer, 224-230
    framing, roof, 249-254
        collar ties, 252
        headers, 251
        lookout, 253
        pitch, 251
        rafters, 249, 251
        ridge board, 252, 253
        sheathing, 249, 254
        slope, 251
        trimmers, 251
    framing, wall, 232-247
        headers, 233, 234, 241-243, 247
        plates, 233, 234, 246
        sheathing, 233, 244, 245
        studs, 233-241, 246
    labor, 257
    laminated, 208, 209
    shingles, 313
    trim, 257
    trusses, 206, 207, 255